2008 IEEE

International Conference on

Microelectronic Test Structures

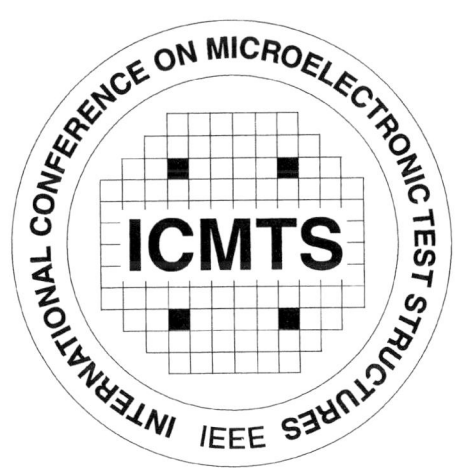

Conference Proceedings

March 24-27

The University of Edinburgh, UK

Copyright Notice:

Copyright and Reprint Permission: Abstracting is permitted with credit to the source. Libraries are permitted to photocopy beyond the limit of U.S. copyright law for private use of patrons those articles in this volume that carry a code at the bottom of the first page, provided the per-copy fee indicated in the code is paid through Copyright Clearance Center, 222 Rosewood Drive, Danvers, MA 01923. For other copying, reprint or republication permission, write to IEEE Copyrights Manager, IEEE Operations Center, 445 Hoes Lane, P.O. Box 1331, Piscataway, NJ 08855-1331. All rights reserved. Copyright ©2008 by the Institute of Electrical and Electronics Engineers.

IEEE Catalog Number: CFP08MTS-PRT
ISBN: 978-1-4244-1800-8
Library of Congress: 2007907607

Additional copies may be ordered from:

IEEE Operations Center

445 Hoes Lane

Piscataway, NJ 08854-4150

USA

+1 800 678 IEEE (+1 800 678 4333)

+1 732 981 1393

+1 732 981 9667 (FAX)

email: customer-service@ieee.org

WELCOME LETTER

Dear Colleagues

The 2008 IEEE International Conference on Microelectronic Test Structures (ICMTS) will be held at the University of Edinburgh, which recently celebrated its 400th anniversary. ICMTS has come of age and this meeting celebrates its 21st anniversary by visiting Edinburgh for the second time. The conference is being run in cooperation with the University of Edinburgh and the Scottish Microelectronics Centre. As was the case for the first ICMTS conference, it is is being sponsored by the IEEE Electron Devices Society, and for the first time the meeting is also being technically co-sponsored by the IEEE Solid State Circuits Society.

The first ICMTS was held in Long Beach in 1988, and since then the conference has cycled between Europe, North America, and Asia. Over the past twenty one years it has brought together designers and users of test structures to discuss recent developments and future directions. As in previous years the conference will be preceded by a one-day Tutorial Short Course on microelectronic test structures and there will also be a related equipment exhibition focused on test structure measurements. We are very fortunate to be holding the conference banquet in the Playfair Library. Its eleven bays of books supporting a coffered vaulted ceiling make it one of Edinburgh's grandest interiors. The conference will be held at the Appleton Tower in the University of Edinburgh, which is located close the heart of Scotland's capital city.

With its stunning Georgian and Victorian architecture, and winding medieval streets, it's easy to see why Edinburgh has been listed as a World Heritage Site. Edinburgh has one of the most beautiful cityscapes in the world with the famous castle perched atop the crags of an ancient volcano dominating the urban skyline. At the opposite end of "The Royal Mile" lies the Palace of Holyrood. This is the Queen's official residence in Scotland and contains historic apartments where Mary, Queen of Scots lived. The capital is bustling with arts, culture, sports and attractions and is famous for playing host to the world's largest arts festival. The city is the birthplace of James Clerk Maxwell, Sir Arthur Conan Doyle, author of the Sherlock Holmes novels and Alexander Graham Bell, inventor of the telephone. After dark, Edinburgh has a lively nightlife with stylish bars and pubs, restaurants, clubs and live entertainment to rival any European City. We are sure that everyone will enjoy attending the various sessions that make up ICMTS and derive great benefit from the numerous networking opportunities. We look forward to welcoming you to Edinburgh.

Sincerely

Anthony Walton, General Chairman

Jurriaan Schmitz, Technical Chair

Stewart Smith, Conference Vice Chair

TABLE OF CONTENTS

Tuesday, March 25

SESSION 1: Matching I – Characterisation

1.1 – Operational Amplifier Based Test Structure For Transistor Threshold Voltage Variation...3

Brian L. Ji, Dale J. Pearson, Isaac Lauer, Franco Stellari, David J. Frank, Leland Chang, and Mark B. Ketchen

IBM T.J. Watson Research Center, NY, USA

1.2 – A Study of Variation in Characteristics and Subthreshold Humps for 65-nm SRAM Using Newly Developed SRAM Cell Array Test Structure............8

A. Mizumura[1], T. Suzuki[1], T. Arima[2], H. Maeda[2], and H. Ammo[1]

[1]*Semiconductor Technology Development Division, Sony Corp.* [2]*Sony LSI Design Incorporated, Kanagawa, Japan*

1.3 – Mismatch Characterization of a High Precision Resistor Array Test Structure...11

Weidong Tian, Philipp Steinmann, Eric Beach, Imran Khan and Praful Madhani

Texas Instruments Inc. USA

1.4 – Rapid Characterization of Parametric Distributions Using a Multi-meter .. 17

Jerry Hayes, Kanak Agarwal, Sani Nassif

IBM Austin Research Labs, TX, USA

SESSION 2: Resistivity

2.1 – Beyond van der Pauw: Sheet Resistance Determination of Arbitrarily Shaped Planar Four-Terminal Devices with Extended Contacts....................23

Martin Cornils and Oliver Paul

Department of Microsystems Engineering (IMTEK), University of Freiburg, Germany

2.2 – Investigation of Electrical and Optical CD Measurement Techniques for the Characterisation of On-Mask GHOST Proximity Corrected Features........29

A. Tsiamis[1], S. Smith[1], M. McCallum[2], A.C. Hourd[3], O. Toublan[4], J.T.M. Stevenson[1] and A.J. Walton[1]

[1]*The University of Edinburgh,* [2]*Nikon Precision Europe GmbH,* [3]*Compugraphics International Ltd., UK* [4]*Mentor Graphics Ltd., France*

2.3 – Comparison of Measurement Techniques for Advanced Photomask Metrology 35

S. Smith[1], A. Tsiamis[1], M. McCallum[2], A.C. Hourd[3], J.T.M. Stevenson[1], A.J. Walton[1], R.G. Dixson[4], R.A. Allen[4], J.E. Potzick[4], M.W. Cresswell[4] and N.G. Orji[4]

[1]*The University of Edinburgh,* [2]*Nikon Precision Europe GmbH,* [3]*Compugraphics International Ltd.,* [4]*National Institute of Standards and Technology*

SESSION 3: Yield and Reliability I

3.1 – Electromigration : from package to wafer level thanks to a heating coil structure 43

C. Chappaz, P. Waltz, L. Castellon

STMicroelectronics, Crolles CEDEX, France

3.2 – Life condition monitoring on smart power devices using a sequence of current and charge-based capacitance measurements 46

Zhenqiu Ning[1], Erwin de Vylder[1], Filip Bauwens[1], Basil Vlachakis[1], H-X Delecourt[1], Renaud Gillon[1], Patrick Van Torre[2] and Dan Hegsted[1]

[1]*AMI Semiconductor Belgium bvba, Belgium,* [2]*Hogeschool Gent, Belgium*

3.3 – Mixed Test Structure for Process Variation and Hard Defect Detection 52

F. Rigaud[1,2], J.M. Portal[1], H. Aziza[1], D. Née[2], J. Vast[2], F. Argoud[2], B. Borot[3]

[1]*L2MP - Laboratoire Matériaux et Microélectronique de Provence,* [2]*STMicroelectronics, Rousset,* [3]*STMicroelectronics, Crolles, France*

3.4 – Short-flow Test Chip Utilizing Fast Testing for Defect Density Monitoring in 45nm 56

Muthu Karthikeyan[1], William Cote[1], Louis Medina[1], Ernesto Shiling[1], Arthur Gasasira[1], Amy Henning[1], William Ferrante[1], Mark Craig[2], and Thomas Merbeth[3]

[1]*IBM Systems and Technology Group, NY, USA,* [2]*AMD Austin, TX, USA,* [3]*AMD Saxony, Dresden, Germany*

SESSION 4: Poster Presentations

4.1 – Conduit Diffusion of Dopants in Tungsten Silicide Layers 65

S. Liao, M. Bain, P. Baine, D.W. McNeill, B.M. Armstrong and H.S. Gamble

Queen's University of Belfast, UK

4.2 – A Novel High Speed Automatic Layout System to Place and Route Test Structures for Parametric Test Capability **71**

Andrew J. West[1], Samrat Mondal[2], Devjyoti Patra[2], Kalyan Goswami[2] and Shamik Sural[2]

[1]*National Semiconductor Corp, Santa Clara, CA, USA,* [2]*Indian Institute of Technology, Kharagpur, India*

4.3 – A Test Structure for Channel Length Engineering of NAND Gates in Standard Cell Library **76**

T. Matsuda[1], Y. Sugiyama[1], J. Takakuwa[1], H. Iwata[1], T. Ohzone[2]

[1]*Department of Information Systems Engineering, Toyama Prefectural University,* [2]*Dawn Enterprise, Japan*

4.4 – Test Structure for Characterising Low Voltage Coplanar EWOD System ... **80**

Yifan Li[1], Yoshio Mita[1,2], Les Haworth[1], William Parkes[1], Masanori Kubota[2], Anthony Walton[1]

[1]*University of Edinburgh, UK,* [2]*University of Tokyo, Japan*

4.5 – Measurement of the MOSFET Drain Current Variation Under High Gate Voltage **86**

Tetsuo Chagawa[1], Kazuo Terada[1], Jianyu Xiang[1], Katsuhiro Tsuji[1], Takaaki Tsunomura[2] and Akio Nishida[2]

[1]*Faculty of Information Sciences, Hiroshima City University, Japan,* [2]*MIRAI-Selete, Tsukuba, Japan*

4.6 – Spacing Impact on MOSFET Mismatch **90**

A. Cathignol[1,3] , S. Mennillo[2], S. Bordez[1], L. Vendrame[2], G. Ghibaudo[3]

[1]*STMicroelectronics, Crolles, France,* [2]*STMicroelectronics, Advanced R&D, Italy,* [3]*IMEP, Minatec, Grenoble, France*

4.7 – Highly Automated Test Chip Layout and Test Plan Development for Parametric Electrical Test **96**

Ann Gabrys, Wendy Greig, Andrew J. West, Philipp Lindorfer, and William French

National Semiconductor Corp., Santa Clara, CA, USA

4.8 – Circular Geometry MOS Transistor Analysis of SOI Substrates for High Energy Physics Particle Detectors **101**

S.L. Suder[1], F.H. Ruddell[1], J.H. Montgomery[1], B.M. Armstrong[1], H.S. Gamble[1], G. Casse[2], T. Bowcock[2], P.P. Allport[2]

[1]*Northern Ireland Semiconductor Research Centre, Queen's University Belfast, UK,* [2]*Liverpool Detector Centre, University of Liverpool, UK*

SESSION 5: 3D Integration

5.1 – Prediction of Stress-Induced Characteristics Changes for Small-scale Analog IC 107

Naohiro Ueda, Eri Nishiyama, Hideyuki Aota, Hirobumi Watanabe

Electronic Devices Company, Ricoh Co. Ltd., Japan

5.2 – Test Structures for the Measurement and Optimisation of Bond Strength for Anodic Bonding of Glass to Dielectric Thin Films 111

G. Cummins, H. Lin and A.J. Walton

The University of Edinburgh, UK

5.3 – Test Structures For The Evaluation Of 3D Chip Interconnection Schemes 117

A. Mathewson[1,2], J. Brun[2], R. Franiatte[2], A. Nowodzinski[2], R. Ancient[2], N. Sillon[2], F. Depoutot[3] and B. Dubois-Bonvalot[3]

[1]*Tyndall National Institute, Cork, Ireland,* [2]*CEA-Leti-Minatec, Grenoble,* [3]*Hardware Research Group, Gemalto, Grenoble, France*

5.4 – An Evaluation of Test Structures for Measuring the Contact Resistance of 3-D Bonded Interconnects 123

H. Lin, S. Smith, J.T.M. Stevenson, A.M. Gundlach, C.C. Dunare, and A.J. Walton

The University of Edinburgh, UK

SESSION 6: Yield and Reliability II

6.1 – High Density Test Structure Array for Accurate Detection and Localization of Soft Fails 131

Christopher Hess[1], Michele Squcciarini[1], Shia Yu[1], Jonathan Burrows[1], Jianjun Cheng[2], Ron Lindley[2], Andrew Swimmer[2], Steven Winters[2]

[1]*PDF Solutions Inc., San Jose, USA,* [2]*PDF Solutions Inc., San Diego, USA*

6.2 – New method for non destructive snap-back characterization in multifinger power MOSFETs 137

François Dieudonné[1], Aurore Constant[2], Julien Rosa[1], Benoit Gautheron[1], Jean-François Revel[1]

[1]*STMicroelectronics, Crolles, France,* [2]*SID/CNM, Bellaterra, Spain*

6.3 – Test Circuit for Measuring Pulse Widths of Single-Event Transients Causing Soft Errors 142

Balaji Narasimham, Matthew J. Gadlage, Bharat L. Bhuva, Ronald D. Schrimpf, Lloyd W. Massengill, W. Timothy Holman, Arthur F. Witulski, Kenneth F. Galloway

Vanderbilt University, Nashville, TN 37235, USA

6.4 – CMOS Latch Metastability Characterization at the 65-nm-Technology Node 147

Manjul Bhushan[1], Mark B. Ketchen[2] and Koushik K. Das[2]

190 [1]*IBM Systems and Technology Group, NY, USA,* [2]*IBM Research, T.J. Watson Research Center, NY, USA*

SESSION 7: MOS Modeling and Characterisation

7.1 – Characterization of MOSFETs Intrinsic Performance using the In-Wafer Advanced Kelvin-Contact Device Structure for High Performance CMOS LSIs 155

R. Kuroda[1], A. Teramoto[2], T. Komuro[3], W. Cheng[2], S. Watabe[1], C.F. Tye[1], S. Sugawa[1] and T. Ohmi[2]

[1]*Graduate School of Engineering, Tohoku University,* [2]*New Industry Creation Hatchery Center, Tohoku University,* [3]*Agilent Technologies International Japan Ltd.*

7.2 – New Y-function-based methodology for accurate extraction of electrical parameters on nano-scaled MOSFETs 160

Dominique Fleury[1,2], Antoine Cros[1], Hugues Brut[1], Gérard Ghibaudo[2]

[1]*STMicroelectronics, Crolles,* [2]*IMEP, Grenoble, France*

7.3 – A Novel Biasing Technique for Addressable Parametric Arrays 166

Brad Smith[1], Uma Annamalai[2], Alexandre Arriordaz[1], Venkat Kolagunta[1], Jeff Schmidt[1] and Mehul Shroff[1]

[1]*Freescale Semiconductor, Austin, TX,* [2]*University of Arkansas, AR, USA*

SESSION 8: RF

8.1 – 2.6GHz RF Inductive Power Delivery for Contactless On-Wafer Characterization 175

Jonathan Tompson, Adam Dolin, and Peter Kinget

Department of Electrical Engineering, Columbia University, NY, USA

viii

8.2 – Advanced Test Structure Design for Dielectric Characterisation of Novel High-K Materials 180

John A. O'Sullivan[1], Wenbin Chen[2], Kevin G. McCarthy[1] and Gabriel M. Crean[1]

[1]*Department of Electrical and Electronic Engineering, University College Cork, Ireland,* [2]*Tyndall National Institute, Ireland*

8.3 – Characterization of T-Shape Terminal Impedances of Differential Short Stubs in Advanced CMOS Technology 185

Chiaki Inui and Minoru Fujishima

University of Tokyo, Japan

8.4 – Identifying dielectric and resistive electrode losses in high-density capacitors at radio frequencies 190

M.P.J. Tiggelman[1], K. Reimann[2], J. Liu[2], M. Klee[3], W. Keur[3], R. Mauczock[3], J. Schmitz[1] and R.J.E. Hueting[1]

[1]*MESA+ Research Institute for Nanotechnology, University of Twente,* [2]*NXP Semiconductors,* [3]*Philips Research, The Netherlands*

SESSION 9: Interconnect

9.1 – A Study of Cross-Bridge Kelvin Resistor Structures for Reliable Measurement of Low Contact Resistances 199

N. Stavitski[1], J.H. Klootwijk[2], H.W. van Zeijl[3], A.Y. Kovalgin[1] and R.A.M. Wolters[1,4]

[1]*MESA+ Institute for Nanotechnology, University of Twente,* [2]*Philips Research,* [3]*DIMES, Delft University of Technology,* [4]*NXP Research Eindhoven, The Netherlands*

9.2 – Comb Capacitor Structures for Measurement of a Post-processed Layer ... 205

D. Roy[1], J.H. Klootwijk[2], N.A.M. Verhaegh[1], H.H.A.J. Roosen[2], and R.A.M. Wolters[1]

[1]*NXP Semiconductors,* [2]*Philips Research, The Netherlands*

9.3 – Test structure for characterizing metal thickness in damascene CMP technology 210

Alain Toffoli, Sylvain Maitrejean, Jean Duport de Pontcharra, Francois de Crecy, David Bouchu, Lucile Arnaud, Fabien Boulanger

CEA-LETI, Grenoble Cedex, France

9.4 – Test Structure Definition for Dummy Metal Filling Strategy Dedicated to Advanced Integrated RF Inductors 214

Carine Pastore[1,2], Frédéric Gianesello[1], Daniel Gloria[1], Emmanuelle Serret[1], Phillippe Benech[2]

[1]*STMicroelectronics, Crolles,* [2]*IMEP, Grenoble Cedex, France*

SESSION 10: Matching II – Mechanisms

10.1 – Fully considered layout variation analysis and compact modeling of MOSFETs and its application to circuit simulation **223**

Takuji Tanaka[1], Akira Satoh[2], Mitsuru Yamaji[1], Osamu Yamasaki[1], Hiroshi Suzuki[2], Tsuyoshi Sakata[1], Yoshio Inoue[3], Masaru Ito[1], Seiichiro Yamaguchi[1] and Hiroshi Arimoto[1]

[1]*FUJITSU Ltd.,* [2]*FUJITSU Laboratories Ltd.,* [3]*FUJITSU VSLI Ltd. Tokyo, Japan,*

10.2 – On-Mask Mismatch Resistor Structures for the Characterisation of Maskmaking Capability **228**

S. Smith[1], A. Tsiamis[1], M. McCallum[2], A.C. Hourd[3], J.T.M. Stevenson[1], A.J. Walton[1] and S. Enderling[1]

[1]*The University of Edinburgh,* [2]*Nikon Precision Europe GmbH,* [3]*Compugraphics International Ltd., UK*

10.3 – Physics and Modeling of Transistor Matching Degradation under Matched External Stress **233**

Xiaoju Wu, Zhenwu Chen, Praful Madhani

Analog Technology Development, Texas Instruments, TX, USA

10.4 – Influence of STI stress on drain current matching in advanced CMOS ... **238**

Nicole Wils[1], Hans Tuinhout[1], Maurice Meijer[2] [1]*NXP-TSMC Research Center,* [2]*NXP Semiconductor Research, The Netherlands*

CONFERENCE OFFICIALS

ICMTS 2008 Conference Committee

General Chairman:
Anthony J. Walton
The University of Edinburgh

Technical Chairman:
Jurriaan Schmitz
University of Twente

Conference Vice Chair:
Stewart Smith
The University of Edinburgh

Treasurer:
Jonathan G. Terry
The University of Edinburgh

Tutorial Chairman:
Johan Klootwijk
Philips Research Europe

Local Arrangements:
Les I. Haworth
The University of Edinburgh

Equipment Exhibition:
Stewart Wilson
Agilent Technologies

USA Representative:
Loren W. Linholm

European Representative:
Anthony J. Walton
The University of Edinburgh

Asian Representative:
Kunihiro Asada
The University of Tokyo

Technical Committee

Richard Allen, NIST, USA
Hugues Brut, STMicroelectronics, France
Rino Choi, USA
Steve S. Chung,
National Chiao Tung University, Taiwan
Michael W. Cresswell, USA
Kelvin Y.Y. Doong, TSMC, Taiwan
Satoshi Habu, Agilent Technologies, Japan
Yoshiaki Hagiwara, Sony Corp., Japan
Christopher Hess, PDF Solutions Inc., USA
Kjell Jeppson,
Chalmers University of Technology, Sweden
Choongho Lee, Samsung Electronics, Korea
Hi-Deok Lee, Chungnam National Univ., Korea
Loren W. Linholm, USA
Emilio Lora-Tamayo,
Universitat Autonoma de Barcelona, Spain
Alan Mathewson,
Tyndall National Institute, Ireland
Colin McAndrew, Freescale Semiconductor, USA
Kevin McCarthy,
University College Cork, Ireland

Yoshio Mita, University of Tokyo, Japan
Tatsuya Ohguro, Toshiba Corp., Japan
Mark Poulter, National Semiconductor, USA
Willy Sansen, Katholieke Univ. Leuven, Belgium
Ulrich Schaper,
Infineon Technologies AG, Germany
Jurriaan Schmitz,
Univ. of Twente, The Netherlands
Dieter Schroder, Arizona State Univ., USA
Franz Sischka, Agilent Home Office, Germany
Giovanni Soncini, IRST, Italy
Lee Stauffer, Keithley Instruments, Inc., USA
Kiyoshi Takeuchi, NEC Corp., Japan
Yoichi Tamaki, Hitachi Ltd., Japan
Hans P. Tuinhout,
NXP-TSMC Research Center, The Netherlands
Bill Verzi, Agilent Technologies, USA
Anthony J. Walton,
University of Edinburgh, UK
Larg H. Weiland, PDF Solutions Inc., USA
Greg M. Yeric, USA

SESSION 1
Matching I – Characterisation

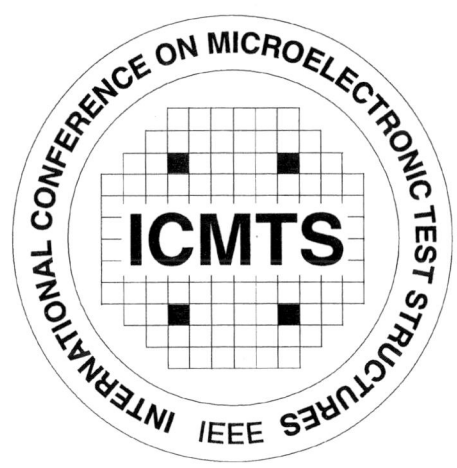

March 25, 09:10–10:30

Co-Chairs: Greg M. Yeric, *USA*

Hans P. Tuinhout, *NXP-TSMC Research Center, The Netherlands*

2008 IEEE Conference on Microelectronic Test Structures, March 24-27, Edinburgh, UK

Operational Amplifier Based Test Structure
For Transistor Threshold Voltage Variation

Brian L. Ji, Dale J. Pearson, Isaac Lauer, Franco Stellari,
David J. Frank, Leland Chang, and Mark B. Ketchen

IBM T. J. Watson Research Center
P.O. Box 218, Yorktown Heights, NY 10598, USA
e-mail: blji@us.ibm.com

ABSTRACT

A new test structure has been developed, which is comprised of MOSFET arrays and an on-chip operational amplifier feedback loop for measuring threshold voltage variation. The test structure also includes an on-chip clock generator and address decoders to scan through the arrays. It can be used in an inline test environment to provide rapid assessment of Vt variation for technology development and chip manufacturing. Hardware results in a 65nm technology are presented.

INTRODUCTION

Silicon CMOS technology faces serious scaling challenges due to increasing variability of device characteristics. Moving into the 65nm node and beyond, device variability plays a major role in chip functionality, yield, and technology optimization. Thorough yet efficient characterization of variability must therefore be implemented in technology development and manufacturing. In particular, FET threshold voltage (Vt) variation is critical to SRAM stability and functionality [1, 2].

Operational amplifier (op-amp) based methods have been previously proposed for simple Vt measurement [3]. Combined with an addressable FET array, such op-amp schemes enable rapid characterization of Vt distributions [4]. To our knowledge, such previously reported works require op-amps on an external test rack and are thus not compatible with inline testers typically available in manufacturing. In this report, we discuss a test structure which has op-amps, clock generation, addressing scheme, and FET arrays all implemented on the same chip. This test structure enables rapid and direct Vt variability measurements suitable for either bench or inline testing.

DESCRIPTION OF TEST STRUCTURE

The gate voltage at which a certain amount of drain-current flows has been used to define the threshold voltage [3]. For example, Vt can be defined as the gate-to-source potential required to drive the threshold drain-to-source current, $Ids(Vt) = Kn*Weff/Leff$ for n-FET, or $Kp*Weff/Leff$ for p-FET, where Kn and Kp are parameters of a given process technology for n-FET

and p-FET, respectively. Fig. 1 shows the circuit schematics for measuring Vt, where the device under test (DUT) is (a) n-FET and (b) p-FET. In Fig. 1(a), V_{set}, V_{ss} and V_{drain} are input terminals; V_{gate} is the op-amp output that is connected to both the gate of DUT and a monitoring digital voltmeter (DVM). The source voltage, V_{source}, is set by and equal to V_{set} via the op-amp loop. The bias current is equal to (V_{set}-V_{ss})/R_{load}, where R_{load} is the resistance of a precision load resistor. This current is set to be equal to the threshold current value and fed to the source terminal of DUT. Finally, V_{gate} is automatically adjusted by the op-amp feedback to maintain the threshold current, thus becomes a direct Vt measure. Fig. 1(b) shows the Vt measurement setup for a p-FET.

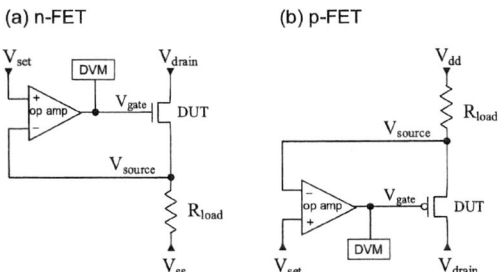

Fig. 1. Circuit schematics for Vt measurement, where the device under test (DUT) is (a) n-FET and (b) p-FET.

A two-dimensional addressable array of nominally identical devices is used for the measurement of Vt variation, as shown in Fig. 2. CMOS transmission gate switches are used to direct the threshold current through one FET at a time. The transmission gates are large enough to have negligible voltage drop at the current level for Vt measurement. As shown in Fig. 2, the gate terminals of FETs in a selected row are connected to the op-amp output V_{gate}, while the gates in unselected rows are turned off. Exemplary implementations of column switches are also shown. Fig. 2(a) shows that the drain terminals of n-FETs in a selected column are connected to V_{drain} through column switches, while Fig. 2(b) shows that the column switches are used at the source terminals of p-FETs. In practice, column switches may be used at either the source or the drain terminals as long as most of the current is steered through the one FET under test. Although this usually is not a problem for a small array, it could be a concern when scaling up the array

978-1-4244-1801-5/08/$25.00 ©2008 IEEE

size. The optimum solution to minimize the unintended current paths will depend on the sub-threshold leakage current, gate-oxide leakage current, transistor sizes and bias conditions for both DUTs and switches.

This test structure was first implemented and demonstrated in 90 nm technology [5]. The current work describes an improved design and hardware results in an experimental 65 nm PDSOI technology.

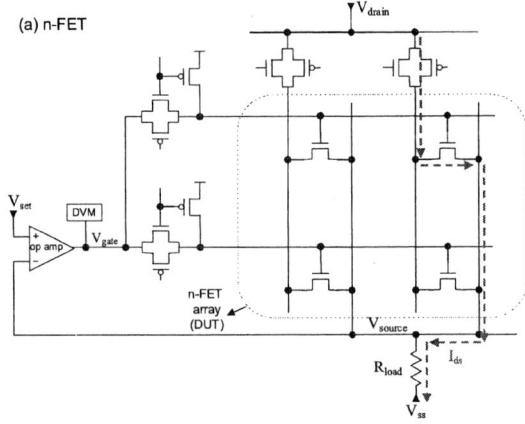

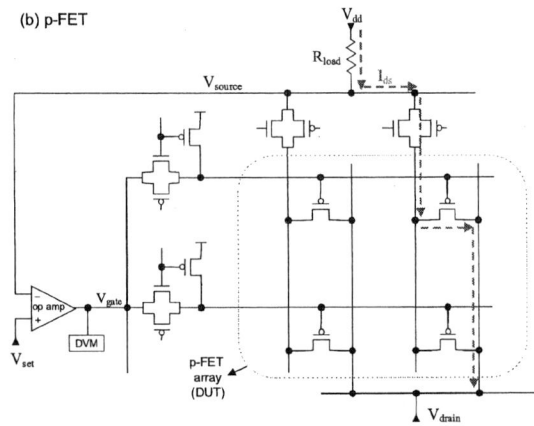

Fig. 2. High level schematic of the op-amp based test structure for local (stochastic) Vt variation measurement. (a) n-FET DUT array. (a) p-FET DUT array.

Fig. 3 shows the circuit schematic of a conventional two-stage op-amp and its bias current generation circuit [6] that is implemented in this test structure. The internal bias block is implemented with current mirror circuitry and a 1 kΩ resistor (designed to carry 40 μA nominal current). This op-amp has a 5 pF compensation capacitor, which eliminates voltage oscillations of the op-amp output during probe test, as it matches to the capacitive load of a probe pad ranging from 2 to 10 pF. From transient response simulations at various process corners, the op-amp settling time is about 10 to 20 ns when a better than 0.3 mV accuracy is achieved. The operating voltage range of this op-amp topology is more suitable for n-FET test when the op-amp shares the voltage supply

and ground with the transistor arrays. For p-FET arrays, therefore, the op-amp uses dedicated voltage supply and ground pads, thus to eliminate voltage constraints at the cost of two additional pads. Alternatively, an op-amp with the complementary topology could be designed for p–FET arrays.

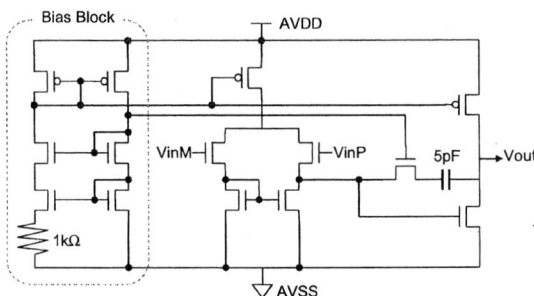

Fig. 3. Schematic showing two-stage op-amp with internal bias control and 5pF matching capacitor.

An on-chip clocking and addressing scheme is implemented for test automation and to minimize the number of probe pads. This addressing scheme has been previously reported by M. Bhushan et al. [7] for use in inline testing. As shown in Fig. 4, an internal clock is generated by a ring oscillator (RO) followed by a frequency divider. This RO clock then drives address counters and row and column decoders. The decoders control the row and column transmission gate switches (as shown in Fig. 2) and enable sequential measurements over an array of nominally identical transistors. Available clock frequencies at about 10 Hz and 40 Hz are designed and selectable using a clock-period-selection pad (not shown). At these low frequencies, transient behavior due to SOI floating body effects is avoided. Alternatively, the CLKSEL signal allows the use of a low frequency external clock (EXCLK) to replace the RO clock. A voltage signal synchronized with the counter output, TRIGOUT, can be used as a trigger to an external digital voltmeter (DVM).

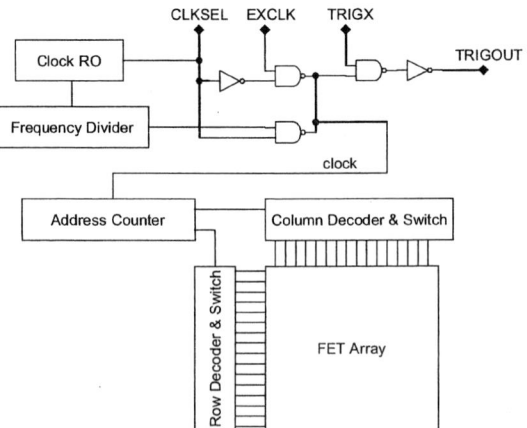

Fig. 4. Block diagram showing clock generation path (with internal and external clock options), address counter, row and column decoders and switches, and a 16x16 FET array.

4

In the hardware reported in this work, the arrays have 256 devices in 16 rows and 16 columns. They are contained in a small area, with surrounding dummy devices to eliminate variation due to the layout environment. Fig. 5 shows the layout of the operational amplifier, load resistor (i.e., Rload in Fig. 1 and Fig. 2), and a 16×16 FET array. Rload is typically designed for a 300 mV voltage drop at the threshold current for a given DUT. The load resistor shown in Fig. 5 is about 90 kΩ to accommodate n-FETs with channel width of 0.38 μm. Larger resistance values are used for p-FET test structures due to the smaller Ids used in the p-FET Vt definition.

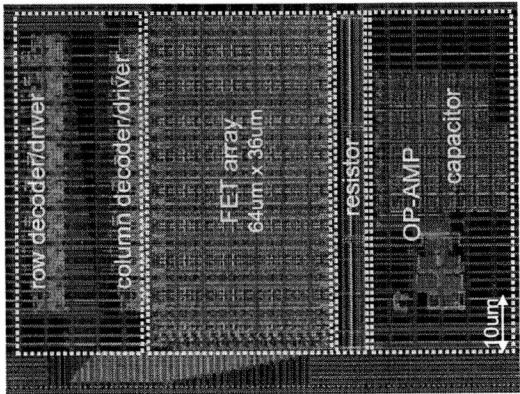

Fig. 5. Layout view of row and column decoders, an FET array, a high precision load resistor, and an op-amp with a 5pF capacitor.

EXPERIMENTAL RESULTS

We have tested hardware manufactured using an experimental 65nm PDSOI technology. Referring to DVM shown in Figs. 1 and 2, the op-amp output is measured using a high speed HP3458A voltmeter. A LabWindows-based program, running on a nearby computer connected to the instrument with a GPIB cable, is used to collect the data with up to the maximum number of samples allowed by the DVM memory (i.e. 4,000) for each data transfer. The DVM could be triggered either by the synchronized output TRIGOUT from the chip, or directly from the TRIG signal of an external clock source when the EXCLK option is used (see Fig. 4). The external clock was supplied by an HP33120A Function/Arbitrary Waveform Generator.

Fig. 6 is the waveform measured using an over-sampling technique (at 600 Hz) to record the detailed waveform shapes when the chip is operated with the internal RO clock of about 44 Hz. Fig. 6 shows the voltage-modulated output signal repeated at time intervals of 256×T as the macro scans through the 256 devices in the array, where T is the clock period for address switching. The repeatability after 256×T and the time stability of voltage levels between transitions are shown. The EXCLK option using an external clock is also verified. Figs. 7 to 11 are measurements

in EXCLK mode, with the HP3458A triggered at the clock rate and the HP33120A providing both EXCLK (at 40 Hz) to the chip and the TRIG signal to the HP3458A.

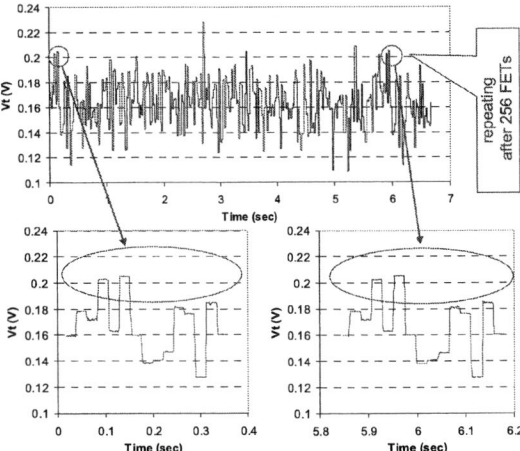

Fig. 6. Voltage modulated output signal repeated at time interval of 256 T, where T is the clock period for addressing.

In addition to the fast measurement time, the op-amp-based method has an advantage that it can be used at any bias condition. Vt and Ids measured as a function of Vds for a single p-FET is shown in Fig. 7. Note that Ids is well regulated at the required constant level by the op-amp. By enabling the addressing clock to switch through all FETs in an array at each Vds step, Vt variation can be easily measured as a function of Vds.

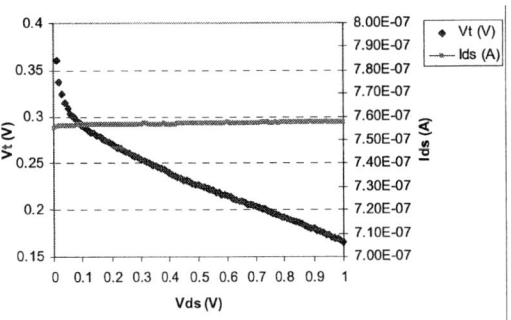

Fig. 7. Vt measured as a function of Vds for 1 PFET. Here the measured Ids is regulated at a constant level by the op-amp loop.

Fig. 8 shows a Vt vs. Vds measurement on a p-FET array consisting of 256 devices. It also shows Vt histograms at linear and saturated regions, with Vds of 0.05V, 0.7V and 1V, respectively.

Exemplary hardware data of Vt mean and standard deviation (σVt) as a function of Vds measured on a p-FET array and an n-FET array are shown in Fig. 9. The measured linear σVt value (i.e., at Vds of 0.05V) of about 15 mV fits well with an estimate based on intrinsic random-dopant-fluctuation. However, significant Vds bias dependence is observed. The measured σVt increases with Vds and reaches a plateau at Vds~0.5V. As shown in Fig. 9, σVt in the saturated region is 40% to 50% higher than that in the linear region. This behavior likely reflects the interplay between DIBL and PDSOI floating body effects. Such Vt bias dependence has a large impact on SRAM stability margins and cell designs. A more detailed analysis of this impact will be published elsewhere.

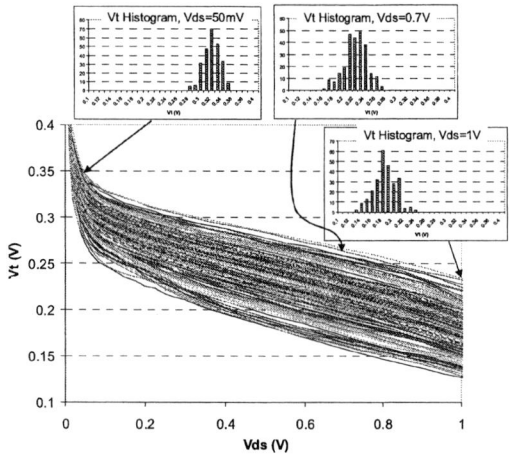

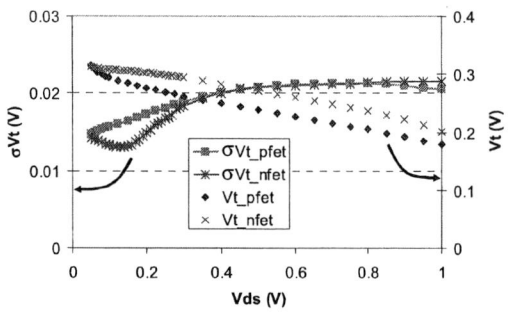

Fig. 8. Measured Vt versus Vds curves for 256 p-FETs and Vt histograms at Vds of 0.05V, 0.7V, and 1V, respectively.

Fig. 9. Exemplary Vt-mean and σVt as a function of Vds measured on a p-FET array and an n-FET array.

The op-amp test structure can be used for both systematic chip-to-chip and local variation of Vt. In Fig. 10, exemplary wafer-level test results show that the chip-mean Vt has a systematic dependence on the wafer location: the center of the wafer has higher Vt. On the other hand, σVt follows a more random pattern across the wafer, which is consistent with the view that local Vt variation is due to random dopant fluctuation. Across the wafer, the mean value of the measured local σVt is 20.8 mV; the standard deviation of this measurement over 89 chips on the wafer is 1.2 mV.

An aggregate statistical analysis of the local component of Vt variation is done for the same wafer and shown in Figs. 11-12, where a normalization method (by subtraction of the median Vt per chip) used to eliminate the systematic chip-to-chip Vt variation. Fig. 11 shows a histogram of 22784 samples from 256 n-FETs per chip and 89 chips per wafer. Fig. 12 is a normal Q-Q plot ("Q" stands for quantile) which shows that the local Vt variation measured on this wafer is normally distributed to 4σ. The tail characteristics of local Vt mismatch are critical for yield analysis of large SRAM arrays.

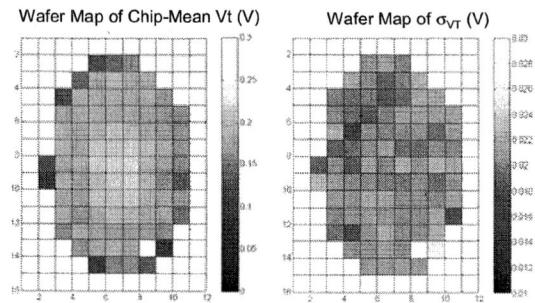

Fig. 10. Wafer map of chip-mean Vt and σVt, measured on n-FET arrays at Vds = 0.7V.

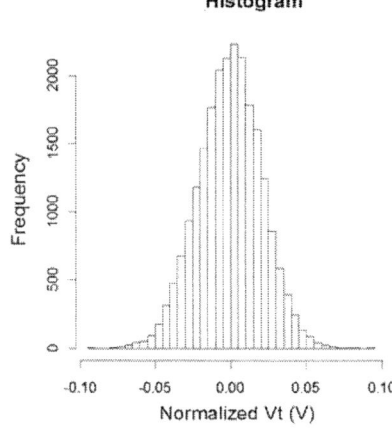

Fig. 11. Histogram of measured Vt for 22784 devices. Vt is normalized by subtraction of the median Vt per chip.

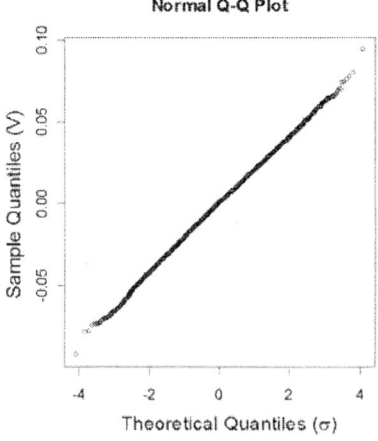

Fig. 12. Normal Q-Q plot. Vt is normalized by subtraction of the median Vt per chip.

SUMMARY

We have designed and demonstrated a test structure for characterization of MOSFET threshold voltage variations. With an on-chip operational amplifier, clock, and addressable array, a method for direct readout of Vt variation is implemented. This leads to simple and efficient bench and inline testing, and can provide a very rapid measurement of Vt variability. 65nm hardware results demonstrate that the scheme can be used to study local Vt statistics and the impact of bias conditions, and across-wafer variation.

ACKNOWLEDGEMENT

We thank Manjul Bhushan and Wilfried Haensch for insightful discussions.

REFERENCES

[1] A. J. Bhavnagarwala, X. Tang, J.D. Meindl, "The impact of intrinsic device fluctuations on CMOS SRAM cell stability", IEEE J. Solid-State Circuits, 36, p658 (2001).

[2] L. Chang, et al, "Stable SRAM cell design for the 32 nm node and beyond", 2005 Symposium on VLSI Technology, Digest of Technical Papers, pp. 128-129.

[3] H. Lee, S. Oh, and G. Fuller, "A simple and accurate method to measure the threshold voltage of an enhancement-mode MOSFET", IEEE Trans. Electron Devices, 29, p346 (1982).

[4] K. Agarwal, S. Nassif, F. Liu, J. Hayes, K. Nowka, "Rapid characterization of threshold voltage fluctuation in MOS devices", Proceedings of 2007 International Conference on Microelectronic Test Structures, ICMTS-2007, pp. 74-77.

[5] D. J. Pearson, unpublished.

[6] D. A. Johns and K. Martin, "Analog Integrated Circuit Design", John Wiley & Sons, New York, 1997.

[7] M. Bhushan, M.B. Ketchen, S. Polonsky, and A. Gattiker, "Ring oscillator based technique for measuring variability statistics", Proceedings of 2006 International Conference on Microelectronic Test Structures, ICMTS-2006, pp. 87-92.

A Study of Variation in Characteristics and Subthreshold Humps for 65-nm SRAM Using Newly Developed SRAM Cell Array Test Structure

A. Mizumura, T. Suzuki, T. Arima*, H. Maeda*, and H. Ammo

Semiconductor Technology Development Division, Semiconductor Business Group, Sony Corp.
*Sony LSI Design Incorporated
4-14-1, Asahi-cho, Atsugi-shi, Kanagawa, 243-0014, Japan
Phone: +81-46-202-4590, Fax: +81-46-230-5572, E-mail: akira.mizumura@jp.sony.com

ABSTRACT

Variation in the characteristics (Vth and Ion) of driver and load MOSFETs in SRAM cells was investigated by using a newly developed SRAM cell array test structure and is reported here for the first time. Subthreshold humps were also monitored in SRAM cells. By extracting the variations in MOSFET characteristics and by estimating the impact of subthreshold humps in SRAM cells, it is possible to design SRAM circuits more accurately.

INTRODUCTION

Continuous down-scaling has made it difficult to design SRAM circuits with sufficient operating margins. Although they have been designed with a large amount of SRAM cell data such as read/write-operations that are measured with SRAM test circuits [1], they have still been designed using the characteristic data (Vth, Ids, etc.) of MOSFETs with just a few cells, which are measured with a small-scale test element group (TEG), and this leads to difficulties in designing SRAM circuits accurately.

However, MOSFET arrays have been used in logic MOSFETs to gain an understanding of the variation in MOSFET characteristics [2-4]. In this paper, we applied this MOSFET array technique to 65-nm SRAM for the first time, and showed the distribution of data of SRAM cell MOSFET characteristics such as Vth and Ion.

With this test structure, we also found subthreshold humps in cell MOSFETs as well as logic MOSFETs [2-3]. We developed an algorithm that can automatically determine whether humps exist from "subthreshold slope - Vg" curves. We showed for the first time the hump occurrence rate in cell MOSFETs and the impact on matching characteristics for cell MOSFETs.

EXPERIMENTS

Driver and load MOSFETs for 65-nm SRAM were measured. Fig. 1 shows an SRAM cell array with an area of about 500 x 1270 um for each driver and load MOSFET. Each cell array contained 125 device under test (DUT) cells, each of which contained a pair of DUT MOSFETs. As a result, the matching characteristics of 125 pairs of MOSFETs in the DUT cells can be measured in addition to the characteristics of 250 individual MOSFETs. The DUT cells are

placed at regular intervals (about 106 um in the horizontal direction and 25 um in the vertical direction). The other areas of the array except for the DUT cells contain dummy cells that have only active and gate patterns.

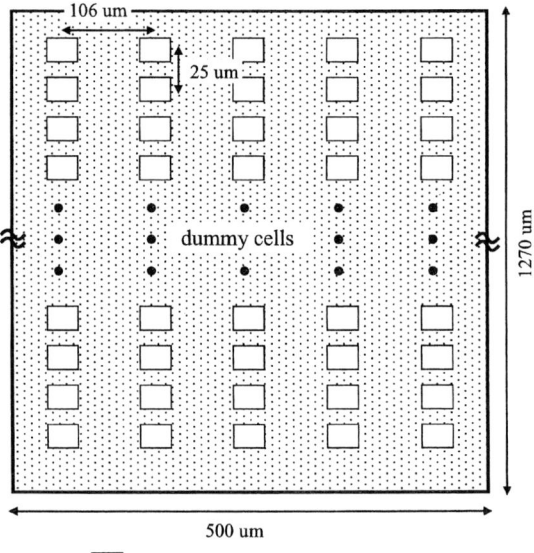

Fig. 1. SRAM cell array area.

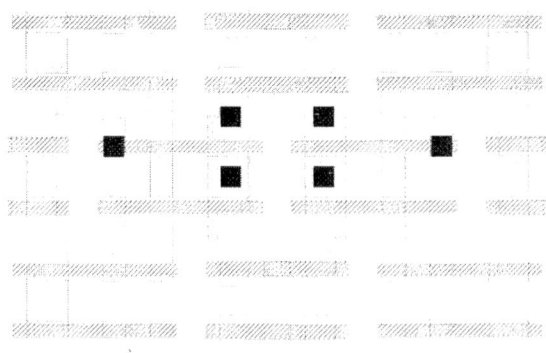

Fig. 2. Pattern layout for driver DUTs in an SRAM cell. Active, poly, and contact patterns are shown.

Fig. 2 shows the pattern layout for two driver DUTs in a cell. The active patterns were partially cut in a DUT cell in order to prevent unexpected leakage currents. The optical simulation was applied when cutting the critical pattern layout. Although shared contacts are used in actual SRAM cells, they were replaced with normal ones in order to measure the DUT MOSFET characteristics exclusively.

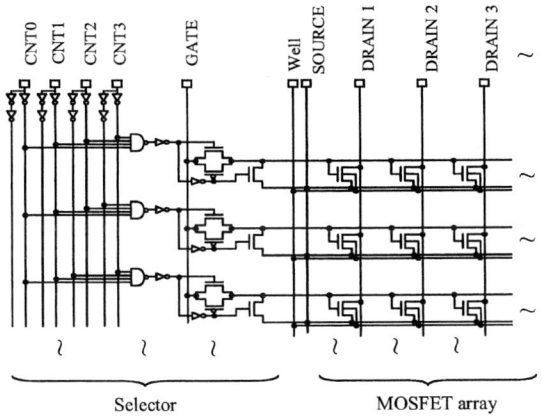

Fig. 3. Circuit diagram for SRAM cell array (driver DUT).

Fig. 3 is a circuit diagram of the test structure [2-3]. The desired gate line is active through control pins, the gate bias is swept (inactive gate lines remain at the source voltage), and the drain current is measured through the required drain pin. The number of DUTs connected to one drain line was limited to five so that subthreshold currents could be observed clearly; this is because the sum of off-leakage currents of DUTs connected in the same drain line disturbs the measured subthreshold currents.

The measurement was carried out in an automatic wafer prober with a parametric tester (Agilent 4073B). The drain current was measured under saturation conditions at Vd = 1 V, Vg = 0 V to 1 V, Vb = 0 V.

RESULTS AND DISCUSSION

Fig. 4 plots examples of the distribution of Vth and Ion for 8000 driver and load MOSFETs in a wafer, which reveals normal distribution. These results can make it possible to more accurately design SRAM circuits despite small operating margins.

The proposed test structure can provide information on subthreshold humps in MOSFETs in SRAM cells because subthreshold currents can be clearly observed.

Fig. 5(a) plots an example of a subthreshold hump that was observed in a driver MOSFET array. Fig. 5(b) plots the subthreshold slope as a function of gate voltage for the sample in Fig. 5(a). A small "hump" appears in the graph, because the subthreshold hump changes the subthreshold slope. In previous studies [2-3], the number of subthreshold humps was determined by manually counting the number of these "humps" in

the graph, i.e. with human eyes. However, Brut and Velghe [5] reported the automatic hump detection using transconductance (Gm). In this study we developed a different algorithm that can automatically recognize the existence of humps as follows. Graphs with a "hump" have inflection points; therefore, we developed an algorithm that can numerically monitor the inflection points in the graph.

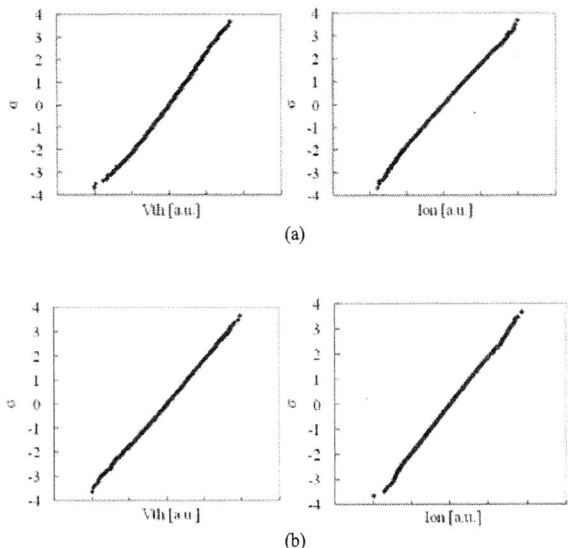

Fig. 4. Distribution of Vth and Ion for SRAM cell array.
(a) Distribution of Vth and Ion for 8000 driver MOSFETs.
(b) Distribution of Vth and Ion for 8000 load MOSFETs.

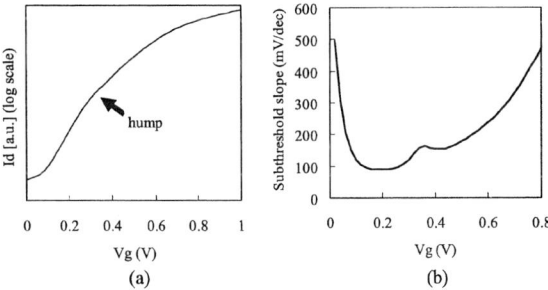

Fig. 5. Example of hump occurrence in driver MOSFET.
(a) Subthreshold hump observed in Id-Vg curve.
(b) Subthreshold slope as a function of gate voltage.

There are two cases in which subthreshold humps occur. One is where both MOSFETs in a MOSFET pair in an SRAM cell have humps, and the other is where only one of them has a hump. We evaluated the rates for the two cases.

Fig. 6(a) shows the hump occurrence rates for 8000 driver MOSFETs. The left graph in Fig. 6(a) shows the rate of MOSFETs with humps out of 8000 MOSFETs, and the graph on the right indicates the rate at which humps were observed in MOSFET pairs

9

in a cell among MOSFETs with humps. Humps were observed in 5.0% of 8000 MOSFETs, but only 3.5% of the MOSFETs with humps were monitored in the same cell, although they were positioned close together. Fig. 6(b) shows the same hump occurrence rates for 8000 load MOSFETs. Humps were observed in 0.9% of 8000 MOSFETs, and all humps were monitored in different cells. These results for driver and load MOSFETs indicate the strong randomness of subthreshold hump occurrence.

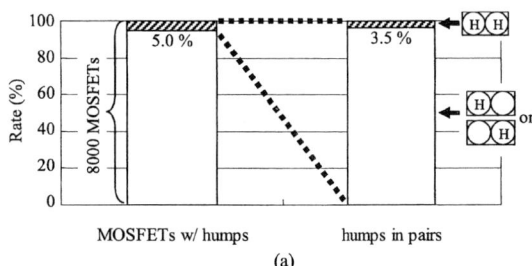

(a)

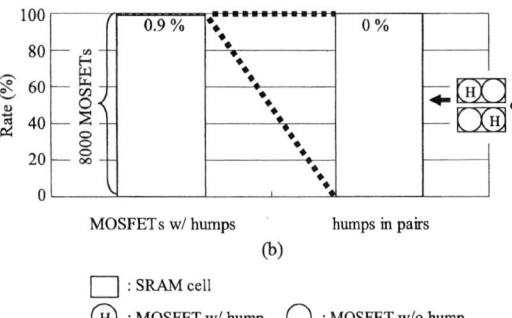

(b)

☐ : SRAM cell

(H) : MOSFET w/ hump   ◯ : MOSFET w/o hump

Fig. 6. Rates at which humps occurred in 8000 MOSFETs and those at which humps occurred in pairs among MOSFETs with humps. (a) Driver MOSFETs (b) Load MOSFETs

Concerning the driver MOSFETs whose rate of subthreshold hump occurrence was larger than the load ones, the impact of the subthreshold humps on matching characteristics was calculated for driver MOSFET pairs in a cell. Table I gives the comparison of σ (ΔVth) for driver MOSFET pairs with and without humps, where ΔVth denotes the Vth difference between the MOSFET pairs in an SRAM cell. It revealed that the subthreshold humps increase

σ (ΔVth) by 21%, which demonstrates that we should pay attention to the impact of humps as well as to the variation in MOSFET characteristics when designing SRAM circuits.

TABLE I

HUMP IMPACT ON VTH MATCHING FOR PAIR DRIVER MOSFETs IN AN SRAM CELL. ΔVTH DENOTES THE VTH DIFFERENCE BETWEEN MOSFET PAIRS IN AN SRAM CELL.

	without humps	with humps
σ (ΔVth)	43 mV	52 mV

CONCLUSION

Distribution of driver and load MOSFET characteristics (Vth and Ion) in SRAM cells was reported for the first time by using a newly developed SRAM cell array test structure. Subthreshold humps were monitored in SRAM cells with an algorithm that can automatically recognize their occurrence. Strong randomness in the occurrence of subthrehshold humps was revealed. The impact of subthrehshold humps on matching characteristics for driver MOSFETs in a cell is not negligible, which increases σ (ΔVth) by 21%.

REFERENCES

[1] M. Khellah, et al., "Effect of power supply noise on SRAM dynamic stability," in Proc. VLSI Circuits, 2007, pp. 76-77.

[2] A. Mizumura, et al., "A study of 90nm MOSFET subthreshold hump characteristics using newly developed MOSFET array test structure," in Proc. ICMTS, 2005, pp. 39-42.

[3] A. Mizumura, et al., "Study of 90-nm MOSFET subthreshold hump characteristics using newly developed MOSFET array test structure," IEEE Trans. Semiconductor Manufacturing, vol. 19, No. 1, pp. 19-26, Feb. 2006.

[4] T. Mizuno, J. Okamura, A. Toriumi, "Experimental study of threshold voltage fluctuations using an 8k MOSFET's array," in Proc. VLSI Tech., 1993, pp. 41-42.

[5] H. Brut, R.M.D.A. Velghe, "Contribution to the characterization of the hump effect in MOSFET submicronic technologies," in Proc. ICMTS, 1999, pp. 188-193.

Mismatch Characterization of a High Precision Resistor Array Test Structure

Weidong Tian, Philipp Steinmann, Eric Beach, Imran Khan and Praful Madhani

Texas Instruments Inc.
Dallas, Texas, USA
wtian@ti.com

ABSTRACT

In this paper we describe a 2x (2xn) pad resistor array test structure. The structure can also be used for analog CMOS, bipolar, capacitor arrays. It provides a flexible and convenient way to study devices in an array. We will focus on various matching properties of the precision thin film resistors, specifically, between any one of the resistors with the average of the whole array, between two resistors (neighbor and far away pair), and between resistor combinations in the array. Both random and systematic mismatch properties are studied. We illustrate key factors for resistor matching from both process and layout perspectives. The principles, analyses and main conclusions apply to other active and passive devices as well. The array layout can be copied directly from and into a circuit design.

INTRODUCTION

Matched precision device arrays are widely used in many high performance analog applications, such as thermometer-code or R-2R ladder resistor arrays in digital-to-analog converters (DACs), binary weighted capacitor arrays in successive approximation analog-to-digital converters (ADCs), and CMOS/bipolar transistor arrays in current-steering high speed DACs [1-3]. For these products, the accurate characterization and modeling of array mismatch are critical.

In this paper we describe a device array test structure for the array matching characterization. The structure is based on the traditional 2xn pads configuration. However, multiple 2xn pads are used to expand necessary array connections. An example of two (2x10) pads structure is illustrated in Fig.1. We designed our structures for high-performance parametric probe stations as well as manual testers. Based on statistical data, both random and systematic mismatch analysis can be performed. These approaches are different from that of H.P. Tuinhout et al [4,5], in which the array was used in bench measurements for an array systematic mismatch study. In addition, we reduced the force current pads from 2xn in reference [4,5] to 2x2. This results in a more compact structure because more pads can be assigned to the devices in the structure.

The structure provides a flexible and convenient way to study matched device arrays used in ADCs and DACs. In this paper we will focus on various matching properties of a precision thin film resistor (TFR) array. CMOS, bipolar, capacitor arrays can be constructed and analyzed similarly. The mismatch properties studied include mismatch between any one of the resistors with the average of the array, between two neighboring resistors, between one resistor and any other one in the structure, and between resistor combinations in the array. Both random and systematic mismatch properties are discussed. In addition, we illustrate key factors for resistor matching, from both process and layout perspectives. Again the principles and main conclusions apply to other active and passive device arrays as well.

TEST STRUCTURE AND CHARACTERIZATION METHOD

2xn pads testing structures are widely used in semiconductor device/process development and manufacturing [6]. By simply sandwiching a standard structure into multiple 2xn pads, the resistors array test structure can be constructed. The resistor array can have different configurations, such as 2x (n-2), 4x (n-6) etc.. In Fig.1, a 2x8 array is shown for a 2x (2x10) pad frame. In the array, each resistor can be tested separately by the conventional 4-terminal (Kelvin) measurement method without a switching circuit. The layout can be copied partially or fully, from or into, an ADC or DAC circuit.

The test structures were fabricated in a high-performance CMOS technology. A SiCr TFR was chosen for our study. Compared to other resistor technologies, this resistor offers a low temperature coefficient (TC) of resistance, tight TC tracking, and excellent long-term stability, all of which are essential for a precision ADC or DAC.

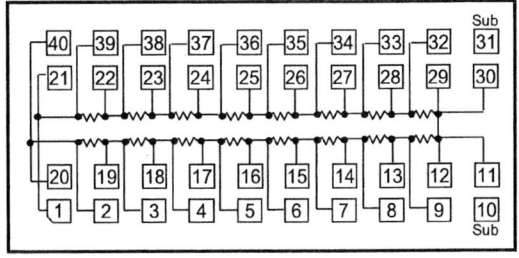

Fig.1 a 2x(2x10) pads resistor array test structure

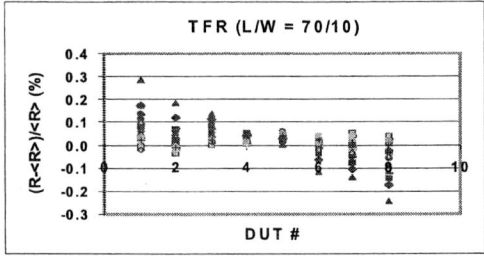

Fig.2 TFR array normalized resistance on a typical wafer

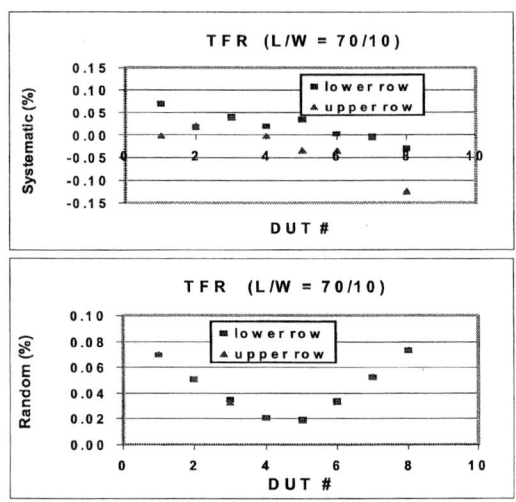

Fig.3 systematic and random mismatch of an array

A parametric test station such as Keithley S630 is used to collect statistical data. Multiple silicon lots with multiple wafers in each lot were manufactured and tested. After taking the data, gross measurement errors and/or defects are analyzed. Once outliers are removed, statistical properties such as average, standard deviation and correlation are calculated. As we can see below, these quantities define all mismatch properties.

A RESISTOR IN AN ARRAY

In many applications such as in a high-precision voltage divider, the absolute value of resistance is not too important as long as the matching of the resistors within an array is good. A normalized random variable $r_m = (R_m - <R_m>_m)/<R_m>_m$ can be defined to evaluate this property by calculating the resistance matching of each individual resistor (R_m, where m is the resistor number index) with respect to the average resistance value of the array it belongs to ($<R_m>_m$).

Although each resistor is laid out in exactly the same way in the array, r_m varies due to a superposition of many interdependent processes operating both at the local and global levels. These include photolithography variations, localized etch rate non-uniformity, long range film thickness and sheet resistance variations often resulting from temperature gradients over a wafer, and contact resistance variations. The effect can be illustrated by Fig.2, in which silicon data from one of our structures in a typical wafer is shown. Each curve in the plot is the resistance of each unit resistor in the array on one die, normalized to the average of the array. As the figure shows, in this technology, the spatial dependence of the array is approximately a liner gradient function $r_m = r_{m0} + a \ast x_m$, where r_{m0} and slope a are random variables.

To evaluate statistical significance, we need to know its systematic and random mismatch. By definition the systematic mismatch is measured by the sample's average representing the deterministic effects in the device, whereas random mismatch is measured by standard deviation representing the stochastic effects.

In Fig 3, they are plotted for a uniform distributed TFR array that we studied. Systematic and random mismatch show approximately a linear dependency of the resistor location. The reason is obvious: if the resistance coordinate dependence due to process gradient effects is approximately $r_m = r_{m0} + a \ast x_m$, then the systematic and random mismatch of r_m will be:

$$<r_m> = <r_{m0}> + <a> \ast x_m \qquad (1)$$
$$\sigma(r_m) = [\sigma^2(r_{m0}) + \sigma^2(a) \ast x_m^2]^{1/2} \qquad (2)$$

Systematic mismatch can be impacted in numerical ways. In our study we found a systematic mismatch difference among different test chips with the same TFR structures. In Fig.4, the systematic mismatch difference between two resistors in a small resistor array is plotted for device A and B. Since the process is same for two devices, we suspected that the difference came from two sets of TFR photo masks used. To confirm, we measured the 5x TFR photo mask geometry. It turns out that the systematic mismatch error on the photo mask and the electrical measurements indeed agree with each other. We conclude that photo masks can cause systematic mismatch device dependence.

Random mismatch, on the other hand, measures random variation of the array spatial linearity in the fabrication process as mentioned early. As Fig.3 shows it goes higher at both ends of the array (i.e. magnified by the factor x_m) just as Equ. 2 described.

In general, normalized random variable r can be written as a function of position in a two dimensional array as $r_0 + r_x \ast x + r_y \ast y + r_{xx} \ast x^2 + r_{xy} \ast x \ast y + r_{yy} \ast y^2$, where the x*y term is due to process gradients other

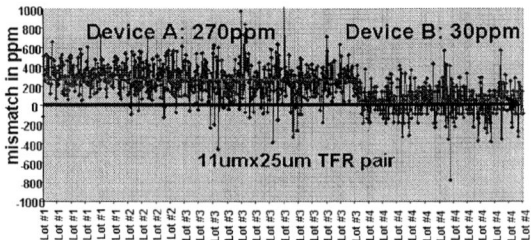

Fig.4 systematic mismatch device dependence

12

than simple parabolic with main axis aligned along the x and y directions. Thus:

$$<r> = <r_0> + <r_x*x> + <r_y*y> + <r_{xx}*x^2> + <r_{xy}*x*y> + <r_{yy}*y^2> + ... \qquad (3)$$

$$\sigma(r) = [\sigma^2(r_0) + \sigma^2(r_x*x) + \sigma^2(r_y*y) + \sigma^2(r_{xx}*x^2) + \sigma^2(r_{xy}*x*y) + \sigma^2(r_{yy}*y^2) + ...]^{1/2} \qquad (4)$$

This information gives designers a feel of how the voltage (or current) is statistically distributed across and sheds some light on circuit design tolerance. The information also helps designers to locate the matching frontier for two or more arrays matching, as discussed later in the last section

MISMATCH BETWEEN A RESISTOR PAIR

Assume the expectation value of a fixed geometry unit resistor in terms of its lateral coordinates due to gradient effects is approximated by:

$$R(x,y)= R_0 + R_x*x + R_y*y + R_{xx}*x^2 + R_{xy}*x*y + R_y*y^2 \qquad (5)$$

where R can be either absolute or normalized resistance value. For any two matched resistors R_1, R_2 in the array, the resistance difference between them is written as:

$$R_1-R_2 = (R_{10}-R_{20}) + R_x*(x_1-x_2) + R_y*(y_1-y_2) + R_{xx}*(x_1^2 - x_2^2) + R_{xy}*(x_1y_1-x_2y_2) + R_{yy}*(y_1^2 - y_2^2) \qquad (6)$$

(x_1, y_1) and (x_2, y_2) are the coordinates of two resistors.

Once resistance data set $\{R_1\}$, $\{R_2\}$ obtained, the resistor pair system and random mismatch can be estimated from the average and standard deviation of $\{R_1-R_2\}$. As Equ. 6 implies, both systematic and random mismatch have a local component, which is not related to spatial gradients, and a global component, which is spatial coordinate dependent. Mismatch of a neighboring pair is local component dominated, whereas a far away pair is global component dominated. In following we will look into both components of the resistor random mismatch. Systematic mismatch can be discussed similarly – as a matter of fact, since systematic mismatch does not involve pair correlation, the analysis is more straight forward.

A Neighboring pair random mismatch

If the standard deviations of R_{10}, R_{20} are σ_1, σ_2, their respective correlation is κ, and spatial gradients between two resistors can be neglected, the random mismatch of a neighbor pair is

$$\sigma(R_1-R_2)= \{\sigma_1^2+\sigma_2^2-2\kappa \sigma_1\sigma_2\}^{1/2} \qquad (7)$$

Usually the matched resistors have a similar variation $\sigma_1 \approx \sigma_2 = \sigma$. Under this condition, Equ. (7) can be simplified to:

$$\sigma(R_1-R_2)= \sqrt{2}*\sigma*\sqrt{(1-\kappa)} \qquad (8)$$

i.e. the resistor random mismatch is proportional to σ and $\sqrt{(1-\kappa)}$. The minimum value of $\sqrt{(1-\kappa}$ [thus $\sigma(R_1-R_2)$] is 0 when $\kappa = 1$ (both resistors vary simultaneously with the same magnitude and direction

thus are perfectly matched). If $\kappa = 0$ (two resistors change independently), $\sigma(R_1-R_2)= \sqrt{2}*\sigma$.

It is well known that the resistance random mismatch is proportional to the standard deviation of each resistor. In this paper, we will address the importance of correlation κ.

Fig. 5 shows the random mismatch of a resistor pair (70um long and 10 um wide) as a function of $\sqrt{(1-\kappa)}$ and σ. All random mismatch, σ and $\sqrt{(1-\kappa)}$ are calculated from statistical data collected on the structure as described in Fig.1. $\sigma(R_1-R_2)$ shows a strong linear dependence of $\sqrt{(1-\kappa)}$ in Fig.5a. From Fig. 5b, it is clear that the resistors' random mismatch $\sigma(R_1-R_2)$ can be reduced by several orders of magnitude than σ due to the correlation factor $\sqrt{(1-\kappa)}$, where $\kappa \neq 0$. Should $\kappa=0$ be assumed, random mismatch $\sigma(R_1-R_2)=\sqrt{2}\sigma$ is orders of magnitude over-estimated, and penalty has to be paid in a design in terms of device area and cost.

B Far away pair random mismatch

As the distance between two matched resistors increases, the mismatch error increases as well. This is confirmed in Fig. 6a, where the random mismatch between the first resistor and each of other resistors in the row is plotted from silicon data. The figure shows that the mismatch error increases linearly with the distance. This is the signature of the technology used for our test chip. In general, the random mismatch between two resistors is:

$$\sigma(R_1-R_2)=\{\sigma_1^2+\sigma_2^2-2\kappa\sigma_1\sigma_2+\sigma^2[R_x*(x_1-x_2)]+\sigma^2[R_y*(y_1-y_2)]+\sigma^2[R_{xx}*(x_1^2-x_2^2)]+\sigma^2[R_{xy}*(x_1y_1-x_2y_2)]+\sigma^2[R_{yy}*(y_1^2 - y_2^2)] \}^{1/2} \qquad (9)$$

Clearly the mismatch distance dependence is partially due to spatial gradients (R_x, R_y, R_{xx}, R_{xy}, R_{yy} terms) as discussed in the last section, partially due to resistor correlation κ distance dependence.

Taking spatial gradient effect into consideration, the distance dependence of correlation can be calculated

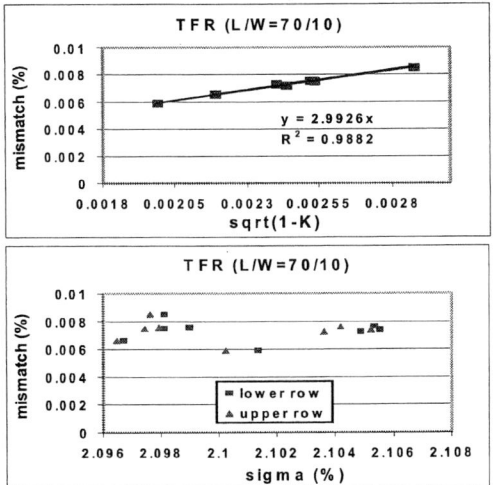

Fig.5 a resistor pair random mismatch as a function of (a) $\sqrt{(1-\kappa)}$ and (b) σ

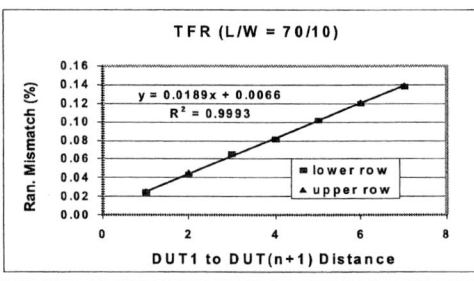

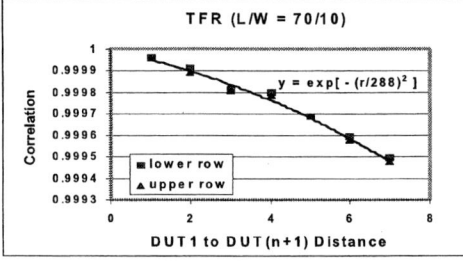

Fig.6. (a) σ vs. distance, (b) κ vs. distance

form Equ.9. The result from our test chip is presented in Fig.6b. As it shows, correlation decreases exponentially with the square of the distance between two resistors:

$$\kappa = \exp[-(r/r_0)^2] \qquad (10)$$

where r_0 can be defined as correlation radius. Essentially correlation between two resistors is the highest at their minimum space, reduces dramatically when resistors are apart, and becomes negligible if the distance is beyond r_0.

Correlation radius r_0 is strongly device dependent. Active devices such as CMOS, bipolar transistors tend to have much smaller r_0 than passive devices such as resistors and capacitors, due to their small sizes and thermal diffusion processes involved. r_0 is also technology dependent. In the actual practice of fabrication, correlation radius r_0 is improved by local variable uniformity control, such as material or equipment induced spatial variation reduction. For a given r_0, layout matching improvement can be done, for example, by adjusting matched device geometry and their arrangement so that they are located within r_0, or better yet, at their minimum design rule space.

RESISTOR SUB-ARRAY MISMATCH

In many high precision small signal applications, matched resistors are often laid out in a form of two or more sub-arrays, with each consisting of two or more identical unit resistors. Without losing generality, we will look at any two sub-arrays first. Starting with an array with the linear spatial gradient function, we will study its systematic and random mismatch. Next two sub-arrays with the general spatial gradient function as Equ.6 described is presented. In the last we will discuss briefly the general case, namely the mismatch of multiple sub-arrays with each consists of multiple unit resistors in an MxN array.

A   Sub-array mismatch: linear gradients

Assume the resistance of each sub-array (i=1,2) is:

$$R_i = \Sigma R_{ij} + R_x * \Sigma x_{ij} + R_y * \Sigma y_{ij} \qquad (11)$$

where summation is over all unit resistors in the sub-array (j=1…N). As above equation implies, a good layout of a resistor sub-array should be able to cancel out spatial process gradients (R_x, R_y terms).

Let us take two matched resistor sub-arrays as an example, which consists of two 8 unit resistors (Fig.7). 180 degree rotational symmetry layout style is often seen under the linear gradient condition, in which for any sub-array, any unit resistor at a location (x_{ij}; y_{ij}) another one exists at location (x_{ik}; y_{ik}) = ($-x_{ij}$; $-y_{ij}$). The layout has an expectation of $\Sigma x_{ij} = \Sigma y_{ij} = 0$, which cancel out spatial process gradients. The systematic mismatch is just $<R_1-R_2> = \Sigma<R_{1j}> - \Sigma<R_{2j}>$.

If unit resistor j in sub-array i has a standard deviation of σ_{ij}, and correlation with another unit resistor (either in the same array or in different array) $\kappa_{ij,i'j'} = \exp[-(r_{ij,i'j'}/r_0)^2]$ where i.i' = 1,2; j,j' =1,2,…N (when i=i', j≠ j'). The random mismatch of two matched components will be

$$\sigma(R_1-R_2) = \{\Sigma\Sigma\sigma_{ij}^2 + 2\Sigma\Sigma^s\kappa_{ij,i'j'}\sigma_{ij}\sigma_{i'j'} - 2\Sigma\Sigma^d\kappa_{ij,i'j'}\sigma_{ij}\sigma_{i'j'}\}^{1/2}$$
$$(12)$$

where Σ^s is the summation for correlated resistors in the same sub-array, whereas Σ^d is the summation for correlated resistors in the different sub-arrays. The second summation is over all sub-arrays.

This result is intriguing. Unlike a pair of two single resistors discussed in the last section where correlation always improves matching, process improvement of correlation can make matching worse here if the array is not laid out properly. The reason behind this is that only the unit resistor correlations between different groups are 'good' correlations, since they reduce the random mismatch as expressed in the last term of Equ.12. The unit resistor correlations within the same groups, on the other hand, will increase random mismatch, as described in the second term of the same

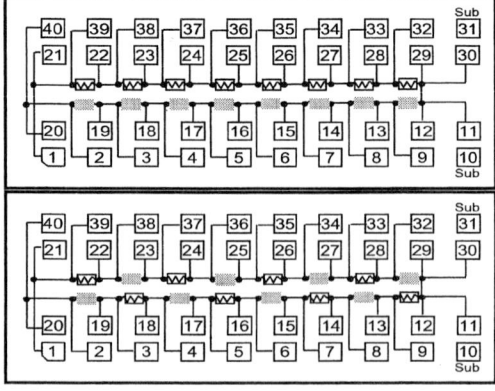

Fig.7 (a) side by side array: systematic mismatch 358ppm, random mismatch 52 ppm; (b) 180 degree rotational symmetry array: systematic mismatch -12 ppm, random mismatch 26 ppm.

equation. The correct layout, therefore, should make the closest neighbors of a unit resistor belong to a different group, as they have the strongest correlations. Based on this, 180 degree rotational symmetry layout style is more preferable.

In Fig. 7, the systematic and the random mismatch of two 180 degree rotational symmetry sub-arrays are compared to that of two side by side sub-arrays. It is clear that 180 degree rotational symmetry layout reduces both systematic and random mismatch significantly.

B   Sub-array mismatch: parabolic gradients

In this case, the resistance of each sub-array can be written as:

$$R_i = \Sigma R_{ij} + R_x * \Sigma x_{ij} + R_y * \Sigma y_{ij} + R_{xx} * \Sigma x_{ij}^2 + R_{xy} * \Sigma x_{ij} * y_{ij} + R_{yy} * \Sigma y_{ij}^2 \qquad (13)$$

where summation is over j, all unit resistors in the sub-array.

For parabolic gradients, a good layout usually has a mirror symmetry style, in which for any sub-array, any unit resistor at a location $(x_{ij}; y_{ij})$ another one exists at location $(x_{ik}; y_{ik}) = (-x_{ij}; y_{ij})$ [or $(x_{ij}; -y_{ij})$]. This arrangement cancels out not only R_x, R_y terms, but also R_{xy} term as well. R_{xx}, R_{yy} terms survive, but make little contribution to systematic mismatch:

$$\langle R_1 - R_2 \rangle = \Sigma \langle R_{1j} \rangle - \Sigma \langle R_{2j} \rangle \qquad (14)$$

R_{xx}, R_{yy} terms do, however, make contributions to random mismatch:

$$\sigma(R_1 - R_2) = \{\sigma^2(\Sigma R_{1j} - \Sigma R_{2j}) + 2\sigma^2(R_{xx} * \Sigma x_{ij}^2) + 2\sigma^2(R_y * \Sigma y_{ij}^2)\}^{1/2} \qquad (15)$$

where the first term can be written as Equ.12. As discussed in last sub-section, since some unit resistors have other unit resistors in the same group as their neighbors, the random mismatch due to correlation in the first term is not minimized. Should 180 degree rotational symmetry layout be used, the matching error from this term is reduced. However, R_{xy} term adds back in both systematic and random mismatch, probably costs more.

C   A general resistor array

In general, p sub-arrays with each consisting of q unit resistors are laid out in an MxN array, with parabolic process gradients [7]. As M and N increase, the number of ways to arrange these p sub-arrays

increases dramatically.

All possible layout configurations, however, can be scatter plotted in a $\langle R_1 - R_2 \rangle$ vs. $\sigma(R_1 - R_2)$ plan, where $\langle R_1 - R_2 \rangle$ and $\sigma(R_1 - R_2)$ are the systematic and random mismatch of any two sub-arrays. An example of this plot is illustrated in Fig.8. The point MRRS on the scattered data border has a minimum random match and a reasonable systematic mismatch; the MSRR point has a minimum systematic mismatch and a reasonable random mismatch. A good layout, therefore, should have all (p-1)! data points located close to the border connecting MRRS and MSRR, where (p-1)! is the total number of all two sub-array pairs in this layout. If we call this part of border 'matching frontier' (defined as the best matching configurations), the challenge for an analog circuit design is to find a layout configuration near matching frontier, adjust by design trade-offs. For example, if random (or systematic) mismatch is the most important for the circuit, the design will be at the west border (or close to x axis) of the data set.

CONCLUSIONS

A TFR array is designed, fabricated, tested and analyzed. We think it is a powerful and flexible tool to characterize resistor mismatch. We discussed both random and systematic mismatch between any one of the resistors with the average of the whole array, between two resistors (neighbor and far away pair), between two resistor sub-arrays. Key factors for resistor matching from both process and design/layout perspectives were illustrated. We concluded that this test structure and its characterization can provide valuable insights for improving resistor array implementations for high precision analog and mixed-signal applications. The array structure, analyses and main conclusions apply to other devices such as CMOS transistors, bipolar transistors and capacitors as well.

ACKNOWLEDGMENT

The authors gratefully acknowledge S. Burich, Z. Chen, C. Spencer, J. Lee for characterization support. Also thank M. Carrier and J. Carroll for collecting optical data on TFR photo masks for the resistor pair systematic mismatch study.

REFERENCES

[1] Paul R. Gray, Paul J. Hurst, Stephen H. Lewis, and Robert G. Meyer, "Analysis and Design of Analog Integrated Circuits (4th Edition)", John Wiley & Sons, 2001

[2] Behzad Razavi, "Principles of Data Conversion System Design", The Institute of Electrical and Electronics Engineers Inc., 1995

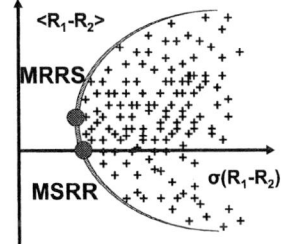

Fig.8 Resistor array matching plot and matching frontier

[3] Rudy J. van de Plassche, "CMOS Integrated Analog-to-Digital and Digital-to-Analog Converters (2nd edition)", Kluwer Academic Publishers, 2003

[4] H.P. Tuinhout et al, "Design and Characterization of a High Precision Resistor Ladder Test Structure", Proceedings of IEEE 2002 ICMTS, Vol.15, April 2002, pp.223-228

[5] H.P. Tuinhout et al, "Design and Characterization of a High Precision Resistor Ladder Test Structure", IEEE Trans. on Semi. Manu. Vol.16, May 2003, pp187-193

[6] A. Walton, "Introduction to Microelectronic Test Structures", 20th IEEE International Conference on Microelectronic Test Structures, Tutorial, March 2007, pp5-50

[7] J. Paul A. van der Wagt, "A Layout Structure for Matching Many Integrated Resistors", IEEE Trans. on Circuits and Systems, Vol. 21, January 2004, pp 186-190

Rapid Characterization of Parametric Distributions Using a Multi-meter

Jerry Hayes, Kanak Agarwal, Sani Nassif

IBM Austin Research Labs
11501 Burnet Rd, Austin, TX 78758, USA

Email: jerrhaye@us.ibm.com

Abstract — We present a technique for fast characterization of the statistical mean and sigma of parametric variations. The technique uses a scan chain to sequentially cycle through a device array, creating a periodic waveform that can be directly measured using a multi-meter. The DC and RMS values of the waveform directly give the mean and sigma of the parameter distribution. We show the technique is sufficiently general and can be applied to wide range of characterization strategies. A V_{TH} characterization array was implemented in a 65nm bulk CMOS process where we compare traditional individual device measurements for calculating statistics with the direct mean and sigma measurement technique using the multi-meter.

Index Terms — Random dopant fluctuation, process variation, test array, MOSFET, threshold voltage

I. INTRODUCTION

Circuit performance sensitivity to manufacturing variations depends upon whether the variations are systematic or random in nature and the general topology of the circuit. Systematic process variations are classified as variations that globally influence circuit performance while random variations affect performance within small regions of the circuit. The sensitivity to systematic process variations is roughly equivalent for circuits with many CMOS devices as compared to circuits with only a few devices. On the other hand, the sensitivity to random variations is inversely proportional to number of devices defining circuit performance. Large differences in sensitivity to random variations can be observed between long data paths and SRAM cells. These differences in sensitivities, coupled with the fact that process variability continues to increase in scaled technologies, is making it increasing more important to properly characterize process variability [1].

To characterize systematic and random process variability, enough samples should be used to produce statistically robust results. For systematic variations only the first moment or mean of the distribution is needed which requires roughly 30 or more samples. One technique is to individually measure each device and then compute the mean by post processing the data [2]. This characterization process is typically quite slow due to the test time required to measure, record, and process the data. A faster technique for acquiring the mean is to use a ring oscillator that averages the individual devices to a single frequency which can be mapped back to the mean of the parameter [3]. Although this technique is capable of quickly obtaining the mean of a distribution, it often suffers from parameter isolation issues that make it difficult to isolate the effects of a single parameter. In addition, ring oscillators are typically not well suited to extract the second moment or variance of distributions that are required to more fully understand the impact variability will have on the design. A test circuit for quickly obtaining distribution statistics on MOSFET channel conductance and threshold voltages has been proposed using two DC measurements [4-6]. Although the technique is fast, possible V_{DS} variations across the DUTs in the matrix can be incorrectly interpreted as voltage threshold variations. In addition, the technique is also not general enough to be extended to other parametric distributions such as capacitance. Using traditional techniques to measure devices may not be economically feasible, particularly in the production environment due to long test times. Faster techniques that leverage speed as illustrated by ring oscillators for acquiring distribution means and detail as illustrated by accessing individual devices for acquiring distribution standard deviations will be required.

By comparing the definition of standard deviation in (1) with the definition of root mean square (RMS) of a periodic signal in (2), we observe that for signals with zero mean the standard deviation is equivalent to the RMS value of the signal. By converting the parametric variations across N devices into a periodic time varying signal, a multi-meter can be used to directly obtain the standard deviation. Figure 1 illustrates a generic structure for

$$\text{Std Deviation} = \sqrt{\frac{1}{N} \sum_{i=1}^{N} (X_i - \mu)^2} \qquad (1)$$

$$\text{RMS} = \sqrt{\frac{1}{N} \sum_{i=1}^{N} (X_i)^2} \qquad (2)$$

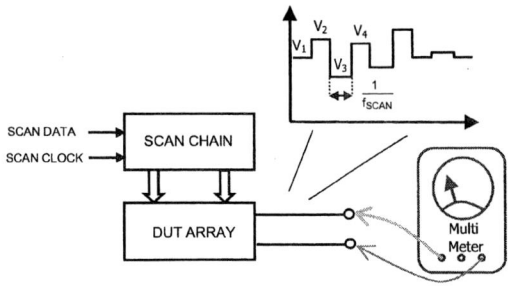

Fig 1: Generalized structure for obtaining the mean and standard deviation of parameter variations using a multi-meter.

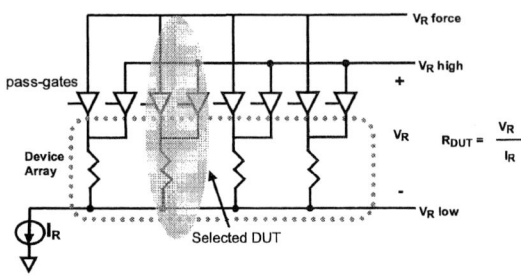

Fig 2: DUT array architecture for measuring resistance distributions.

producing and measuring periodic time varying signals that can be correlated back to parametric distributions. A scan chain is used to sequentially access each device under test (DUT) in the array at a rate determined by the scan clock frequency f_{SCAN}. By periodically sequencing though all N devices, a periodic waveform is produced at the output of array where the magnitude of the waveform at any given time corresponds to the parametric value of a single device. The mean is easily obtained using the DC mode of the multi-meter while the standard deviation is obtained using the AC mode which inserts a series capacitor to block the DC average, thus producing a zero mean signal, and measures the RMS value that is equivalent to the standard deviation. To produce stable multi-meter readings, the integration time of the multi-meter should be long enough to include enough samples as determined by f_{SCAN} that results in statistical robustness. Typical integration times are a few seconds or less with scan clock frequencies of 1 kHz or more. The relationship between the magnitude of a periodic voltage waveform and parametric values can either be a direct relationship, such as V_{TH} measurements using source modulation techniques [7], or an indirect relationship as in the case of resistance measurements where the mean and standard deviation of the voltage waveform needs to be normalized by a constant current. In the next section we give examples of both direct and indirect DUT array architectures.

II. DUT ARRAYS

Figure 2 illustrates the array architecture for measuring resistance distributions. A single device is selected by enabling the V_R force and V_R high pass-gates for the column containing the device under test. A fixed current I_R and voltage V_R-force are applied to the device. The characterization process uses the scan chain to sequentially step through each column DUT, (via, metal, OP resistor, etc.,) while sensing V_R-high and V_R-low directly with a high impedance multi-meter to obtain the voltage V_R across the DUT. The variation in DUT resistance across the array can be quickly found by dividing the measured mean and sigma of V_R by I_R. Note

the structure is free of IR drop issues because current forcing and voltage sensing occur at opposite ends of the array. Unwanted waveform spikes can occur during DUT switching which causes measurement error that is proportional to the scan frequency. Common techniques to eliminate this source of error include sample and hold circuits and synchronizing the trigger of the multi-meter with the scan chain clock.

Figure 3 illustrates the array architecture for measuring capacitance distributions. The signals CLKC and CLKB are high speed non-overlapping signals that control the charge and discharge rate of node X. When CLKC is on, node X ramps to V_X, then CLKB turns on and ramps node X back to ground potential. Typically the capacitance at node X is calculated from the measured average current, I_{AVE}, delivered by the supply V_X over many charge and discharge cycles. In this scenario the measurement averaging is done for an individual DUT rather than across the array as required for obtaining distributions. To enable averaging across the array, a resistor R is inserted between the V_X supply and node X as shown. Low pass filters (LPF) are connected to each end of R to directly obtain V_{AVE} with one multi-meter sample. The cutoff frequency (f_C) of the low pass filters are chosen to be much less than f_{CLKC} to filter out high frequency transients, and much greater than f_{SCAN} to allow V_{AVE} to

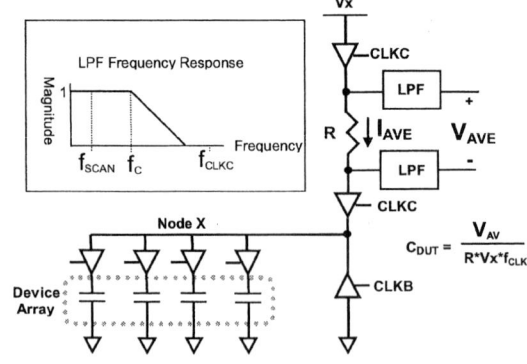

Fig 3: DUT array architecture for measuring capacitance distributions.

$$\mu_{CDUT} = \frac{\mu V_{AVE} - V_{AVE}(cal)}{(R*Vx*f_{CLKC})} \qquad (3)$$

$$\sigma_{CDUT} = \frac{\sigma V_{AVE}}{(R*Vx*f_{CLKC})} \qquad (4)$$

vary from DUT to DUT. The resulting configuration allows direct measurement of the mean and sigma of V_{AVE} that can be used to extract the mean and sigma capacitance variations as shown in (3) and (4) respectively. The $V_{AVE}(cal)$ term is the voltage obtained during a calibration run with no DUTs selected in the array. Another calibration run is used to obtain the value of R with CLKC held low. The pass-gates associated with charge and discharge paths are typically dual oxide devices to minimize gate leakage and coupling. The capacitive parasitics associated with the DUT pass-gates in figure 3 is a source of error that can limit the accuracy of the measurements. To eliminate this error, crosstalk based capacitance measurement techniques can be used as shown in figure 4 [8]. One side of the DUT is connected directly to node X while the other side is connected to either ground or an aggressor signal. When the select signal is low, one end of the DUT is connected to $V_{AGGERSSOR}$ that transitions from ground to V_x during the time CLKC is active. The net change in voltage across the DUT is zero, thus removing the effect of this capacitance from node X. The total node X capacitance when select is low is $C_{PARASITICS}$. When select is high, C_{DUT} and $C_{PARASITCS}$ are in parallel, resulting in a total node X capacitance of $C_{DUT} + C_{PARASITICS}$. The difference between the two measurements gives C_{DUT}.

Figure 5 shows the DUT array architecture for measuring FET voltage threshold (V_{TH}) distributions [7]. A constant current source I_{DS} is used to fix the current through the DUT with a fixed applied gate voltage. A unity gain source-follower stage senses the source voltage and adjusts the drain voltage accordingly to keep V_{DS} of the DUT constant. The drain-to-source voltage can be controlled by changing the bias current I_P of the source-follower. If the threshold voltage of the device changes, it causes the source voltage of the DUT to change by an equal and opposite value such that the voltage overdrive ($V_{GS} - V_{TH}$) is maintained at a constant level. A V_{TH} shift can be measured by sensing the change in the source voltage.

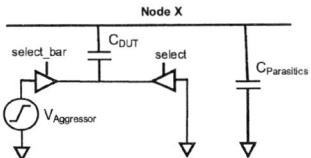

Fig 4: Crosstalk based capacitance measurement structure for improved DUT isolation

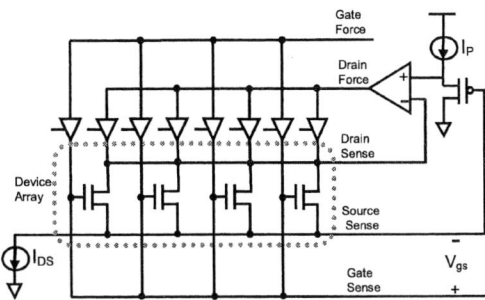

Fig 5: DUT array architecture for measuring FET voltage threshold (V_{TH}) distributions.

Different values for I_{DS} and V_{DS} are used to define V_{TH} in the saturation (V_{Tsat}) and linear (V_{Tlin}) operating regions. By using the architecture in figure 5, a direct measurement of V_{Tsat} and V_{Tlin} is obtained by appropriately adjusting I_{DS} and V_{DS} and measuring V_{GS}. The array architecture in figure 5 was implemented in a 65nm bulk process for three different PFET device widths. In the next section we compare hardware measurement results between traditional V_{TH} measurement techniques and a multi-meter for directly obtaining distribution statistics for each PFET device width.

III. MEASUREMENT RESULTS

The three PFET device widths, 120nm, 500nm, and 1um, all used the minimum L_{DRAWN} channel length of 60nm. A triple well process was used to enable local well ties to the source in order to null out V_{TH} variations due to V_{BS}. Each DUT array consists of 1000 identical devices. Figure 6 illustrates a portion of the source periodic waveform for approximately 30 devices of the array. Variations in V_{TH} are directly observable by the varying magnitude of the waveform. The maximum scan frequency f_{SCAN} used to sequence the DUTs is proportional to the device size. Minimum size devices increase both the dynamic range of the variance as well as increase the time

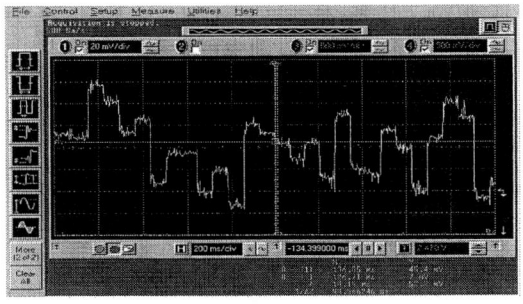

Figure 6: Source sense output waveform illustrating variations of V_{TH} across the DUTS.

constant associated with driving the source sense line. For minimum size PFET devices, the maximum scan frequency was around 500Hz due to tester parasitic capacitances. To increase the scan frequency, on chip compensation would be required to isolate tester parasitic capacitances from the source sense line. Figure 7 shows the measured ΔV_{Tlin} for an array containing 1um width PFET devices where each device was measured using traditional gate sweeping techniques. Figure 8 plots PFET V_{Tlin} sigma versus device gate area for multiple dies. The salient features of figure 8 is that sigma is inversely proportional the square root of gate area and the spread in sigma from array to array increases with decreasing device size [9]. Table 1 compares the individual device measurement technique with the direct multi-meter

Table 1: PFET V_{Tlin} comparison between individual device measurements and direct distribution measurement.

W/L	Individual DUTs		Multi-meter	
	Mean	Sigma	DC mode	AC mode
1.00/0.06	562mV	11.8mV	554mV	12.1mV
0.50/0.06	559mV	16.7mV	561mV	16.5mV
0.12/0.06	526mV	29.0mV	530mV	29.2mV

As illustrated by table 1, direct measurement of the mean and sigma is achievable with a multi-meter and the test time is extremely small, a few seconds, as compared to indirect methods. Similar results were observed for the NFET arrays.

IV. CONCLUSION

We demonstrated a method for rapid characterization of parameter distributions using a multi-meter for directly obtaining both the mean and sigma of the statistics. The proposed method is sufficiently general in that it and can be applied to a wide range of characterization strategies as shown for resistance, capacitance, and device V_{TH} characterization.

REFERENCES

[1] S. Nassif, "Modeling and Forecasting of Manufacturing Variations", Statistical Metrology, 2000 5th International Workshop, pp. 2 -10.

[2] A. Keshavarzi et al., "Measurements and Modeling of Intrinsic Fluctuations in MOSFET Threshold Voltage", Intl Symp on Low Power Electronic Design, pp. 26-29, 2005

[3] M. Bhushan, M. Ketchen, S. Polonsky, A. Gattiker, "Ring oscillator based technique for measuring variability statistics", Intl Conf on Microelectronic test structures, 2006, pp. 87-92.

[4] K. Terada et al., "A Test Circuit for Measuring Standard Deviations of MOSFET Channel Conductance and Threshold Voltage", Intl Conference on Microelectronic Test Structures, 2002, pp. 61-66

[5] K. Terada et al., "Physical Meaning of σ Value Estimated with V_{TH}-Mismatch Evaluation Circuit.", Intl Conference on Microelectronic Test Structures, 2005, pp. 165-16

[6] K. Terada et al., "Further Study of V_{TH}-Mismatch Evaluation Circuit", Intl Conference on Microelectronic Test Structures, 2004, pp. 155-159

[7] K. Agarwal et al., "Rapid Characterization of Threshold Voltage Fluctuation in MOS Devices", Intl Conference on Microelectronic Test Structures, 2007, pp. 74-77

[8] L. Vendrame et al., "Crosstalk-Based Capacitance Measurements: Theory and Applications", IEEE Trans on Semiconductor Manufacturing, vol. 19, no. 1. February 2006.

[9] A. Asenov and S. Saini, "Suppression of Random Dopant-Induced Threshold Voltage Fluctuations in Sub-0.1um MOSFETs with Expitaxial and δ-Doped Channels", IEEE Trans on Electron Devices, vol. 46, pp. 1718-1724, 1999

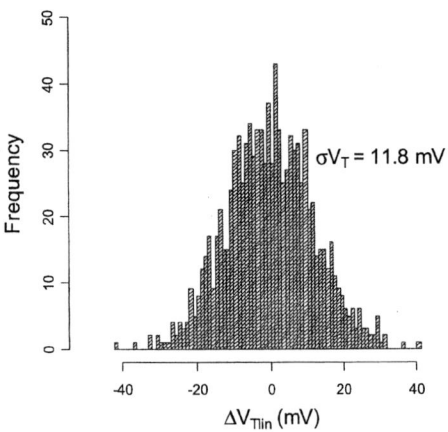

Fig 7: Measured histogram of PFET V_{Tlin} fluctuations in a 65nm bulk technology.

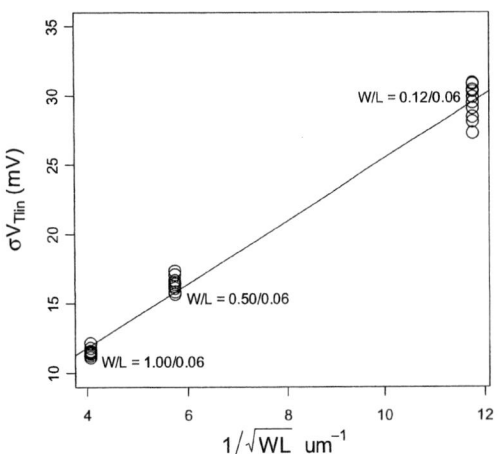

Fig 8: PFET V_{Tlin} sigma versus device gate area for multiple arrays.

SESSION 2

Resistivity

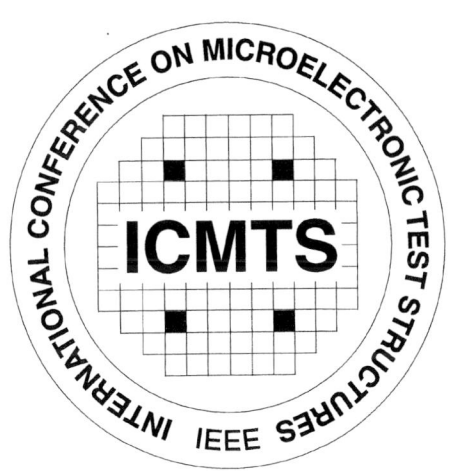

March 25, 11:00–12:00

Co-Chairs: Satoshi Habu, *Agilent Technologies, Japan*

Richard A. Allen, *NIST, USA*

Beyond van der Pauw: Sheet Resistance Determination from Arbitrarily Shaped Planar Four-Terminal Devices with Extended Contacts

M. Cornils and O. Paul

Department of Microsystems Engineering (IMTEK)
University of Freiburg, Germany

ABSTRACT

An extension of van der Pauw's method to determine the sheet resistance R_{sq} from arbitrarily shaped devices with four peripheral contacts of any size is presented. Extracting R_{sq} requires the measurement of six electrical resistances on these devices and the simultaneous solution of a system of six algebraic equations. The results are obtained using the method of conformal mapping in combination with the method of images. The computational effort to extract R_{sq} is modest. Using finite-element analysis, our findings were verified within a relative accuracy of better than 0.1%. Experimental tests with differently shaped structures are fully consistent with the novel theory.

INTRODUCTION

The calibration and in particular the electrical characterization of planar four-contact sensors such as integrated Hall plates, temperature sensors, conductometric chemical sensors, and sheet resistance test structures plays an important role in the semiconductor industry and in microelectromechanical systems engineering. One property of interest in this context is the sheet resistance R_{sq} of conducting layers. However, methods to determine R_{sq} usually require either dedicated measurement setups or special test structures satisfying the requirements of van der Pauw's method [1], namely, that their contacts be effectively point-like. In fact, such test structures make it possible to extract the sheet resistance from just two resistance measurements as shown in Figs. 1 (a) and (b) using

$$\exp\left(-\pi\frac{R_{AB,DC}}{R_{sq}}\right) + \exp\left(-\pi\frac{R_{BC,AD}}{R_{sq}}\right) = 1. \quad (1)$$

Here, A, B, C, and D denote the four consecutive peripheral contacts of the device and $R_{ij,kl} = V_{kl}/I_{ij}$ denotes the resistance determined by the voltage drop V_{kl} between contacts k and l when a current I_{ij} is applied between contacts i and j.

In contrast to the ideal case, microsensors such as Hall plates [2], [3] and stress sensors [4]-[6] have contacts usually covering a significant fraction of their periphery and thus rarely satisfy the requirement of being point-like. Recently, extensions of van der Pauw's method holding for

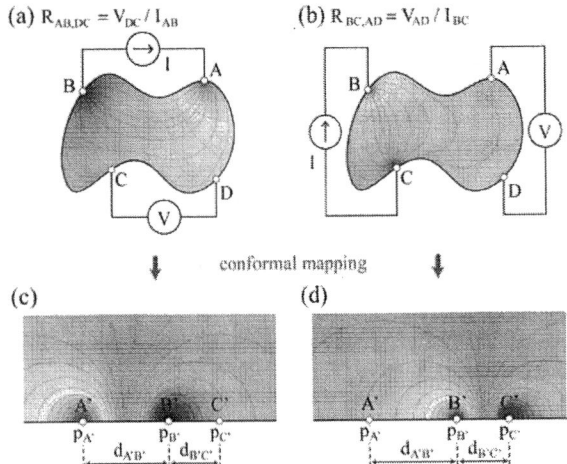

Fig. 1. (a) and (b) Ideal van der Pauw structure and two resistance measurement configurations required to extract the sheet resistance R_{sq}. (c) and (d) Equivalent half plane representations simplifying the calculations of the corresponding resistances.

symmetric devices with contacts of finite size [7] as well as for arbitrarily shaped devices with one and two extended contacts [8] were presented. Up to now, the extraction of R_{sq} from arbitrarily shaped devices with four extended contacts has required the use of finite-element modeling [9]. Thus, our goal was to extend the method of van der Pauw to devices with arbitrarily sized contacts with the perspective of an R_{sq} extraction of similar simplicity as in van der Pauw's original method based purely on electrical measurements and some algebra. As a consequence this would make it possible to monitor and possibly recalibrate sensors in-situ.

For this purpose, in the next section, we will review the procedure to determine the sheet resistance of arbitrarily shaped planar four-terminal devices with up to two extended contacts. This will pave the way to a better understanding of the subsequent theory for devices with four extended contacts. Finally, the theoretical results are checked with finite-element simulations. In addition, experimental tests with differently shaped micro structures fabricated using the physical vapor deposition (PVD) of thin metal films corroborate the novel theory.

DEVICES WITH UP TO TWO EXTENDED CONTACTS

A. Devices with four point-like contacts

The extraction of the sheet resistance R_{sq} of arbitrarily shaped planar four-terminal devices with point-like contacts from two resistance measurements using (1) was introduced by van der Pauw [1]. The derivation of (1) was based on the technique of conformal mapping [10]. In the following, we will give a derivation similar to, but slightly simpler than van der Pauw's original approach.

We start by conformally mapping the arbitrarily shaped device in Figs. 1 (a) and (b) onto the upper half plane in such a way that one of the contacts, say D, is projected to infinity, while the other contacts are projected onto the corresponding contacts A', B', and C' located at $p_{A'}$, $p_{B'}$ and $p_{C'}$ on the real axis. Up to linear scaling, two parameters characterize the novel geometry, i.e., the distances $d_{A'B'}$ between contacts A' and B' and $d_{B'C'}$ between contacts B' and C', as illustrated in Figs. 1 (c) and (d).

The solution of the potential problem with two current contacts can be obtained by superposition of the solutions for the potential problems with a single current contact. Injecting the current I_0 at any point-like contact X located at p_X on the real axis involves the potential

$$\Phi_X(z) = -\frac{R_{sq}I_0}{\pi}\ln|z - p_X| + \phi \qquad (2)$$

at any point z in the upper half plane, where ϕ denotes an integration constant that can be omitted in the following development. Extracting I_0 changes the sign of the first term on the right hand side of (2). In the case of Fig. 1 (c), the overall potential is $\Phi(z) = \Phi_{A'}(z) - \Phi_{B'}(z) = (R_{sq}I_0/\pi) \times \ln|(z - p_{B'})/(z - p_{A'})|$. Evaluation between contact C' and infinity finally yields the desired expression for the resistance $R_{AB,DC} = [\Phi(\infty) - \Phi(p_{C'})]/I_0$, namely,

$$R_{AB,DC} = \frac{R_{sq}}{\pi}\ln\left(\frac{d_{A'B'} + d_{B'C'}}{d_{B'C'}}\right). \qquad (3)$$

Similarly, $R_{BC,AD}$ is found to be

$$R_{BC,AD} = \frac{R_{sq}}{\pi}\ln\left(\frac{d_{A'B'} + d_{B'C'}}{d_{A'B'}}\right). \qquad (4)$$

Finally, noticing that the arguments of the logarithms of $R_{AB,DC}$ and $R_{BC,AD}$ fulfill the identity $[(d_{A'B'} + d_{B'C'})/d_{B'C'}]^{-1} + [(d_{A'B'} + d_{B'C'})/d_{A'B'}]^{-1} = 1$, both expressions for $R_{AB,DC}$ and $R_{BC,AD}$ can be combined to yield (1) by elimination of the geometry parameters $d_{A'B'}$ and $d_{B'C'}$. Van der Pauw's original result thus relates R_{sq} with two resistance measurements, irrespective of the device geometry.

B. Devices with one extended contact

Consider now the arbitrarily shaped planar four-contact device with three point-like contacts A, C, and D and the extended contact B shown in Fig. 2 (a). It has been shown that the electrical response of such a device is fully specified by three parameters, i.e. two geometry parameters and the sheet resistance R_{sq} [8]. It was concluded that three independent resistance measurements are required to extract these parameters from the device. The derivation of analytic expressions for one possible set of resistances, e.g. $R_{DB,CB}$, $R_{DB,AB}$, and $R_{AB,CB}$, requires an extension of the procedure for devices with point-like contacts only. First, to simplify the potential problems, the arbitrarily shaped device is mapped onto the upper half plane so that the point-like contact D opposite to the extended contact B is projected to infinity. Further, the remaining point-like contacts A and C are projected onto point-like contacts A' and C' located at $p_{A'}$ and $p_{C'}$, respectively, on the real axis. Finally the image B' of the extended contact B forms a semi-circle with radius $r_{B'}$ centered at the origin of the complex plane as illustrated in Fig. 2 (b).

In a second step, using the method of images [11], this geometry is extended to a full half plane with point-like contacts only located on the real axis as shown in Fig. 2 (c). In fact, the potential distribution in the upper half plane due to the current contacts A' and B' can be constructed by replacing B' by an additional point-like contact A^* located at

$$p_{A^*}(p_{A'}, r_{B'}, p_{B'}) = \frac{r_{B'}^2}{p_{A'} - p_{B'}} + p_{B'} \qquad (5)$$

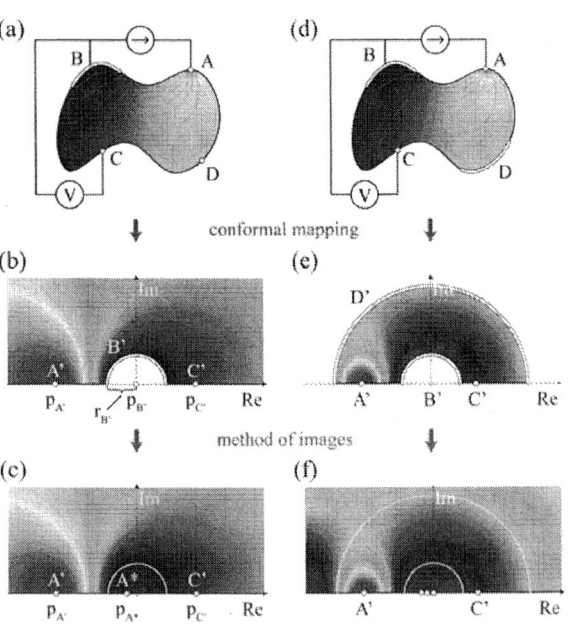

Fig. 2. (a)-(c) One among three independent resistance measurement configurations required to extract R_{sq} from a device with one extended contact and equivalent half plane representations simplifying the calculation of the corresponding potential problem. (d)-(f) One among four independent resistance measurement configurations required to extract R_{sq} from a device with two extended contacts and corresponding half-annulus and half-plane representations.

on the real axis, where $p_{B'}$ denotes the position of the center of B'. In other words, A^* is the mirror image of A with respect to B' guaranteeing that an equipotential line in the upper half plane coincides with the semi-circular contact B'. As a consequence, this configuration is equivalent to the situation shown in Figs. 1 (c) and (d), i.e., a potential distribution in the upper half plane caused by a pair of point-like current contacts on the real axis. Now, using appropriate analogies of (2), the calculation of the sought resistances is straightforward according to the procedure for devices with van der Pauw geometry. Again, the three expressions for the three resistances can be combined to eliminate the geometry parameters, finally yielding

$$\exp\left(\pi\frac{R_{AB,CB}}{R_{sq}}\right) = \frac{\cosh\left[\frac{\pi}{2R_{sq}}\left(R_{DB,CB} + R_{DB,AB}\right)\right]}{\cosh\left[\frac{\pi}{2R_{sq}}\left(R_{DB,CB} - R_{DB,AB}\right)\right]}. \quad (6)$$

Equation (6) relates the three resistance measurements with the sheet resistance in the sense of van der Pauw's original result (1).

C. Devices with two extended contacts

Let us now turn to arbitrarily shaped planar four-contact devices with two point-like contacts A and C and two extended contacts B and D, as illustrated in Fig. 2 (d). It has been previously shown that the electrical response of such a device is fully specified by four independent parameters, namely three geometry parameters and R_{sq} [8]. As a consequence, four independent resistance measurements are needed to extract these parameters. Again, the derivation of analytic expressions for one possible set of four resistances, e.g. $R_{DB,CB}$, $R_{DB,DB}$, $R_{DB,AB}$, and $R_{AB,CB}$, is carried out using a conformal mapping. Here, the arbitrarily shaped device is mapped onto an equivalent half-annulus in the upper-half plane so that the extended contacts B and D are projected onto semi-circles B' and D', respectively, centered at the origin. Further, the remaining point-like contacts A and C are projected onto point-like contacts A' and C', respectively, located on the real axis, as can be seen in Fig. 2 (e). In analogy to the structures with one extended contact, this geometry can be extended to a full half plane with exclusively point-like contacts located on the real axis. However, in this case, an infinite set of point-like current contacts obtained from the iterative application of the method of images with mirroring on both extended contacts B' and D' is required to construct the potential distribution in the half plane and thus to calculate the four resistances as illustrated in Fig. 2 (f). Again from these expressions the geometrical parameters are eliminated and the relationship

$$e^{\pi\frac{R_{AB,CB}}{R_{sq}}} = \prod_{n=-\infty}^{\infty} \frac{\cosh\left[\frac{\pi}{2R_{sq}}\left(R_{DB,CB} + R_{DB,AB} - 2nR_{DB,DB}\right)\right]}{\cosh\left[\frac{\pi}{2R_{sq}}\left(R_{DB,CB} - R_{DB,AB} + 2nR_{DB,DB}\right)\right]} \quad (7)$$

is obtained. Equation (7) relates the sheet resistance with the four measured resistance values, thus naturally extending van der Pauw's original result for devices with two extended contacts.

DEVICES WITH FOUR EXTENDED CONTACTS

Taking advantage of these techniques, it is now possible to address the arbitrarily shaped planar four-contact device with its four extended contacts A, B, C, and D shown in Fig. 3 (a). By virtue of Riemann's mapping theorem [10], such a device can be mapped conformally onto an equivalent unit disk and corresponding contacts A' to D' illustrated in Fig. 3 (b). Using a second conformal mapping, the contacts can be shifted on the circumference of the unit disk so that the images B'' and D'' of the original contacts B and D are located exactly opposite each other and have identical opening angles, as shown in Fig. 3 (c). Thus, it is concluded that a resistive device with four extended contacts has six degrees of freedom. These are the contact opening

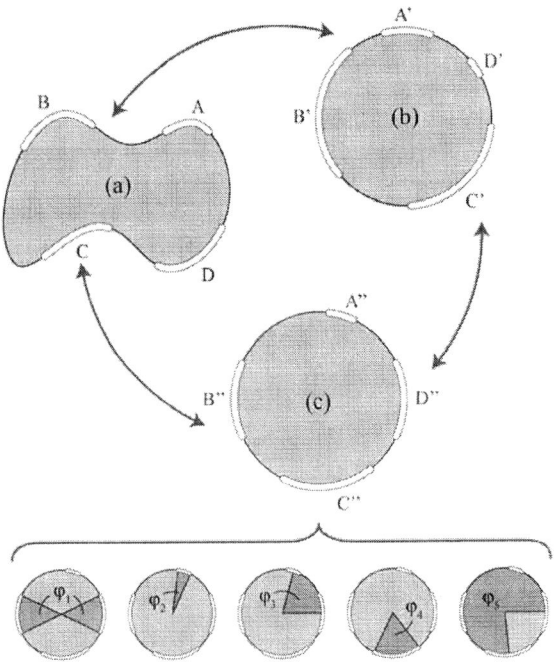

Fig. 3. (a) Arbitrarily shaped four-contact device with extended contacts. (b) Equivalent unit disk. (c) Equivalent unit disk with aligned contacts B'' and D'' adjusted to equal length. The corresponding five geometric degrees of freedom are φ_1 to φ_5, as indicated; R_{sq} is the sixth parameter specifying the device.

angle φ_1 of the two opposite contacts B'' and D'', the opening angles φ_2 and φ_4 and positions φ_3 and φ_5 of the two remaining contacts A'' and C'', and the sheet resistance R_{sq}. Consequently six independent resistance measurements of the form $R_{ij,kl} = f_{ij,kl}(\varphi_1, \varphi_2, \varphi_3, \varphi_4, \varphi_5, R_{sq})$ are expected to be sufficient to extract these parameters. Among them is the desired R_{sq}. One set of possible measurement configurations, i.e., $R_{DB,AB}$, $R_{DB,DB}$, $R_{DB,CB}$, $R_{AB,AB}$, $R_{AB,CB}$, and $R_{CB,CB}$, is presented in Fig. 4 (a)-(f).

To derive analytic expressions for these six resistances the corresponding potential problems have to be solved. Similar to the procedure for devices with up to two extended contacts, a conformal mapping is used to map the unit disk in Fig. 3 (c) onto an equivalent half annulus so that the contacts B'' and D'' are projected onto concentric semi-circular contacts B''' and D''' with respective radii $r_{B'''}$ and $r_{D'''}$ centered on $p_{B'''} = p_{D'''}$. Without loss of generality, the structure can be scaled and shifted so that $r_{D'''} = 1$ and $p_{B'''} = p_{D'''} = 0$. Further, the two remaining contacts A'' and C'' are projected onto semi-circular contacts A''' and C''' with respective radii $r_{A'''}$ and $r_{C'''}$ centered on $p_{A'''}$ and $p_{C'''}$ on the real axis. The resulting geometry is illustrated in Fig. 4 (g).

In analogy to the procedure for devices with one and two extended contacts, using the method of images, this geometry is now extended to the entire upper half-plane by replacing the semi-circular contacts for each type of resistance measurement by a set of appropriately located point-like current sources and current sinks on the real axis, guaranteeing the contacts to be equipotential lines.

The construction of this set is illustrated for the case of $R_{CB,CB}$. We start with a point-like current source $c+$ and a point-like current sink $c-$ ensuring equipotential lines at the semi-circular contacts B''' and C'''. In the case of $R_{CB,CB}$ this requires the positions $p_{c-}(p_{c+}, r_{B'''}, p_{B'''})$ and $p_{c+}(p_{c-}, r_{C'''}, p_{C'''})$ of $c+$ and $c-$ to be

$$p_{c+} = p_{C'} + \frac{2r_{C'}^2 (p_{B'} - p_{C'})}{(p_{B'} - p_{C'})^2 - r_{B'}^2 + r_{C'}^2 + \sqrt{\Theta}} \qquad (6)$$

and

$$p_{c-} = \frac{p_{B'}^2 - p_{C'}^2 - r_{B'}^2 + r_{C'}^2 + \sqrt{\Theta}}{2(p_{B'} - p_{C'})}, \qquad (7)$$

respectively, where

$$\Theta = (p_{B'} - p_{C'} - r_{B'} - r_{C'}) \times (p_{B'} - p_{C'} + r_{B'} - r_{C'})$$
$$\times (p_{B'} - p_{C'} - r_{B'} + r_{C'}) \times (p_{B'} - p_{C'} + r_{B'} + r_{C'}). \qquad (8)$$

Using appropriate analogies of (5), both point-like contacts are mirrored successively with respect to all four semi-circular contacts yielding a first generation of six point-like current sources and sinks on the real axis approximating the potential distribution in the upper half plane. This procedure is repeated, creating more and more current sources and

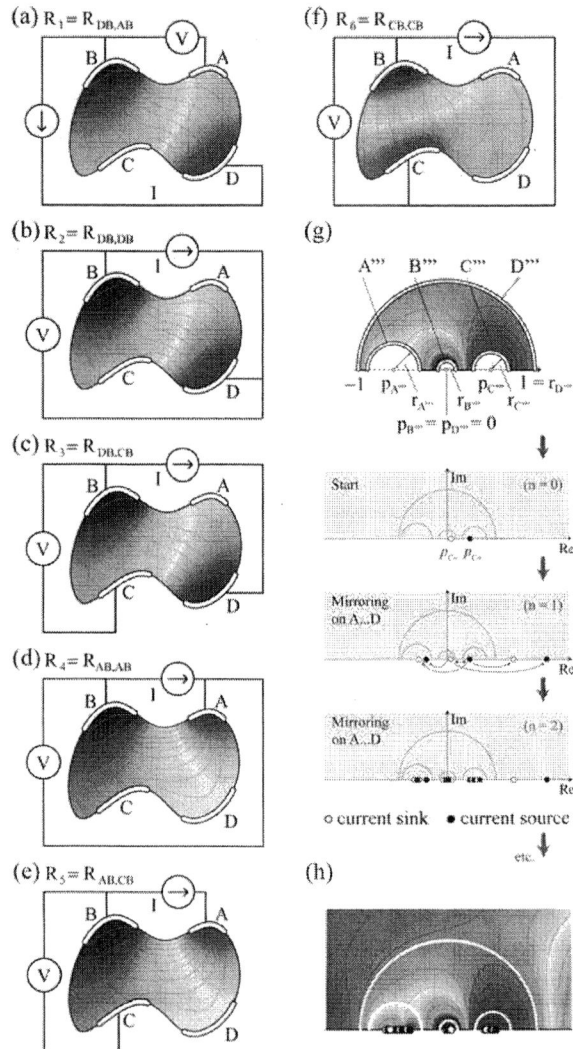

Fig. 4. Structure with four extended contacts and six resistance measurement procedures required to extract R_{sq} from the structure. The principle to represent the potential distribution of such structures in the upper half plane using an infinite set of current sources and sinks is sketched for the case of $R_1 = R_{CB,CB}$ in (g). In the case of $n = 2$, some of the sources and sinks lie outside the shown area of the half plane.

sinks from iteration n to iteration $n + 1$. After each iteration, the number of sources and sinks increases by a factor of 3 yielding a total number of 2×3^n after the n-th iteration. As n tends towards infinity, since the resulting distribution of sources and sinks is its own mirror image with respect to mirroring at all four semi-circular contacts, the contacts are equipotential lines, and thus the potential problem within the structures shown in Figs. 4 (g) and thus (f) is solved. By construction, the locations of all point-like current sources and current sinks in the six measurement situations are unambiguously defined by the three radii $r_{A'''}$, $r_{B'''}$, and $r_{C'''}$, and the two center positions $p_{A'''}$ and $p_{C'''}$. Superposition of the potentials due to all these current sources and sinks for the six measurement configurations establishes six analytical, although cumbersome relations $R_i = f_i(r_{A'''}, r_{B'''}, r_{C'''}, p_{A'''}, p_{C'''}, R_{sq})$, with $i = 1, \ldots, 6$, between the five geometry parameters and R_{sq} on one hand and the six resistance values R_i represented in Figs. 4 (a) to (f) on the other hand.

When extracting R_{sq}, one proceeds in the opposite direction, since the six resistance values are known by experiment. Then, the above relations can be simultaneously solved for the six common arguments $r_{A'''}$, $r_{B'''}$, $r_{C'''}$, $p_{A'''}$, $p_{C'''}$, and R_{sq} of the six functions f_i. One of the results is the desired R_{sq} value.

Obviously, an infinite number of sources and sinks cannot be handled realistically. The iterative construction of the set of sources and sinks therefore has to be cut off at a maximum iteration number. The approximation of the potential distribution within the geometry in Fig. 4 (g) by only six current contacts is rather poor. However, the quality is improved from iteration to iteration. In practice, a series of values $R_{sq,n}$ is computed, based on the successively expanded sets of sources and sinks. Once the relative change $(R_{sq,n} - R_{sq,n-1}) / R_{sq,n}$ is smaller than a cutoff value ε_{conv}, the iteration is stopped and the last value is taken as a reasonable approximation of the true R_{sq}. Numerical experiments have shown that in general decent convergence is achieved with $n = 3$, involving 54 current sources and sinks. The procedure is outlined in Fig. 5.

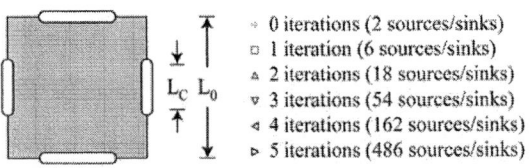

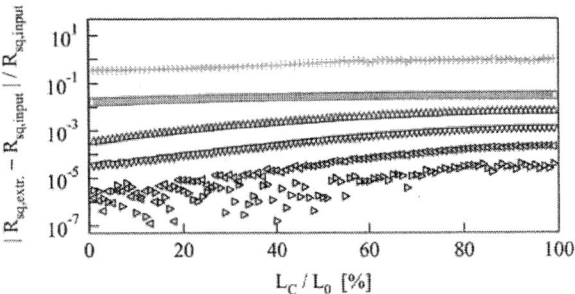

Fig. 6. Relative accuracy of the sheet resistance R_{sq} of square devices for different contact length to side length ratios L_C / L_0 extracted using the new procedure with up to five iterations. The six required resistance values were computed using finite-element simulations.

SIMULATIONS

The procedure was analyzed by numerical simulations using the finite-element tool Comsol Multiphysics (COMSOL AB, Sweden) with the square device shown in Fig. 6. Three contacts were set to a fixed length, while the length L_C of the fourth contact was varied between 1% and 99% of the edge length L_0 in steps of 1%. The six resistance values $R_1, \ldots, R_6$ were extracted as a function of the ratio L_C / L_0, based on an input value $R_{sq,input} = 1\ \Omega$. These values were in turn used to extract $R_{sq,extr.}$ with the new method using source/sink sets of increasing iteration levels. The result is shown in Fig. 6. After three iterations only, i.e. with 54 sources and sinks, the recovered $R_{sq,extr.}$ value already agrees to within 0.1% with the input value $R_{sq,input}$.

EXPERIMENTAL RESULTS

Experimental tests with seven differently shaped test structures shown in Fig. 7 consisting of a 45-nm-thick Cr layer with Au contacts fabricated using physical vapor deposition (PVD) on an oxidized 4"-Si wafer corroborate the theoretical results. Table I lists measured resistance values for representative devices as well as the sheet resistances extracted using the new procedure. With 24 devices with four extended contacts, a mean sheet resistance $\langle R_{sq} \rangle = 30.72\ \Omega$ and a standard deviation $\sigma_{Rsq} = 1.62\ \Omega$ are obtained. These results are consistent with $R_{sq} = 32.81 \pm 0.82\ \Omega$ obtained by measurements on four ideal van der Pauw test structures [device (g) in Fig. 7].

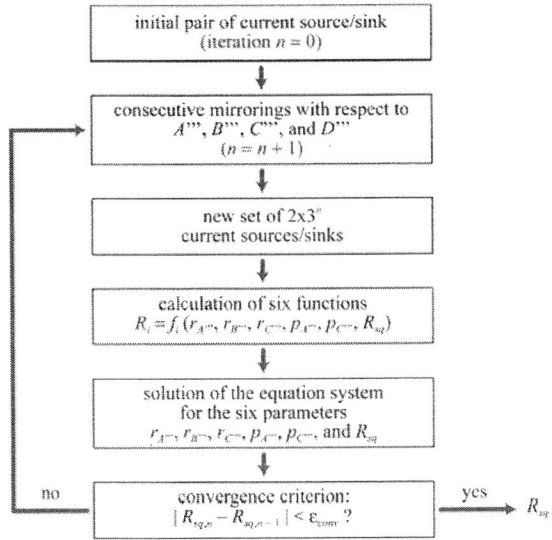

Fig. 5. Schematic procedure to determine the sheet resistance with the desired accuracy by iterative mirrorings of all current sources and current sinks with respect to A''', B''', C''', and D'''.

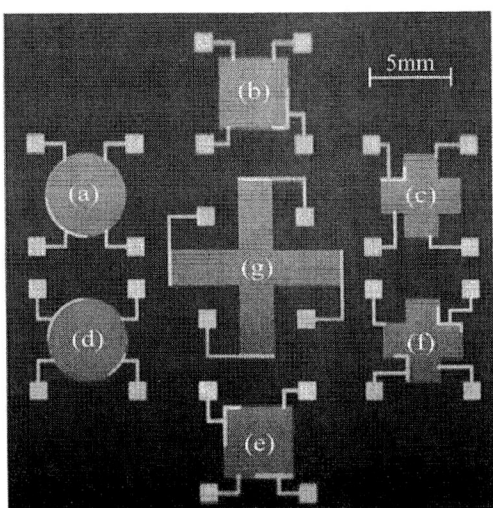

Fig. 7. Optical micrograph of test structures consisting of 45-nm-thick chromium layer with gold contacts fabricated using physical vapor deposition (PVD).

CONCLUSION

In summary, a new method to extract the sheet resistance R_{sq} of arbitrarily shaped planar four-terminal devices with contacts of finite size was developed. Fully overcoming the requirement of point-like contacts of van der Pauw's original method, this approach enables the determination of R_{sq} of any planar test structure without holes. This is achieved by measuring six resistance values on the device. No prior knowledge is required about the geometry of the structure, which can indeed be considered as an electrical black box with four contacts.

ACKNOWLEDGMENT

This work has been supported by the Ph.D. research training program "Embedded Microsystems" funded by the Deutsche Forschungsgemeinschaft DFG through grant no. GRK 1103/1.

REFERENCES

[1] L. J. van der Pauw, "A method of measuring specific resistivity and Hall effect of discs of arbitrary shape", Philips Res. Rep., vol. 13, no. 1, pp. 1-9, 1958.

Table I: Measurement results (i) for different four-terminal devices with extended contacts [Figs. 7 (a)-(f)] and corresponding sheet resistance R_{sq} extracted using the new method, and (ii) for the van der Pauw structure [v.d.P., Fig. 7 (g)] with R_{sq} extracted using van der Pauw's formula. All resistances are in units of Ω.

(i)	$R_{DB,AB}$	$R_{DB,DB}$	$R_{DB,CB}$	$R_{AB,AB}$	$R_{AB,CB}$	$R_{CB,CB}$	R_{sq}
Disk (a)	15.68	61.01	24.89	46.30	12.51	56.04	31.42
Square (b)	21.88	64.94	22.56	53.46	15.32	65.09	31.85
Cross (c)	18.66	73.30	14.12	47.35	10.01	67.83	28.82
Disk (d)	23.94	47.93	31.79	55.19	20.53	62.33	30.64
Square (e)	24.58	45.32	20.70	68.84	15.78	49.20	31.19
Cross (f)	16.87	45.87	15.40	47.21	10.12	67.85	28.32
v.d.P. (g)	83.87	167.7	83.87	160.8	76.73	160.8	32.33

(ii)	$R_{AB,DC}$						R_{sq}
v.d.P. (g)	7.24						32.81

[2] R. S. Popovic, "Hall-Effect Devices", Sensors and Actuators, vol. 17, no. 1, pp. 39-53, May 1989.

[3] R. Steiner, Ch. Maier, A. Häberli, F.-P. Steiner, and H. Baltes, "Offset reduction in Hall devices by continuous spinning current method", Sensors and Actuators A, vol. 66, no. 1-3, pp. 167-172, April 1998.

[4] J. Bartholomeyczik, S. Brugger, P. Ruther, and O. Paul, "Multidimensional CMOS in-plane stress sensor", IEEE Sensors J., vol. 5, no. 5, pp. 872-882, October 2005.

[5] M. Doelle, D. Mager, P. Ruther, and O. Paul, "Geometry optimization for planar piezoresistive stress sensors based on the pseudo-Hall effect", Sensors and Actuators A, vol. 127, no. 2, pp. 261-269, March 2006.

[6] A. Mian, J. C. Suhling, and R. C. Jaeger, "The van der Pauw stress sensor", IEEE Sensors J., vol. 6, no. 2, pp. 340-356, April 2006.

[7] M. Cornils and O. Paul, "Sheet Resistance Determination Using Symmetric Structures with Contacts of Finite Size", IEEE Trans. Electron Devices, vol. 54, no. 10, pp. 2756-2761, October 2007.

[8] M. Cornils and O. Paul, "Sensor Calibration of Planar Four-Contact Devices with up to Two Extended Contacts", in Proc. of the Sixth IEEE Sensors 2007 Conference, Atlanta, USA, pp. 1259-1262.

[9] R. Chwang, B. J. Smith, and C. R. Crowell, "Contact size effects on the van der Pauw method for resistivity and Hall coefficient measurement", Solid State Electron., vol. 17, no. 12, pp. 1217–1227, December 1974.

[10] S. Lang, "Complex Analysis", 4th ed., Springer-Verlag, New York 1999, Chapters VIII and X.

[11] J. D. Jackson, "Classical Electrodynamics", 3rd ed., John Wiley & Sons, Inc., New York 1999, Chapter 2.

Investigation of Electrical and Optical CD Measurement Techniques for the Characterisation of On-Mask GHOST Proximity Corrected Features

A. Tsiamis*, S. Smith*, M. McCallum[†], A.C. Hourd[‡], O. Toublan[§], J.T.M. Stevenson* and A.J. Walton*

*Institute of Integrated Micro and Nano Systems, School of Engineering and Electronics,
Scottish Microelectronics Centre, The University of Edinburgh, Kings Buildings, Edinburgh, EH9 3JF, U.K.
Email: A.Tsiamis@ed.ac.uk

[†]Nikon Precision Europe GmbH, Appleton Place, Appleton Parkway, Livingston, West Lothian, EH54 7EZ, U.K.

[‡]Compugraphics International Ltd., Eastfield Industrial Estate, Glenrothes, Fife, KY7 4NT, U.K.

[§]Mentor Graphics Ltd., French Branch, 180 Av de l'Europe,
ZIRST de Montbonnot, 38334 ST ISMIER CEDEX, France

Abstract—This paper reports the measurement results from a set of electrical, on-mask test structures based on industry standard test feature layouts normally used to investigate process proximity effects and improve optical proximity correction (OPC) models. The electrical test structures were fabricated on a binary photomask using the GHOST proximity correction technique to compensate for typical e-beam induced proximity errors. This is one of the first times that electrical test structures have been used to evaluate GHOST exposure. The test structures were measured electrically and optically with a dedicated photomask metrology tool and the results from the two techniques are presented.

I. INTRODUCTION AND BACKGROUND

The work presented in this paper is part of an ongoing project which is developing electrical test structures and metrology techniques for photomask characterisation. Published work so far has demonstrated that on-mask test structures are capable of sheet resistance, critical dimension (CD) and pitch extraction. Initial work concentrating on alternating aperture phase shifting masks (Alt-PSMs) showed that electrical CD results obtained from such structures compared favourably with measurements made on the same test structures using CD-SEM [1], [2], AFM [3]–[6] and optical mask metrology tools [7].

This work was extended by adapting industry standard optical metrology patterns into electrically measured, on-mask test structures. A binary plate, written without any corrections for e-beam proximity effects, was electrically and optically measured, to investigate a fundamental photomask fabrication process [8]. This present work describes the design and fabrication of binary, on-mask electrical test structures on a chrome-on-quartz plate that was written using the GHOST proximity correction technique [9]. Results from the electrical and optical measurements made on the test structures, have been used to examine the effectiveness of this method of e-beam proximity correction and to further evaluate the performance of the two mask metrology methods.

II. TEST STRUCTURES AND PHOTOMASK FABRICATION

The design of the electrical test structures which are based on the cross-bridge [10] and split-cross-bridge resistors [11], has been adapted from a set of industry standard optical test sites supplied by Mentor Graphics (MG). A full description of these optical test features can be found in [8]. In brief, the MG patterns consist of a wide range of isolated and dense, line and space metrology features. These are normally used by the mask making industry to investigate proximity effects or other fabrication artifacts, to improve the current proximity correction models and thus increase the dimensional agreement between rendered and designed features.

A GHOSTed binary mask (MSN6659) was fabricated, in an attempt to correct the proximity effects normally introduced in a standard e-beam lithography exposure [8]. The GHOST technique works by performing a second exposure where the inverse of the main pattern is written with a defocused electron beam of lower dose. The beam conditions for the correction exposure are adjusted to mimic the shape of the backscatter energy distribution produced by the main pattern exposure [9]. The backscattered electron energy dose received by all points in the pattern is equalised, resulting in a pattern that will have an energy distribution within the resist that is roughly uniform. This is intended to produce features which are largely free from proximity induced CD variations.

Mask MSN6659 includes 9 identical blocks (A1-C3) of on-mask, electrical test structures and two blocks of printable versions that can be used for on-wafer metrology when printed with a 4X photolithography system. Fig. 1(a) is an image of the mask, which also includes on-mask mismatch resistor structures for [12], while the layout of a block of on-mask structures is presented in Fig. 1(b). This design consists of 120 structures split into 10 sets of 12 structures. Nine of the sets consist of isolated and dense, dark and clear line combinations, which are identical to the test structure designs used in mask MSN5757 [8].

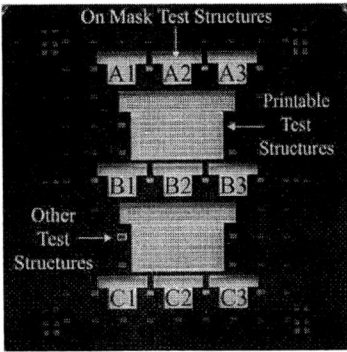

(a) Layout of binary mask (MSN6659).

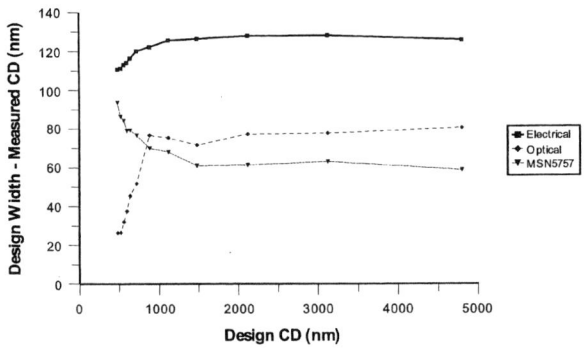

(b) Expanded view of a block of on-mask test structures.

Fig. 1. Layout of binary mask (MSN6659) with a close up view of one block of on-mask test structures.

the two measurement techniques track each other well for nominal CDs >1500nm. These results show that the fabrication process is independent of the feature dimension. For linewidths <1500nm both measurement techniques suggest that the lines appear wider as dimensions reduce. However, the optical method indicates that lines become wider at a much faster rate. This again shows how ambiguities in the optical measurement may affect the interpretation of the capability of the mask making process.

Fig. 2. Comparison of ECD and optical results for a set of isolated cross-bridge structures (GHOSTed MSN6659) with reference ECD measurements from a standard exposure mask (MSN5757).

However, unlike MSN5757 each set of split-bridge structures on MSN6659 is contained within a single row to minimise any CD variation due to the position of the test structure on the mask. As these structures are used to characterise fabrication effects that are proximity or dimension dependent, it is consequently very important to minimise any other CD variations, which may be introduced during the writing process. Hence, another alteration from the original design is that on MSN6659 the width of the bridge voltage taps (W_t) is always equal to the minimum CD (480nm for this mask). This was not the case for MSN5757 where $W_t = W_b$ (i.e., the designed linewidth of the measured line). By keeping the width of the voltage taps at the smallest possible dimension the electrical measurement error caused by current shunting at the taps is minimised [13].

III. MEASUREMENTS AND RESULTS

Electrical and optical measurements were performed using the same tools as those used in reference [8]. Detailed information on the procedures employed to measure the cross-bridge and split-bridge structures can be found in [7] and [8].

A. Isolated Lines

The linewidths for a set of isolated structures were measured electrically and optically. The measured widths have been subtracted from the nominal widths and the results are presented in Fig. 2. The electrical results from similar structures on MSN5757 are also included for reference and

For CDs >1500nm the ECD results from the GHOSTed lines and those on MSN5757 track each other well. The offset between them indicates that fabrication process parameters, such as resist development time, had to be changed to accommodate the GHOST exposure. For linewidths <1500nm, and in particular for sub-micron features, the fabricated dimensions become non-linear. On MSN5757 forward and backward electron scattering through the resist causes narrow sub-micron lines located between large clear areas to receive more electrons (i.e. a greater dose is delivered to the resist by the e-beam tool) and thus become overexposed. For isolated lines the inverse image in the GHOST process is an isolated space. Narrow isolated space features would normally lose most of their electron energy due to scattering and will develop incompletely. Therefore, superimposing the correction exposure on the pattern exposure will result in a reduced energy dose and incomplete development. This effect appears to be causing sub-micron isolated lines on MSN6659 to become wider as the nominal dimensions reduce.

B. Dense Lines

Measurements were made on dense line structures and the results from the set of 520nm wide lines are presented in Fig. 3. For all features the nominal width of the measured line is 520nm, while the line-spacing (S) between the measured and surrounding dummy lines varies. For reference purposes the results from the same structures on MSN5757 are included.

For $S > 1500$nm the electrical and optical results agree and, although the optical measurements appear more noisy,

30

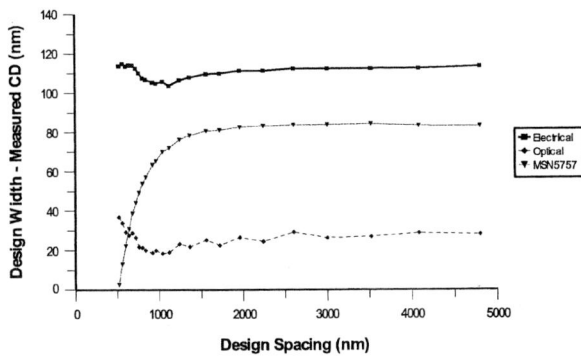

Fig. 3. Comparison of ECD and optical results for a set of 520nm wide dense cross-bridge structures (GHOSTed MSN6659) with reference ECD measurements from a standard exposure mask (MSN5757).

there is an almost constant offset between the two techniques. In addition the electrical trends from the GHOSTed lines and those on MSN5757 are similar. This is expected as there are no effects associated with line-to-line proximity for the above spacing dimensions.

While for $S < 1500$nm proximity has a very clear effect on the fabricated lines of MSN5757, the results from MSN6659 indicate that the GHOST correction appears to be effective, particularly above one micron. For sub-micron S, although there are still proximity related effects with the GHOSTed results, the CD range of the fabricated lines has reduced significantly from that of MSN5757. The effect seen on the lines of MSN6659 is most likely related to the secondary exposure of the GHOST technique. The inverse of a dense line pattern is a dense space pattern that produces an energy distribution approximating that caused by backscattering on the primary pattern. The sum of the exposures would ideally result in a constant energy distribution across the dense pattern. However, there is obviously a dimensional variation for the narrowest spaces which appears to be density dependent. Measurement results, not presented here, for other CDs have have shown that there is also a clear dependance on design linewidth. One explanation for this is that the total energy dose delivered to the resist is above or below the ideal uniform dose and this is dependent on the nominal linewidth and line spacing.

The dense set consists of four groups of structures, each with features of equal design widths and varying line-to-space ratios. A correlation plot between electrical and optical measurement results can be used to visualise the data from all groups of structures. This is presented in Fig. 4(a) for MSN6659 and suggests that the variation of the ECD measurement results is smaller than the variation of the OCD results. To confirm this, the standard deviations (σ) of the measured CDs for each group of dimensions have been calculated for the electrical and optical measurements and the results are plotted in Fig. 4(b).

It would be expected that the linewidth variability and

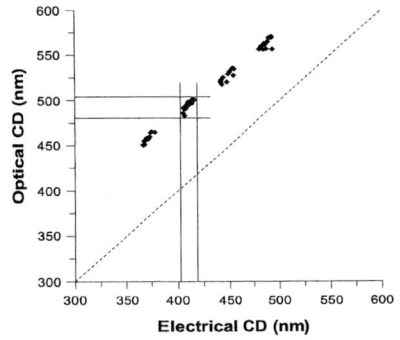

(a) Correlation plot between ECD and OCD measurement results.

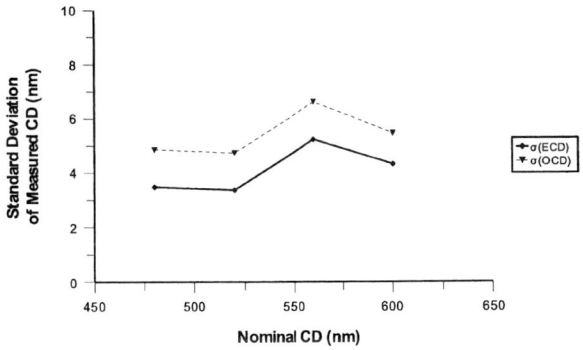

(b) Standard deviation of ECD and OCD measurement results against design linewidth.

Fig. 4. Comparison of electrical and optical measurement results for four groups of dense features with similarly sized CDs on different duty cycles, on MSN6659.

therefore the standard deviation values would remain nearly the same for both measurement techniques. However the standard deviation for each group of OCD measurements is higher, by over 1nm, than that of the ECD measurements. This suggests that there is an error associated with the OCD measurement, which is caused by optical proximity effects between the measured and surrounding dummy patterns of the dense feature. Although the GHOST correction scheme reduces proximity effects significantly, it cannot yield ideal features and the fabrication process will be characterised by a proximity based non-linear CD transfer. This can be seen in Fig. 4(a) as linewidth variations between dense features of nominally equal widths and different design densities.

C. Isolated Spaces

In order to electrically measure the line-spacing associated with a split-bridge structure, firstly the widths of the solid line (W_B) and the split-line (W_S) have to be measured. The line-spacing is then $S = W_B - 2W_S$. Fig. 5(a) shows the measurement results of the, nominally 16µm wide, solid line sections of an inverse isolated line (i.e. isolated space) set on MSN6659; reference ECD results from a similar set on MSN5757 are also presented. It can be seen that the ECD variation between each set is significantly reduced for the corrected mask with $\sigma = 2.3$nm on MSN6659 compared with 6.6nm on MSN5757. There is also no change in the measured CD with position, unlike MSN5757. These effects are most likely related to the fact that the structures on MSN6659 are located in single a row, while those on MSN5757 are located across two rows. Finally, the width of the taps has been reduced from 16μm on MSN5757, to the minimum feature dimension on the GHOSTed mask. This is at least part of the reason that the lines on MSN5757 appear electrically wider than those on MSN6659.

Fig. 5(b) presents the ECD results from measurements made on the split-line sections of the same two sets of inverse isolated structures. Note that as the design CD of the split-lines increases, the line-spacing between them decreases. The results from the non-GHOSTed lines show that due to proximity effects there is an increase in the ECD of the split-lines whose nominal width is > 14.5μm (i.e. $S < 1500$nm). The GHOSTed results show that the lines appear to have been corrected for proximity induced effects and this is the case for split-lines up to 15μm. Hence, when S is greater than 1000nm, the GHOST technique is at its most effective and the width of the fabricated lines is similar to features which are largely free from proximity induced CD variations.

the uncorrected lines increases significantly as S reduces. In addition, for the GHOSTed lines the correction exposure has partially compensated for proximity. However, the linewidth variation across these features is greatly reduced. The inverse pattern of a two-line and space combination (i.e. an isolated split-bridge) is an isolated line. For the dimensions discussed here the inverse pattern consists of narrow sub-micron lines which suffer from scattering effects and receive a higher energy dose than the ideal. Therefore the total energy dissipated in the patterns is still not uniform, however it is a better approximation to the nominal than the uncorrected version.

Once the two bridge sections have been measured, the electrical line-spacing can be calculated. Fig. 6 shows the line-spacing results for the structures on MSN6659 and the reference electrical results from MSN5757. The results from MSN5757 illustrate that for two closely spaced lines (nominal $S < 1500$nm) the space between them is underexposed and appears narrower than normal. Although these spaces are approaching their design target this is misleading as they are actually moving away from the mean fabrication process target. To investigate the dimensional offset from design targets in a fabrication process, features that do not suffer from proximity effects should be considered. In this case the wider line-spacings (>1500nm), which also appear to have the highest offset, should be the point for comparison.

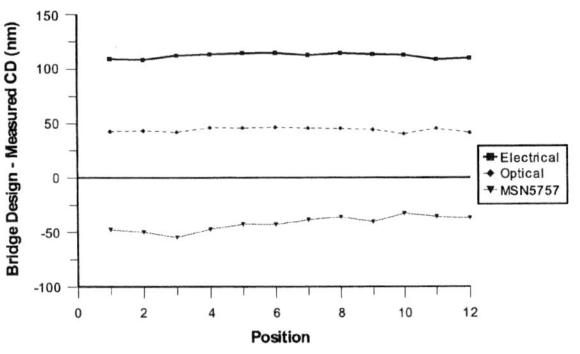

(a) Isolated bridge section

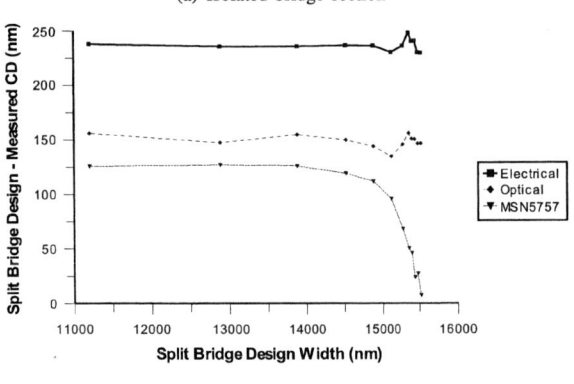

(b) Isolated split-bridge section

Fig. 5. Comparison of ECD and optical results from a set of inverse isolated (space) structures (GHOSTed MSN6659) with reference ECD measurements from a standard exposure mask (MSN5757).

For sub-micron line-spacings the rate of change of ECD on

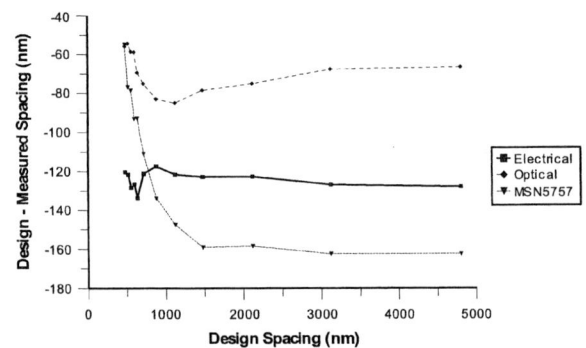

Fig. 6. Comparison of electrical and optical line-spacing results from a set of inverse isolated (space) structures (GHOSTed MSN6659) with reference ECD measurements from a standard exposure mask (MSN5757).

The line-spacing results from MSN6659 show that GHOSTing is most effective for spaces >1000nm. These dimensions are normally affected by proximity, although not as severely as narrower dimensions. In this case the correction process can achieve a relatively uniform energy distribution across the pattern. Although the technique has not completely corrected the features for sub-micron dimensions, the range of line-spacing results is greatly reduced. The trend of the data in the sub-micron region is most likely related to a feature size dependent energy distribution achieved from the combination of the primary and inverse exposures. That is, the sum of the two exposures produces an overall energy dose that is either below or above the nominal distribution.

D. Dense Spaces

Fig. 7(a) presents the results from electrical measurements made on the bridge sections of an inverse dense (space) set (reference results from a similar set on MSN5757 are also included). Although the structures on both masks investigate dense spaces, their solid bridge sections not only differ in their fabrication process but also at their design level. The features on MSN5757 are isolated lines. However, as this was a design error the bridges on the new mask are all dense, with a 520nm nominal spacing between them. The lines on MSN5757 show a decrease in the offset between design and measured ECD, which appears to be feature size dependent and this would be expected for narrower dimensions. However, for dimensions > 1µm the CD offset should remain nearly constant. This is most likely related to the design width of the bridge voltage taps which are wide enough to cause a line to appear electrically wider than is the case. For the dense lines on MSN6659 the tap effect has been eliminated, but the trend of the curve suggests that the width of the fabricated lines depends on their base CD and line-spacing. Due to GHOSTing the widths of these lines also depend on the inverse pattern exposure, which in this case is narrow (520nm) lines with wide spaces between them.

Fig. 7(b) shows the ECD results from measurements made on the split-lines of the same sets of structures. For design linewidths >2240nm, the ECD offset between the two sets remains nearly constant. Although there are proximity related effects for these split-lines, their design CD is sufficiently wide to keep this effect minimal. By design, when the CD increases so does the line to space ratio. When the line to space ratio increases the lines get wider, which is also the case here although at a very slow rate. For a base CD <1600nm the results from the two masks do not agree. The uncorrected split-lines appear narrower as their design width decreases. In addition, line to space proximity reduces significantly for narrow dimensions making the effects more apparent. This means that the fabricated split-lines get narrower at a faster rate, the line-spacing between them becomes also wider at a faster rate. This effect is reversed for the split-line results on MSN6659 where the lines become wider when their nominal CD decreases. Again this is related to GHOST and the exposure of the inverse pattern, which is used to equalise the energy distribution across the pattern. The inverse pattern of a dense split-bridge is a dense space pattern and superimposing this inverse exposure on the main exposure appears to result in an energy distribution which resembles that of the inverse pattern. Therefore the results of the GHOSTed split-lines are similar to the line-spacing results for the uncorrected fabrication process.

Fig. 8 compares the electrical and optical results from measurements on the spacings between the split-lines on MSN6659 and similar structures on MSN5757. The first thing to notice, is that due to the design error with the structures on MSN5757 the assumption that the outer widths of the bridge and split-bridge sections are equal was not valid. This led to errors and misleading trends with the electrical line-spacing

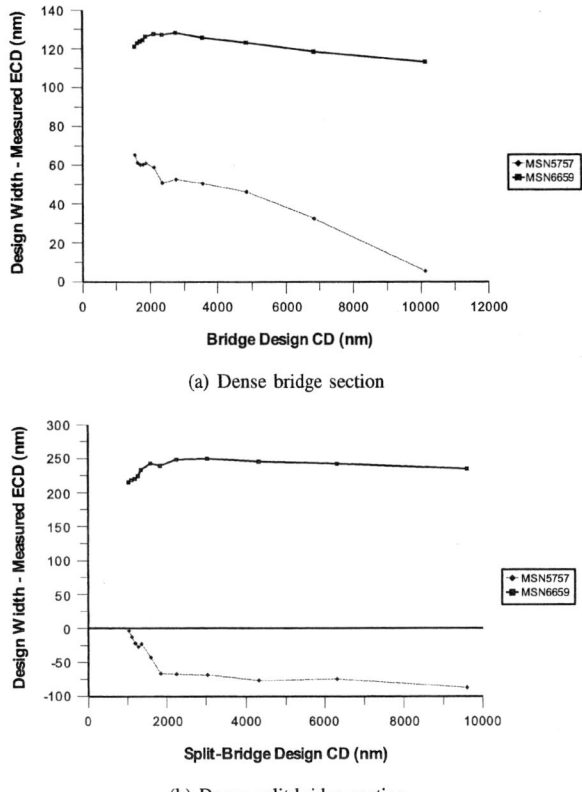

(a) Dense bridge section

(b) Dense split-bridge section

Fig. 7. Comparison of ECD results from a set of inverse dense (space) structures on a GHOSTed (MSN6659) and a standard exposure mask (MSN5757).

offset for the widest split-lines, which has been corrected for the GHOSTed mask. For split-line CDs >2240nm the results of the spacing curves track each other well with a constant offset (both measurement techniques and fabrication processes with the exception of the electrical results on MSN5757). This is expected since the split-lines are wide enough that any further increase in their linewidth does not alter the proximity effects they have on the dimensions of the spacing between them.

For split linewidths <1840nm the line-spacing results of the two masks exhibit different trends. On MSN5757 the measured spacing gets wider as the nominal width of the split lines decreases. Since the nominal line-spacing (520nm) remains constant for all structures, the split-line to space ratio decreases as the width of the split-line decreases. Split-lines with smaller split-line to space ratios appear more narrow (from the nominal) than the lines with a higher ratio and this causes the spacing between them to appear wider. On the other hand the line-spacing results from MSN6659 appear to get narrower as the design CD of the split-lines is reduced. In fact this is the same trend that the split-line results of the uncorrected mask follow (see Fig. 7(b)). In addition a split-line is the inverse pattern of the spacings under investigation. This suggests that when the energy dose from the inverse pattern exposure is superimposed with that of the main pattern exposure it does not create a uniform energy distribution across the pattern. Instead it creates an energy signature that resembles that of the inverse pattern. Finally the range of CD and spacing offset for the proximity affected results has only been reduced by a small fraction. This suggests that the GHOST technique has encountered problems with the correction of the dense features in the inverse dense set.

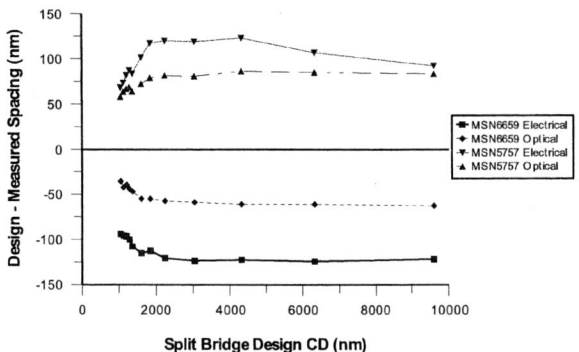

Fig. 8. Comparison of electrical line-spacing results from inverse dense structures on masks MSN5757 (optical results included for reference) and MSN6659.

IV. CONCLUSIONS AND FURTHER WORK

On-mask, electrical test structures based on industry standard OPC layouts have been designed and fabricated on a photomask using the GHOST proximity correction technique. These structures are direct electrical equivalents of optical metrology patterns, which are specifically designed to characterise OPC. Both electrical and optical measurements have been performed to evaluate the effectiveness of the GHOST correction strategy and to further examine the capability of two different metrology techniques.

The analysis, presented in this paper, has shown that GHOST correction has a significant, and positive, effect on the mask manufacturing process. However, the test structures have highlighted that there are limitations associated with the procedure which depend on the density and dimensions of the features being fabricated. The conditions for the correction exposure must be carefully considered, or the CD trends of the fabricated features may resemble those of their uncorrected inverse pattern and in some cases exhibit unexpected CD variations across a range of dimensions.

The next step to this work is to fabricate a GHOST proximity corrected photomask, which is designed for a 90nm lithography process. This will be used to print test structures and will therefore allow comparison between dimensions at the mask and wafer level. This will provide a great deal of information on how the e-beam mask writer and photolithographic tools behave when they are operating near the limit of their capabilities.

ACKNOWLEDGEMENTS

The authors would like to acknowledge the support of the Edinburgh Research Partnership in Engineering and Mathematics and the associated Institute of Integrated Systems, Nikon Precision Europe GmbH, Compugraphics International Ltd and Mentor Graphics.

REFERENCES

[1] S. Smith, M. McCallum, A. Walton, and J. Stevenson, "Electrical CD Characterisation of Binary and Alternating Aperture Phase Shifting Masks," in *Proceedings of IEEE International Conference on Microelectronic Test Structures*, Cork, Ireland, April 2002, pp. 7–12.

[2] S. Smith, M. McCallum, A. Walton, J. Stevenson, and A. Lissimore, "Comparison of Electrical and SEM CD Measurements on Binary and Alternating Aperture Phase-Shifting Masks," *IEEE Transactions on Semiconductor Manufacturing*, vol. 16, no. 2, pp. 266–272, May 2003.

[3] S. Smith, M. McCallum, A. Walton, J. Stevenson, P. Harris, A. Ross, A. Hourd, and L. Jiang, "Test Structures for CD and Overlay Metrology on Alternating Aperture Phase-Shifting Masks," in *Proceedings of IEEE International Conference on Microelectronic Test Structures*, Awaji Yumebutai International Conf. Center, Japan, March 2004, pp. 29–34.

[4] M. McCallum, S. Smith, A. Hourd, A. Walton, and J. Stevenson, "Cost Effective Overlay and CD Metrology of Phase-Shifting Masks," in *Proceedings of the SPIE, Vol. 5567: 24th Annual BACUS Symposium on Photomask Technology*, 2004, pp. 596–603.

[5] S. Smith, M. McCallum, A. Walton, J. Stevenson, P. Harris, A. Ross, and L. Jiang, "On-Mask CD and Overlay Test Structures for Alternating Aperture Phase Shift Lithography," *IEEE Transactions on Semiconductor Manufacturing*, vol. 18, no. 2, pp. 238–245, May 2005.

[6] S. Smith, A. Walton, M. McCallum, A. Hourd, J. Stevenson, A. Ross, and L.Jiang, "Improved Test Structures for the Electrical Measurement of Feature Size on an Alternating Aperture Phase-Shifting Mask," in *Proceedings of IEEE International Conference on Microelectronic Test Structures*, K.U. Leuven, Leuven, Belgium, April 2005, pp. 17–22.

[7] S. Smith, A. Tsiamis, M. McCallum, A. Hourd, J. Stevenson, and A. Walton, "Comparison of Optical and Electrical Measurement Techniques for CD Metrology on Alternating Aperture Phase-Shifting Masks," in *Proceedings of IEEE International Conference on Microelectronic Test Structures*, Austin, Texas, USA, March 2006, pp. 119–123.

[8] A. Tsiamis, S. Smith, M. McCallum, A. Hourd, O. Toublan, J. Stevenson, and A. Walton, "Development of Electrical On-Mask CD Test Structures Based on Optical Metrology Features," in *Proceedings of IEEE International Conference on Microelectronic Test Structures*, Tokyo, Japan, March 2007, pp. 171–176.

[9] G. Owen and P. Rissman, "Proximity Effect Correction for Electron Beam Lithography by Equalization of Background Dose," *Journal of Applied Physics*, vol. 54, no. 6, pp. 3573–3581, 1983.

[10] M. Buehler, S. Grant, and W. Thurber, "Bridge and van der Pauw Sheet Resistors For Characterizing the Line Width of Conducting Layers," *J. Electrochemical Soc - Solid State Technology*, vol. 125, no. 4, pp. 650–654, April 1978.

[11] M. Buehler and C. Hershey, "The Split-Cross-Bridge Resistor for Measuring the Sheet Resistance, Linewidth, and Line Spacing of Conducting Layers," *IEEE Transactions on Electron Devices*, vol. ED-33, no. 10, pp. 1572–9, Oct 1986.

[12] S. Smith, A. Tsiamis, M. McCallum, A. Hourd, J. Stevenson, and A. Walton, "Electrical Measurement of On-Mask Mismatch Resistor Structures," in *Proceedings of IEEE International Conference on Microelectronic Test Structures*, Tokyo, Japan, March 2007, pp. 3–8.

[13] G. Carver, R. Mattis, and M. Buehler, "Design Considerations for the Cross-Bridge Sheet Resistor," National Bureau of Standards, Washington, DC, Tech. Rep. NBSIR 82-2548, 1982.

Comparison of Measurement Techniques for Advanced Photomask Metrology

S. Smith*, A.Tsiamis*, M. McCallum[†], A.C. Hourd[‡], J.T.M. Stevenson*, A.J. Walton*, R.G. Dixson[§],
R.A. Allen[§], J.E. Potzick[§], M.W. Cresswell[§] and N.G. Orji[§]

*Institute of Integrated Micro and Nano Systems, School of Engineering and Electronics,
Scottish Microelectronics Centre, The University of Edinburgh, EH9 3JF, UK   Email: Stewart.Smith@ed.ac.uk

[†]Nikon Precision Europe GmbH, Appleton Place, Appleton Parkway, Livingston, West Lothian, EH54 7EZ, UK

[‡]Compugraphics International Ltd., Eastfield Industrial Estate, Glenrothes, Fife, KY7 4NT, UK

[§] National Institute of Standards and Technology, Gaithersburg, Maryland, USA

Abstract—**This paper compares electrical, optical and AFM measurements of critical dimension (CD) made on a chrome on quartz photomask. Test structures suitable for direct, on-mask electrical probing have been measured using the three techniques and the results show very good agreement between the electrical measurements and those made with a calibrated CD-AFM system.**

I. INTRODUCTION

The use of direct electrical measurement of critical dimension (CD) on advanced photomask plates has been presented in a number of previous publications [1]–[7]. These have described the design, fabrication and testing of sheet resistance and electrical linewidth test structures capable of being electrically probed on-mask. Most recently a set of electrical test structures based on optical metrology features were used to measure iso-dense proximity effects on binary photomasks with designed CDs as low as 480nm (4X) [8]. This mask design also included, for the first time, test structures designed to investigate dimensional mismatch between closely spaced chrome features [9].

In addition to the on-mask, electrical measurements, more traditional metrology techniques such as CD-SEM (CD Scanning Electron Microscopy) and optical CD measurements have been evaluated [1], [3], [8]. These results have demonstrated that there are serious issues with the extraction of linewidth from SEM or optical images of photomask features. This is especially true for alternating aperture phase shifting masks and where optical proximity effects dominate imaging. Regardless of these results, persuading the mask making community to integrate on-mask electrical measurements into their manufacturing process has proved to be difficult. One problem is the issue of probe needles coming into contact with the mask surface, even though the electrical structures would be located outside the exposure area, and the fact that delicate ICs are routinely probed during test. Another concern is the non-physical nature of the electrical measurement, which could mean that effects observed do not transfer to dimensions of features on wafer. In an attempt to demonstrate the power of on-mask electrical measurements, structures on one of the

masks described in [8] and [9] have been measured using one of the few CD-AFM (CD Atomic Force Microscope) tools in the world that is fully calibrated to a CD reference standard [10]. This paper presents a preliminary comparison of CD-AFM measurements to electrical and optical metrology results.

II. TEST STRUCTURES

A. Cross-Bridge Linewidth Structures

The mask used in the present work (MSN6659) was fabricated to the same design as mask MSN5757, as described in [8]. The test structures are based on an optical/SEM metrology feature set from Mentor Graphics who use it to investigate iso-dense bias and other optical proximity effects in photolithography. The basic electrical linewidth test structure is the cross-bridge resistor [11]. This is made up of two parts, a Greek cross sheet resistance structure and a Kelvin connected bridge resistor. The layout of an isolated cross-bridge structure is shown in Fig. 1. The standard chrome on quartz photomask includes an anti-reflective coating (ARC) of insulating chromium oxynitride and so a second level of patterning is required to open the ARC over the pads so that good electrical contacts can be made.

B. Mismatch Test Structures

The x-y mismatch test structures on mask MSN6659 are pairs of Kelvin connected bridge resistors, 600µm long, 0.5µm wide and separated by 30µm. There are two different arrangements with lines running either vertically or horizontally as shown in Fig. 2.

The mask features an array of 54 sets of test structures, placed around and between the blocks of cross-bridge linewidth test structures.

III. MEASUREMENTS

A. Electrical Measurements

1) Cross-Bridge Structures: The sheet resistance R_S of the chrome layer of the mask is measured using the Greek cross structure and the method described in [12]. Current is forced

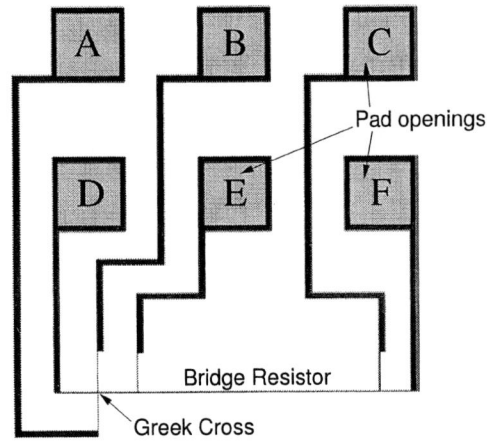

Fig. 1. Cross-bridge linewidth test structure

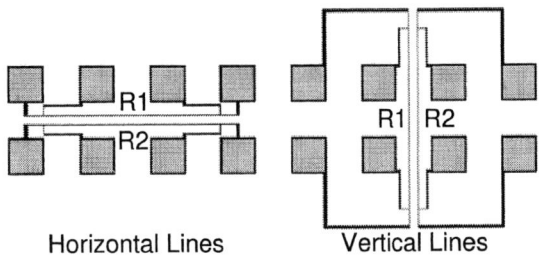

Fig. 2. Mismatch bridge resistor test structures

between two adjacent arms of the cross (pads A and D in Fig. 1) and the voltage between the other arms (pads B and E) is measured. This is repeated with the current reversed and then again with the connection rotated through 90° (I_{DB}, V_{EA} etc.). The results of the four Kelvin V/I measurements are averaged (R_{avg}) and the sheet resistance is calculated using

$$R_S = \frac{\pi R_{avg}}{\ln 2} \qquad (1)$$

The resistance R_B of the bridge resistor section is then measured by forcing a current between pads D and F and measuring the voltage drop between pads E and C. The average electrical linewidth (ECD) of the bridge section can then be calculated using

$$W_B = \frac{R_S L_B}{R_B} \qquad (2)$$

where L_B is the length of the bridge section (400μm).

2) Mismatch Test Structures: The resistance of each Kelvin bridge resistor in the mismatch test block is measured in a similar way to the bridge section of the cross-bridge structure. The approximate linewidth is then calculated using an average value of sheet resistance extracted from the many Greek cross structures on the mask. This is 22.45Ω/□ with a variation of about 1% across the mask. The estimated electrical CDs are then used to calculate the dimensional mismatch, $\Delta W/W(\%)$,

in X (from the vertical lines) and Y (from the horizontal lines). As this is a relative figure the results are unaffected by small errors in the sheet resistance used to calculate the electrical linewidth.

B. Optical Measurements

Optical CD (OCD) measurements have been made using a MueTec <M5K> mask metrology system with 248nm ultraviolet illumination. This captures an image of features on the mask, always in transmission at 248nm, and determines the CD [13]. The system extracts an intensity profile from the image and applies a threshold in order to determine the position of the feature edges. This is a subjective measurement requiring careful calibration and it has been demonstrated in previous publications that this measurement has problems when measuring phase shifted masks [7] or isolated features below ~700nm [8], as a consequence of the calibration methodology used in these references.

C. CD-AFM Measurements

CD-AFM measurements were performed using an SXM320 at the National Institute of Standards and Technology (NIST). Because the tip width is calibrated using a single crystal, critical dimension reference material (SCCDRM), which was developed at NIST, this instrument is capable of performing linewidth measurements with expanded uncertainties as low as 1.5nm (k = 2) [10], [14]. The AFM measurements were obtained near the center of the bridge resistor on both types of test structures. However, this positioning is only approximate due to the length of the structures and the absence of nearby navigation markers. 20 AFM scan lines are made over a 1μm length of track and the average width is calculated. The typical standard deviation of the 20 measurements is 5-7nm. The expanded uncertainties of the values as measured by the AFM ranged between 1.7 nm and 3.8 nm, with tip wear driving the larger uncertainties. However, it should be noted that these estimates do not include the uncertainty resulting from line width roughness (LWR). In order to investigate any longer range changes in linewidth one of the structures was remeasured at 5 different positions, about 100μm apart, along the length of the bridge.

IV. RESULTS

A. Cross Bridge Structures

Due to the length of time required for CD-AFM measurements these initial results only cover the smallest of the isolated linewidth structures. There are also difficulties with the measurement of dense features using the CD-AFM due to the shape of the tip. As a result there are measurements from 6 isolated cross bridge structures with designed linewidths between 480nm and 720nm. These results are plotted along with ECD and OCD results in Fig. 3. It should be noted that this graph shows the measured CD subtracted from the designed linewidth.

These results show excellent agreement between the CD-AFM and ECD measurements, but a significant offset between

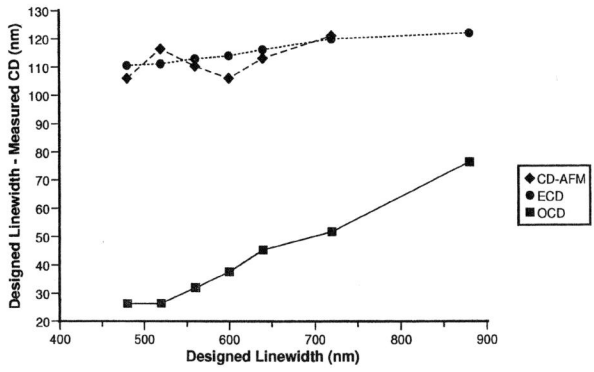

Fig. 3. Comparison of optical, electrical and AFM metrology

them and the optical CD metrology results. The level of agreement between the electrical and standards calibrated CD-AFM results is surprisingly good. It is expected that there would normally be a systematic offset between any two measurement techniques, associated with the type of measurement used [15]. The variation of the AFM results away from the smooth trend of the ECD measurement can be explained by noting that the AFM measurement is looking at a relatively short (1µm) length of the bridge while the electrical results give the average width of a 400µm line.

In order to investigate this, further measurements were made at 5 different positions along a bridge structure with a nominal width of 520nm. This is the second point in Fig. 3 with a measured width of 403.5nm. These measurements were made using an AFM tip with significant wear and so they should be considered as indicative of the variation of width along the line rather then of absolute CD. The results have been normalised to the value for the centre of the line given above (i.e., 403.5nm) and are presented in table I, along with the standard deviation of the 20 individual measurements made at each position. Note that the "Distance" column gives the approximate position of the AFM measurement along the measured line. These results suggest that the variation of linewidth at this long range is very small and is of a similar scale to the standard deviation of the individual measurements at each point.

TABLE I
CD-AFM MEASUREMENTS MADE AT FIVE DIFFERENT POSITIONS

Distance (µm)	Width (nm)	Std. Dev. (nm)
60	404.92	7.95
130	402.15	5.92
200	403.51	7.88
270	400.55	7.22
340	400.28	4.96

Fig. 4 is a more direct comparison of the three different measurement methods. It shows the differences between the measurement results plotted against the CD measured with the NIST CD-AFM. The offset between the ECD and AFM results

is less than 10nm for each of the test structures and does not display an obvious dependence on the width. This is not the case for the optical results where the offset is significantly larger and also seems to reduce as the dimensions increase.

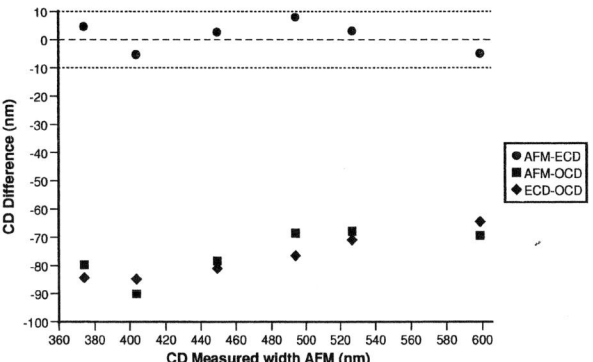

Fig. 4. Measurement offsets for isolated linewidth structures

B. Mismatch Test Structures

Three blocks of mismatch structures from MSN6659 have been measured with the NIST CD-AFM. Each block has 4 individual Kelvin bridge resistors and so this provides measurements of 12 features with a nominal width of 500nm. The structures were chosen on the basis of the initial electrical measurements which either showed larger or smaller offsets in the vertical or horizontal directions than the average. Fig. 5 shows the positions of the three blocks of structures that were measured while Fig. 6 presents the offsets between each of the different measurement methods for these structures.

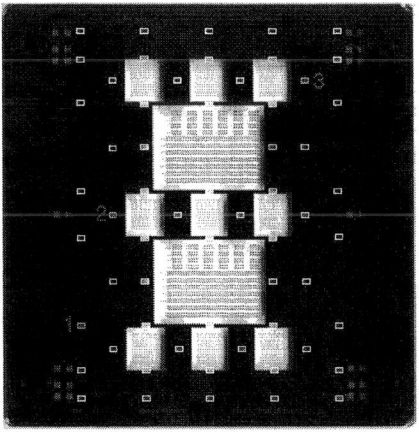

Fig. 5. Scanned image of photomask with measurement sites highlighted

As with the isolated cross-bridge structures there is good agreement between the electrical and CD-AFM measurements where the difference is always less than 10nm. The optical results show a significant difference when compared to the other measurement techniques, probably due to calibration issues, but there is no significant trend with width unlike the

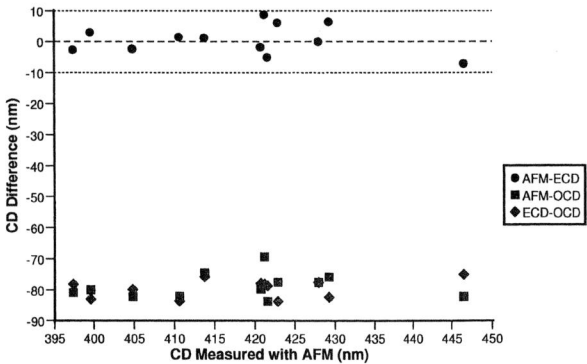

Fig. 6.   Measurement offsets for mismatch test structures

results in Fig. 4. This is probably due to the much smaller range of dimensions in this data set.

Values for the dimensional mismatch between each pair of lines have also been calculated for each measurement technique and the results are shown in Fig. 7. The results from the electrical measurements are closer to those obtained with the CD-AFM for every set of structures except for the vertical lines in block 2. In this structure the offset is close to zero and the observed result may be explained as an artefact of the AFM measurement variability.

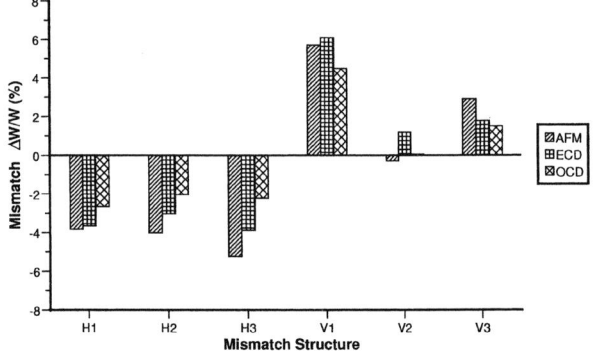

Fig. 7.   Comparison of percentage mismatch $\Delta W/W$ extracted using different metrology techniques

V. CONCLUSION

Three different techniques: electrical, optical and CD-AFM, have been used to measure the linewidths of metal features on a standard chrome on quartz photomask. ECD measurements are made by direct probing onto the mask, while optical measurements are made using a mask metrology and verification tool. The CD-AFM measurements are made using a state-of-the-art system, which is calibrated using a traceable reference standard and has an uncertainty of less than 4nm.

Measurements of isolated cross-bridge linewidth structures with nominal widths between 0.48 and 0.72µm show good agreement between ECD and CD-AFM measurements. The offset is less than ±10nm and shows no obvious dependence

upon nominal size over the range of dimensions measured. This is not the case of the optical measurements which are offset by more than 60nm from the electrical and AFM measurements. The optical results also show a dependence on the nominal width with the offset reducing as dimensions increase. The AFM measurement is taken over a very short (1µm) distance while the ECD is an average over the length of a 400µm line (600µm for mismatch structures). For this reason, further AFM measurements were taken at ∼70µm steps along a bridge structure. These demonstrated a surprisingly small variation in width of less than 5nm.

In addition to the cross-bridge structures a number of Kelvin resistors designed for use as dimensional mismatch structures were also measured with the three different techniques. These 12 structures are all isolated metal features with a nominal width of 0.5µm but the measured dimensions varied by as much as 50nm. As with the cross-bridge structures, the AFM and ECD measurements are within 10nm of each other while the average offset of the optical measurements is 80nm.

Overall, the ECD and CD-AFM measurements show very good agreement with no obvious systematic offset while the optical measurements overestimate the width of these narrow isolated features by as much as 90nm. These results demonstrate the capability of the on-mask electrical measurement technique, especially when compared to optical tools that struggle to measure small dimensions.

ACKNOWLEDGMENT

The authors would like to acknowledge the support of the Edinburgh Research Partnership in Engineering and Mathematics and the associated Institute of Integrated Systems, Nikon Precision Europe GmbH, Compugraphics International Ltd and Mentor Graphics.

The NIST authors acknowledge support from the NIST Office of Microelectronics Programs, the NIST Advanced Technology Program, and the Nanomanufacturing Program of the Manufacturing Engineering Laboratory.

REFERENCES

[1] S. Smith, M. McCallum, A. J. Walton, and J. T. M. Stevenson, "Electrical CD Characterisation of Binary and Alternating Aperture Phase Shifting Masks," in *Proceedings of IEEE International Conference on Microelectronic Test Structures*, Cork, Ireland, April 2002, pp. 7–12.

[2] S. Smith, M. McCallum, A. J. Walton, J. T. M. Stevenson, and A. Lissimore, "Comparison of Electrical and SEM CD Measurements on Binary and Alternating Aperture Phase-Shifting Masks," *IEEE Transactions on Semiconductor Manufacturing*, vol. 16, no. 2, pp. 266–272, May 2003.

[3] S. Smith, M. McCallum, A. J. Walton, J. T. M. Stevenson, P. D. Harris, A. W. S. Ross, A. C. Hourd, and L. Jiang, "Test Structures for CD and Overlay Metrology on Alternating Aperture Phase-Shifting Masks," in *Proceedings of IEEE International Conference on Microelectronic Test Structures*, Awaji Yumebutai International Conf. Center, Japan, March 2004, pp. 29–34.

[4] M. McCallum, S. Smith, A. C. Hourd, A. J. Walton, and J. T. M. Stevenson, "Cost Effective Overlay and CD Metrology of Phase-Shifting Masks," in *Proceedings of the SPIE, Vol. 5567: 24th Annual BACUS Symposium on Photomask Technology*, 2004, pp. 596–603.

[5] S. Smith, M. McCallum, A. J. Walton, J. T. M. Stevenson, P. D. Harris, A. W. S. Ross, and L. Jiang, "On-Mask CD and Overlay Test Structures for Alternating Aperture Phase Shift Lithography," *IEEE Transactions on Semiconductor Manufacturing*, vol. 18, no. 2, pp. 238–245, May 2005.

[6] S. Smith, A. Walton, M. McCallum, A. C. Hourd, J. T. M. Stevenson, A. W. S. Ross, and L. Jiang, "Improved Test Structures for the Electrical Measurement of Feature Size on an Alternating Aperture Phase-Shifting Mask," in *Proceedings of IEEE International Conference on Microelectronic Test Structures*, K.U. Leuven, Leuven, Belgium, April 2005, pp. 17–22.

[7] S. Smith, A. Tsiamis, M. McCallum, A. C. Hourd, J. T. M. Stevenson, and A. Walton, "Comparison of Optical and Electrical Measurement Techniques for CD Metrology on Alternating Aperture Phase-Shifting Masks," in *Proceedings of IEEE International Conference on Microelectronic Test Structures*, Austin, Texas, USA, March 2006, pp. 119–123.

[8] A. Tsiamis, S. Smith, M. McCallum, A. C. Hourd, O. Toublan, J. T. M. Stevenson, and A. J. Walton, "Development of Electrical On-Mask CD Test Structures Based on Optical Metrology Features," in *Proceedings of IEEE International Conference on Microelectronic Test Structures*, Tokyo, Japan, March 2007, pp. 171–176.

[9] S. Smith, A. Tsiamis, M. McCallum, A. C. Hourd, J. T. M. Stevenson, and A. J. Walton, "Electrical Measurement of On-Mask Mismatch Resistor Structures," in *Proceedings of IEEE International Conference on Microelectronic Test Structures*, Tokyo, Japan, March 2007, pp. 3–8.

[10] R. G. Dixson, R. A. Allen, W. F. Guthrie, and M. W. Cresswell, "Traceable Calibration of Critical-Dimension Atomic Force Microscope Linewidth Measurements with Nanometer Uncertainty," *Journal of Vacuum Science and Technology*, vol. B 23, no. 6, pp. 3028–3032, 2005.

[11] M. G. Buehler, S. D. Grant, and W. R. Thurber, "Bridge and van der Pauw Sheet Resistors For Characterizing the Line Width of Conducting Layers," *J. Electrochemical Soc - Solid State Technology*, vol. 125, no. 4, pp. 650–654, April 1978.

[12] M. G. Buehler and W. R. Thurber, "An Experimental Study of Various Cross Sheet Resistor Test Structures," *J. Electrochemical Soc - Solid State Technology*, vol. 125, no. 4, pp. 645–650, April 1978.

[13] A. C. Hourd, A. Grimshaw, G. Scheuring, C. Gittinger, S. Döbereiner, F. Hillman, H.-J. Brück, S.-B. Chen, P. Chen, R. Jonckheere, V. Philipsen, H. Hartmann, V. Ordynskyy, K. Peter, T. Schätz, and K. Sommer, "Reliable sub-nanometer repeatability for CD metrology in a reticle production environment," in *Proceedings of the SPIE, Vol. 4889: 22nd Annual BACUS Symposium on Photomask Technology*, 2002, pp. 319–327.

[14] M. W. Cresswell, W. F. Guthrie, R. G. Dixson, R. A. Allen, C. E. Murabito, and J. V. Martinez De Pinillos, "RM 8111: Development of a Prototype Linewidth Standard," *Journal of Research of the National Institute of Standards and Technology*, vol. 111, no. 3, pp. 187–203, May-June 2006.

[15] R. A. Allen, P. Troccolo, J. C. Owen, J. E. Potzick, and L. W. Linholm, "Comparisons of measured linewidths of sub-micrometer lines using optical, electrical, and SEM metrologies," *Proceedings of the SPIE*, vol. 1926, pp. 34–43, 1993.

SESSION 3
Yield and Reliability I

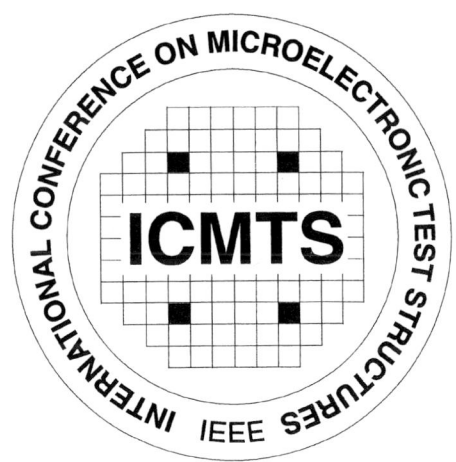

March 25, 13:30–14:50

Co-Chairs: Christopher Hess, *PDF Solutions, USA*

Kelvin Yih-Yuh Doong, *TSMC, Taiwan*

Electromigration : from package to wafer level thanks to a heating coil structure

C. Chappaz, P. Waltz, L. Castellon

STMicroelectronics, 850 rue Jean Monnet 38950 Crolles CEDEX

Tel : + 33 (0)4 76 38 21 08, Fax : + 33 (0)4 76 92 57 32

cedrick.chappaz@st.com

Abstract— **An efficient and precise method for Black's parameters extraction at wafer level is proposed. Using an innovative device including a standard Electromigration Metal Line (EML) – NIST architecture - [1] with a dedicated heating coil, electromigration tests can be successfully leaded at wafer level for activation energy (Ea) and current acceleration factor (n) extraction. Comparison with package results emphasizes a very good behavior of the structure.**

Index: electromigration, heating coil, wafer level, Ea extraction

I. INTRODUCTION

It's mostly admitted that electromigration parameters can only be precisely determined at package level. Indeed, the well known Black's equation [2] implicates to separate current and temperature effects. Usual electromigration at wafer level on standard structures are done using the self-heating brought by a high current in the line. This direct correlation between current and temperature of test doesn't allow extracting Ea and n properly [3].

Previous studies proposed a polysilicon resistor below the metal line to heat the structure [4]. They emphasize good correlation with package level tests. Limitation in this case is that efficiency is lower for high metal level because of dissipation in bulk silicon.

To avoid this problem, we designed a new structure that locally heats the tested metal line separately with the current effect.

II. DESCRIPTION OF THE INNOVATIVE DEVICE

A. A new design

This new architecture is the combination of a standard NIST test line with a heating coil. The coil is a three metal levels device which embraces the EML. This configuration maintains the heat radiated by the coil in the metal line.

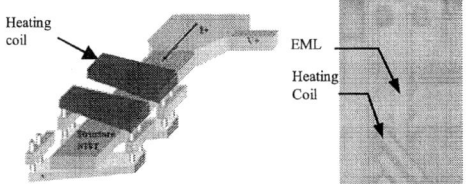

Fig. 1: Schematic view of the heating coil surrounding the metal line tested (left). Picture of current leading and coils (right).

More precisely, three coils surround the EML and fit into each other. Only two of them are dedicated for the heating. The third one can heat and give precisely the temperature near the EML with a four-probe measurement. The metallization stack employed to fabricate this structure is a standard CMOS copper back-end. The only requirement is to have a three metal levels configuration to realize the coils. So, it can be easily adapted to any standard copper process.

B. Heating capabilities

The first step of the test structure study was to check if the temperatures obtained thanks to the heating coils are compatible with standard electromigration at package level tests (300-350°C). Fig. 2 shows the temperature obtained on the EML versus the power injected in the heating coils.

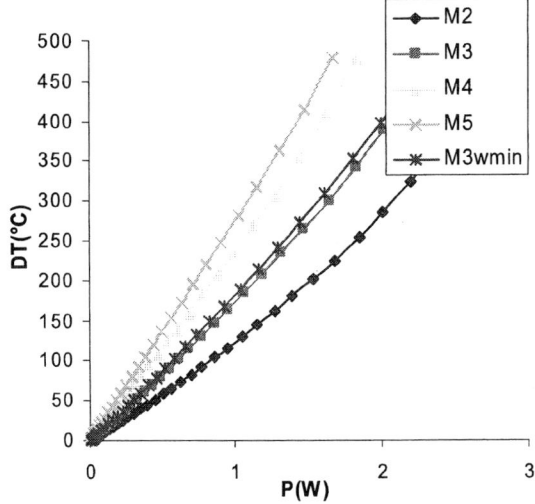

Fig. 2: Temperature measured on EML vs. injected power for different metal levels.

As expected, the efficient heat leads to very high temperature (over 300°C for 2W). Indeed, the architecture limits the heat flow through the silicon bulk.

III. ELECTROMIGRATION TEST PROCEDURE

A. Temperature control

The "temperature coil" allows measuring precisely the temperature of the EML. Thanks to this, we can control the current in the coils heating with a feedback loop. Users define the temperature regulation precision (for example ± 1°C). This is in agreement with package level tests requirements.

B. Starting procedure

Different parameters are required to use the heating coil structure. The user has to previously measure the temperature coefficient of resistance (TCRs) of the EML as well as the heating coils. It will be useful to link the resistance value with the temperature. Before starting the electromigration test, it is necessary to evaluate the initial current to put in the heating coils to insure a good regulation loop and avoid over-heating the EML.

C. Electromigration test

The target temperature (T_t) is reached by adding the heating coil contribution (T_C) and the Joule effect inside the EML (T_N). So, current inside coils is reduced to take into account the self-heating in the metal line.

Coil resistance is measured periodically to regulate current inside the heating coils. Moreover, as it is not correlated with the evolution of the EML, it avoids reducing heating because of a resistance increase due to void formation inside the metal line. In the mean time, the EML resistance is monitored: we define a stop criterion on the resistance variation. When this limit is reached, the test is stopped. Time to failure, EML and coils resistances and all required parameters are stored in a data file.

IV. RESULTS COMPARISON WITH PACKAGE LEVEL TESTS

A. MTF versus metal level

We leaded at the same time standard package level tests and innovative wafer level tests to compare the Ea and n value in both cases.

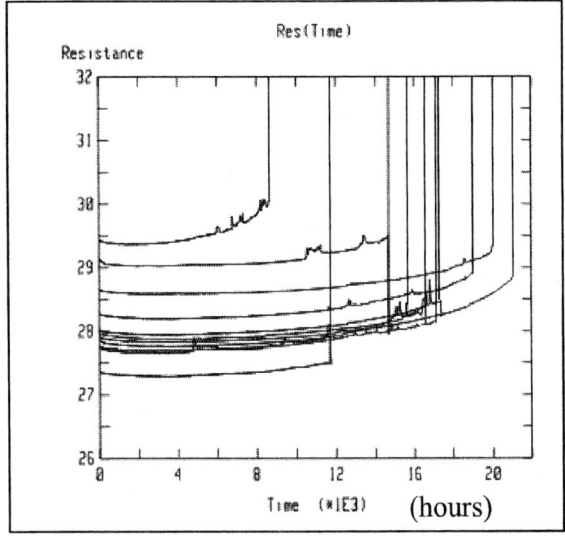

Fig. 3: Resistance variation of EML for different dice at T_T=300°C and I_{EML}=20mA.

For each temperature and current stress, 10 dice were tested to have enough data for statistics.

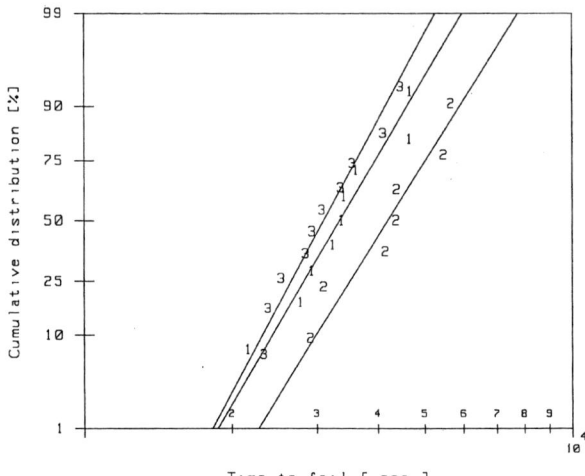

Fig. 4: Time To Failure (TTF) distributions at metal 2 levels for 3 different J_{EML} conditions.

We can observe on Fig. 4 that distributions are parallel together which indicates the same electromigration degradation mode.

TABLE I
MTF VS. TARGET TEMPERATURE AND J_{EML} ON METAL 2

T_T	J_{EML}	I_{coil}	MTF
[°C]	[mA/µm²]	[mA]	[s]
300	19.05	235	6132
320	19.05	240	4951
350	19.05	250	3384
300	181	170	3541
320	181	180	3121
350	181	185	2590
300	200	165	2836
320	200	170	2431
350	200	175	2191

Table 1 resumes some Mean Time to Failure (MTF) results for different target temperatures and stress current densities in the EML. Due to self-heating, we can reduce drastically the test time.

B. Ea and n parameters extraction

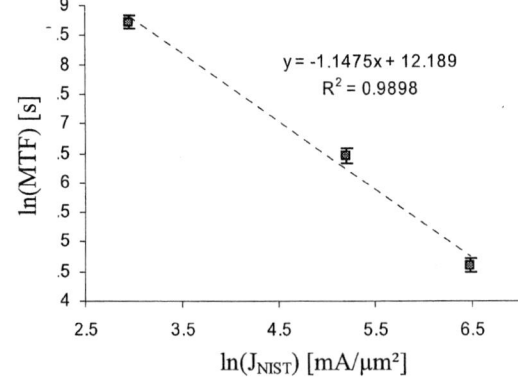

Fig. 5: Ea extraction with 3 current stress conditions.

Fig. 5 clearly shows a good correlation with MTF values

versus J_{EML}.

TABLE II
COMPARISON WAFER/PACKAGE

Metal level	Wafer level test		Package level test	
	Ea (eV)	n	Ea (eV)	n
Metal 2	1.03 [0.94-1.18]	1.08 [0.85-1.91]	1.09 [0.92-1.14]	1.07 [0.86-1.19]
Metal 5	0.75 [0.69-0.78]	1.20 [1.17-1.36]	0.71 [0.54-0.91]	1.31 [1.16-1.54]

These results allow computing Ea and n parameters for each device. Table II summarizes the values obtained for structures with EML at metal 2 and metal 5 levels.

First of all, we can notice that these values are, in both cases, in agreement with those for copper metallization. Secondly, package and wafer level results are very close. This emphasizes the ability of the dedicated structure to extract Black's parameters at wafer level.

V. CONCLUSION

We demonstrated the feasibility of electromigration tests at wafer level and Black's parameters extraction with a novel structure merging a standard EML with heating coils. Using a thermometer coil, temperature of EML is controlled and a loop allows limiting its variations as for package level tests. Ea and n are very close to those obtained at package level. At the same time, we can reduce test time for each die.

REFERENCES

[1] Schafft, H.A., Staton, T.C., Mandel, J., Jschott,D., "Reproducibility of électromigration measurements", IEEE Transactions on Electron Devices, 1987, Vol. ED-34, N°3,p. 673-681.

[2] J.R Black, "Electromigration failure modes in Aluminum metallization for semiconductor devices" Proc IEEE, 1969, Vol. 57, n°9, pp. 1587-1594.

[3] M. Sakimoto, T. Itoo, T. Fujii, H. Yamaguchi, K. Eguchi, "Temperature measurement of Al metallization and the study of Black's model in high current density", Reliability Physics Symposium, 1995. 33rd Annual Proceedings., IEEE International 4-6 April 1995, pp.333 – 341

[4] Hin-Kiong Yap; Kin-Leong Yap; Yew-Chee Tan; Keng-Foo Lo; "Wafer level electromigration testing on via/line structure with a poly-heated method in comparison to standard package level tests", Physical and Failure Analysis of Integrated Circuits, 2003. IPFA 2003. Proceedings of the 10th International Symposium on the 7-11 July 2003 pp. 75 - 79

Life condition monitoring on smart power devices using a sequence of current and charge-based capacitance measurements

Zhenqiu Ning*, Erwin de Vylder*, Filip Bauwens*, Basil Vlachakis*, H-X Delecourt*, Renaud Gillon*, Patrick Van Torre** and Dan Hegsted*

* AMI Semiconductor Belgium bvba
Westerring 15, 9700 Oudenaarde, Belgium
** Hogeschool GENT,
Schoonmeersstraat 52, 9000 GENT, Belgium
zhenqiu_ning@amis.com

ABSTRACT

The precise life condition monitoring on smart power devices is important for smart power circuit design. To perform such a monitoring, a technique based on charge-based capacitance measurement has been developed, by which all DC, AC measurements and stress are implemented in sequence on a device. Measurement, stress and re-measurement are repeated on a logarithmic time sequence, and key parameters for the life condition monitoring are extracted from the measured curves. The degradation of the key parameters is characterized as a measure for the life condition monitoring. The technique is accurate, flexible and with lower cost.

I. INTRODUCTION

Smart power circuits, devices and technologies contribute to the realization of the system-on-chip concept by combining digital logic with analogue signal processing and power and high voltage switching. Design and fabrication of highly reliable and efficient smart power circuits is one of the most important tasks in the design of power systems. To enable a robust design of smart power circuits leading to a first-time-right design with built-in reliability, compact models will be built that accurately describe device operations, including extensions to life condition monitoring (safe-operating area SOA), matching properties and ESD. The precise life condition monitoring on smart power devices is important for smart power circuit design [1][2].

In the CMOS based smart power processes [3], there are many power devices available, e.g. floating nMOS/pMOS devices, floating medium voltage NDMOS (Vds up to 14 V) and floating high voltage lateral PDMOS, etc. Degradation of these devices, e.g. due to hot-carrier injection and radiation damage, etc. has to be characterized for both the understanding of the degradation mechanisms in the device for the improvement of the process for better device, and the life condition monitoring for smart power circuit design [1][4][5]. For the degradation mechanisms, many studies have been done using the charge pumping technique. For the life condition monitoring, the characterization can be done by the measurement of the variation of key DC parameters and capacitance with respect to stress. The degradation of the key DC parameters has been well characterized and presented [5][6]. The degradation of DC parameters together with gate capacitance has not been well characterized yet.

In a smart power circuit, the gate-drain capacitance Cgd of a power device dominates the drain switching waveform due to the Miller effect [7]. The gate-source and gate-drain capacitances, i.e. Cgs and Cgd must be charged and discharged by the gate drive circuit to enable the switching. Due to the creation of interface traps and the injection of charge in the drift region, capacitance variation with respect to electrical stress will occur. Characterization of the capacitance variation (degradation) is important since the capacitances determine device switching speed.

To perform such a characterization for life condition monitoring, a technique based on charge-based capacitance measurement has been developed, by which all DC, AC measurements and stress can be implemented in sequence on a device. Based on the measured Id/Vd, Id/Vg and C/V curves of the device, the key parameters for the life condition monitoring will be extracted, e.g. Vt, Beta, Ron and Idlin from the Id/Vg curve, Idsat from the Id/Vd curve and the gate capacitance at accumulation and strong inversion from the C/V curve. Measurement, stress and re-measurement are repeated on a logarithmic time sequence. Then the degradation of the key parameters can be characterized as a measure for the life condition monitoring.

Comparison with the existing DC and capacitance measurements shows that the newly developed technique has many advantages. It reaches the same accuracy as that of a commercial Analyzer used for DC measurement. Capacitance can be measured on a regular sized device using the charge-based capacitance measurement method, eliminating the use of an LCR meter. The DC and capacitance measurements can be combined together and implemented on the same device. The measurement system is reduced to the use of only a few SMU's. The measurements are accurate, more flexible and with lower cost.

This paper presents the technique in detail. Here a floating n-type DMOS device is selected as an example to show the technique. The DC current and capacitances of the DMOS device will be measured with very high resolution at different bias voltages. The key parameters of the devices, e.g. Vt, Beta, Ron, Idsat, and gate capacitance will be extracted for the characterization of device degradation for the life condition monitoring for smart power circuit design.

In section 2, the device under study will be introduced. The principle of the technique and the design of the test structure are presented in Section 3. The measurement results and their application to the life condition monitoring are shown in Section 4. Conclusions are given in Section 5.

II. DEVICE DESCRIPTION

The cross-section of the device under study is shown in Figure 1, which is a floating Lateral n-type DMOS (FLNDMOS) processed using the smart power technology [3], based on a 0.35 μm CMOS process. Different from the standard CMOS technology, the NDMOS device is built into an n-type epitaxial layer instead of being positioned into a p-epi layer on a p-type substrate. As a consequence, the Pwell will be floating and the Nwell will be common. To provide the high voltage features, some extra masks and implants have been implemented. If punch-through currents are not generated between the p-substrate and the p-regions implanted inside the N-epi at the top surface, the N-epi layer creates a top region virtually floating compared to the underlying p-substrate. From Figure 1 one can see that the FLNDMOS transistor is an asymmetric device, with the drift region on the drain-side located in a lightly doped N-epi. The channel is created by the reversion of the P-well under the gate. High breakdown voltage is achieved due to the depletion region that extends from the P-well into the drift region, which allows for a higher voltage drop between the highly doped drain contact region and the channel-region. A thick field oxide (LOCOS) protects the thin gate oxide from the drain voltage. As a consequence, it can be applied as medium voltage switch and medium voltage current source, where drain terminal can stand up to 14V with respect to the bulk of the device. The whole device can float with respect to the p-substrate due to the epitaxial zone.

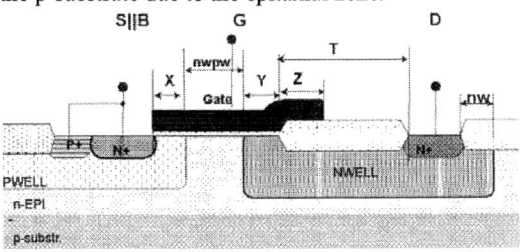

Fig. 1: Cross section of the floating lateral NDMOS for MV application.

The gate capacitance at different bias conditions can be approximated. If Vgs = -Vdd, device channel (X region) is at accumulation and drift region (nwpw and Y) at strong inversion. A continuous hole sheet will be formed over the drift region and the channel. The inner fringing capacitor will be gone due to the shield of field by the continuous hole sheet layer. Let Cof denote the outer fringing capacitor, Col the overlay capacitor, Cox the unit capacitance of the gate oxide, and Cox-field the unit capacitance of the field oxide. The gate capacitance is approximated as:

$$C = (C_{of} + C_{ol}) + C_{ox} + C_{ox-field} \quad (1)$$

with

$$C_{ox} = \frac{\varepsilon_0 \varepsilon_r}{T_{ox}} \quad (2)$$

and

$$C_{ox-field} = \frac{\varepsilon_0 \varepsilon_r}{T_{ox-field}} \quad (3)$$

Where Tox denotes the gate oxide thickness, ε_0 and ε_r denote the vacuum permittivity and the relative permittivity of the gate oxide.

If Vgs >> Vt, e.g. Vgs=Vdd, device channel (X region) is at strong inversion and drift region (nwpw and Y) at accumulation. A continuous electron sheet will created across the surface underneath the gate. The LDD region is deeply accumulated (Col = Cox) and the inner fringing capacitance is gone due to the shield of field by the electron sheet layer. The gate capacitance can be written as

$$C = C_{of} + C_{ox} + C_{ox-field} \cdot \quad (4)$$

If the outer fringing capacitance can be negligible, the gate capacitance at accumulation and at strong inversion can be calculated approximately by:

$$C_g = C_{ox}W(X + NWPW + Y) + C_{ox-field}WZ . \quad (5)$$

with W the width of the device. Equation (5) will be used for a manual calculation of the gate capacitance for LFNDMOS in this paper.

III. DEVELOPMENT OF TEST STRUCTURE

A Design of test structures

For precise analogue characterization of the capacitance of DMOS device at different bias voltages, a technique based on CBCM (Charge-Based Capacitance Measurement) has been developed. The principle of the test structure is shown in Fig.2 (a). The target capacitance C1 is the gate capacitance of the FLNDMOS device discussed in section II, as shown in Fig.2 (b). It is connected to the output node of a "pseudo" inverter, which is different from a normal inverter, since its nMOS and pMOS do not share the same input. Each one has its own gate input to allow them to be driven by two non-overlapping signals to suppress the short circuit current when the inverter is operated. In order to suppress the parasitic effects of the basic "pseudo" inverter, the test structure has a reference structure, which is a duplicate of the basic "pseudo" inverter except that it has no target capacitor at its output node. Here, C2 represents only parasitic capacitance. Both "pseudo" inverters are driven by the same non-overlapping signals.

The design of the CBCM structures has been optimized to reduce the mismatch of both the charge injections and switch parasitics between two pseudo inverters [8]. Here a pass-gate scheme using compensating charge injection has been developed. When the switch is turned off, the charge in the inversion channel of the MOS devices will be dispersed, forcing current to flow into load capacitors at the MOS drain or source. The current due to the charge injection induces an over-estimation of the capacitance. The use of a transmission-gate switch reduces charge injection due to compensation of electrons and holes channel carriers. Since the charge injection error is load dependent, smaller MOS devices will result in smaller charge injection error. However, the smaller devices will result in other mismatches, e.g. the mismatch of drain current and threshold voltage. A design trade-off has been done to meet the different requirements. Within the switches, the open MOS device connected to capacitors introduces capacitance, and the closed MOS device introduces series resistance along the signal path. The cross-quad configuration has been taken for the design of all switches to minimize the mismatch of the capacitance and resistance. The CBCM accuracy also depends on the pulse rising time of the non-overlap clock signals. The faster the clock signals the larger the error due to charge injections. Slow clock signal reduces the measured DC current, consequently reducing the resolution. Optimum design for clock generator also has been done to make the balance between the possible charge injection error and the error due to the DC current reduction.

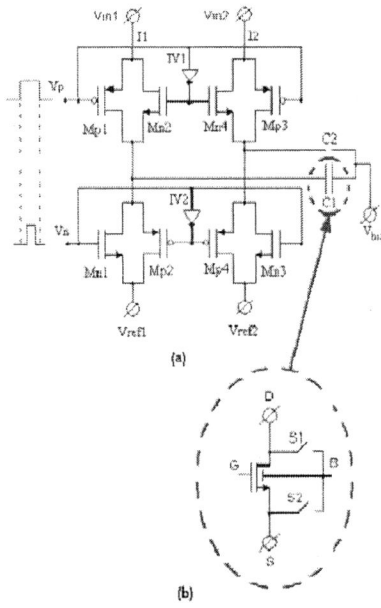

Fig. 2: Principle of the current and capacitance degradation test structure. (a) CBCM test structure, (b) a DMOS device with some switches.

With a transmission gate implementation of CBCM as shown in Fig. 2 (a), capacitance is determined by:

$$C = C_1 - C_2 = \frac{I_1 - I_2}{(V_{in} - V_{ref})F} . \qquad (6)$$

Where I1 and I2 are the average DC currents measured at node Vin1 and Vin2, (Vin – Vref) is the voltage drop for pumping charge and F is the frequency of the non-overlap signal. Here we select:

$$V_{in} = V_{in1} = V_{in2}$$
$$V_{ref} = V_{ref1} = V_{ref2}$$

Let voltages Vin, Vref and Vbias be all with respect to

ground. The effective voltages drop across the capacitor C1 or C2 can be calculated as:

$$V_C = \frac{V_{in} + V_{ref}}{2} - V_{bias} . \qquad (7)$$

From (7), one can see that capacitance as a function of bias voltage can be measured if a scan of Vbias is applied.

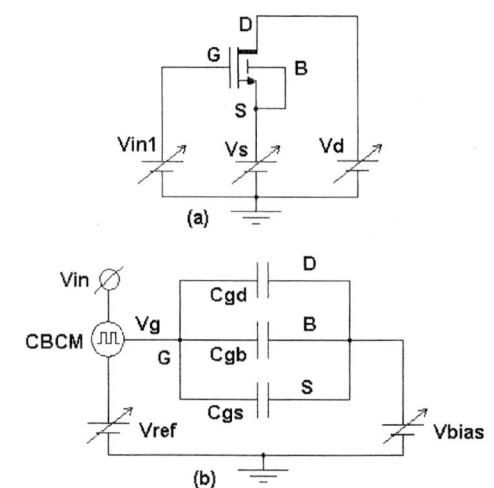

Fig. 3: Equivalent circuits. Measurement of DC current is shown in (a), and measurement of capacitance is presented in (b).

Using this test structure, the measurement of DC currents, capacitances and stress can be done in sequence. Let Vp = Vn = low and with certain options of switches, e.g. switch S1 is off and switch S2 is on, the measurement is converted to DC mode, as shown in Fig. 3 (a). For this setting, devices Mp1 and Mn2 will be on and devices Mn1 and Mp2 will be off. As a result, voltage Vin1 will be copied to the gate, i.e. Vg=Vin1, and the drain, source and bulk voltages will be Vd=VD, Vs=VS and Vb=VS. The drain current as a function of gate and drain voltage, i.e. Id/Vg and Id/Vd curves can be measured automatically by scan of Vg or Vd. The device capacitance is measured by CBCM. Gate is a terminal. The drain, source and bulk of the DMOS device are connected together as another terminal and linked to Vbias, as shown in Fig. 3 (b). The gate capacitance Cgg as a function of Vgs (CV curve) can be measured by a scan of the bias voltages. Stress is done in DC mode. The only difference is that the stress voltage is higher. After stress, measurement of DC curves and capacitance Cgg are repeated again. The Measurement, stress and re-measurement will be implemented on a logarithmic time sequence. Then the degradation of the key parameters can be characterized as a measure for the life condition monitoring.

B. Layout generations

The layout of the test structures is shown in Fig. 4, which consists of a 26-bondpad configuration with a test structure for FNDMOS device inside. The size of the DMOS device is W = 60 μm and the number of fingers is 1. The test structure was designed for the characterization of the gate capacitance

48

and current degradation of the DMOS device. With this test structure, one can measure DC current and gate capacitance Cgg in a stress sequence for degradation study. In order to minimize the mismatch between the two pseudo inverters of CBCM and to reduce leakage current when open-circuit, circuit design avoided minimum device dimensions and gave special care to the layout of the test structure.

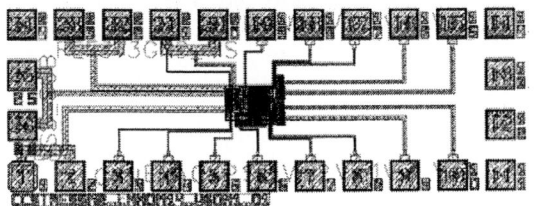

Fig. 4: Layout of the test structure.

C. Development of the automatic measurement setups

The test chip allows for measurement of DC and AC parameters, and DC stressing of the device under test. Here the aim is to develop a measurement setup which allows us to do fully automatic measurement and stress routines: all parameters of the device are measured, the device gets stressed for a predefined time, and parameters are measured again.

To switch between different measurements modes, a few control signals on the chip have to be changed. A test PCB was developed, allowing the PC to control these signals. The schematic of the PCB is given in Fig. 5. It contains a voltage regulator (IC5), four relays and a few buffers, etc. It is a double sided board, with ground planes on both sides. The analog and digital parts have been separated as much as possible; their ground planes are isolated and connected in only one point.

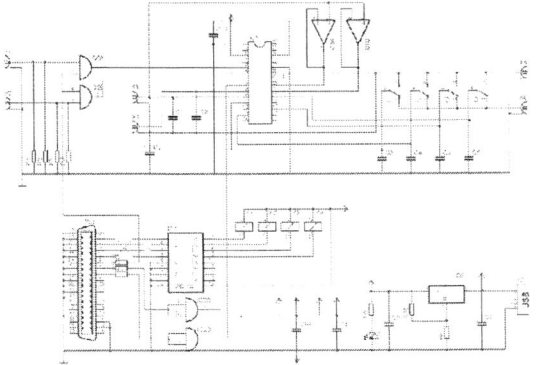

Fig. 5: Schematic of the CC-degradation test PCB.

To allow for efficient measurements, a new software program was developed. It had to be easy to use, to control the whole measurement setup, and to enable automatic measurements without operator interaction. The software was programmed from scratch in Visual Basic 6. It is written in a modular way to allow for easy adaptation or expansion.

Let Id denote device drain current, gm the transconductance, Vt the threshold voltage, Idsat the saturation current, Ron the on resistance, Vc the voltage drop across the capacitor, Cgg-

acc, Cgg-depl, and Cgg-inv the gate capacitance at accumulation, depletion and strong inversion regions. Following functionality has been implemented:

- measuring DC curves (Id and gm vs. Vg, Id vs. Vd)
- extracting DC parameters (Vt, gm, Idsat, Ron)
- measuring AC curves (Cg vs. Vc)
- extracting AC parameters (Cgg-acc, Cgg-depl, Cgg-inv)
- stressing the device during a predefined time
- perform an automatic stress/measurement routine
- save results in CSV format (MS Excel, OpenOffice Calc compatible).

D. Measurement, stress and re-measurement in sequences

Before automation measurement and stress, a calibration between measurement and calculation has been done for gate capacitance, e.g. for a NMOS device with width W=60 μm and length L = 0.35 μm, the difference is 0.9% (99.0 fF to 98.1 fF).

Measurement of DC curves is pretty straightforward: Vg and Vd are scanned in a predefined range, and Id is measured. However, to improve accuracy, the measurement is pulsed, to prevent device heating. The duty cycle is about 5%, so the device can cool down 95% of the time.

Vt is extracted as the gate voltage (Vgs) which is needed to obtain a drain current (Id) of 4μA at a drain-source voltage of 0.1 V. An approximation algorithm was implemented to measure Vt with an accuracy better than 0.01mV.

Parameter GMmax is the maximum of the gm curve obtained at Vds = 0.1 V. An algorithm was also implemented to measure GM as a function of Vg and extract GMmax. The standard deviation of repeated gm measurements is better than 0.03%.

The aim of this project is to stress the device, and measure a set of parameters at logarithmic time intervals. The stress was done at Vds=14V and Vgs=4.5V. The time intervals were chosen as follows: 1s – 2s – 5s – 10s – 20s – 50s – 100s - ... – 10000s.

The measurement/stress/measurement routine has been automated. It contains the following steps:

1. The fresh, unstressed device is measured.

2. The device is stressed for a certain time interval.

3. The device is allowed to cool down.

4. The device is measured again.

5. If desired, return to step 2.

IV. RESULTS AND THEIR APPLICATION TO LIFE CONDITION MONITORING

The test chip has been processed in AMIS 0.35 μm smart power technology [3] and measured automatically.

The measured Id and gm vs. Vgs curves are shown in Fig. 6, in which -b and -a denote data before and after stress. Both Id and gm have decreased significantly.

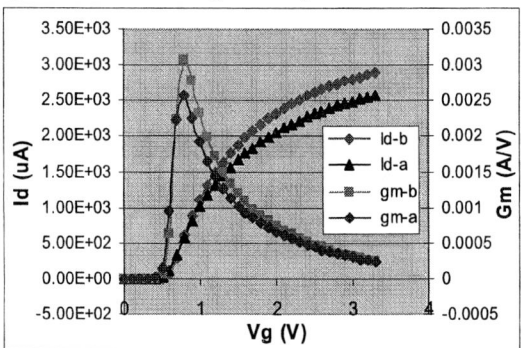

Fig. 6: Id/Vg and Gm/Vg curves, before and after stress.

The measured Id vs. Vds curve are presented in Fig.7, -b and -a also denote data before and after stress. Again, Id significantly decreased due to the stress.

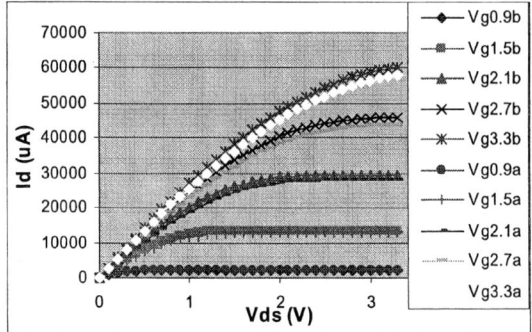

Fig. 7: The measured Id/Vd curves, before and after stress.

In figure 8, the measured Cgg vs. Vc curves are shown. The differences are small, but it is clear that in the depletion region, the capacitance decreased.

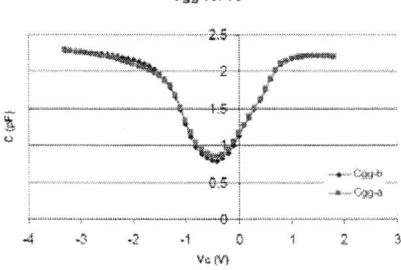

Fig. 8: Cgg vs. Vc, before and after stress.

For lifetime condition monitoring, typically the worst degrading parameters are considered as key parameters, and lifetime is defined as the time where a 10% shift of the key parameters is reached. By varying the stress conditions (Vds, Vgs), the lifetime in the (Vds,Vgs) space can be extrapolated from the stress conditions using the Takeda model [9]: $\log(\text{lft}) = A + B/V(\text{d or g})$s. Here the parameters Ron, Idsat, GMmax, Vt and gate capacitance Cgg are chosen as the key parameters.

Fig. 9 presents the degradation of the on-resistance vs. stress time. Ron1, Ron2, Ros3 and Ron4 denote the on-resistance of the FNDMOS device under different stress conditions. Due to the "hot carrier" effect, the oxide/silicon interface above channel region is damaged by the more energetic hole and electronics. The damage to the interface results in a lower mobility. Additionally, due to the charge injection in the drift region, the channel current is forced into a route that is located somewhat deeper under the gate. These result in an increased on-resistance. The on-resistance is increased about 6% after 10000 seconds of accelerated stress.

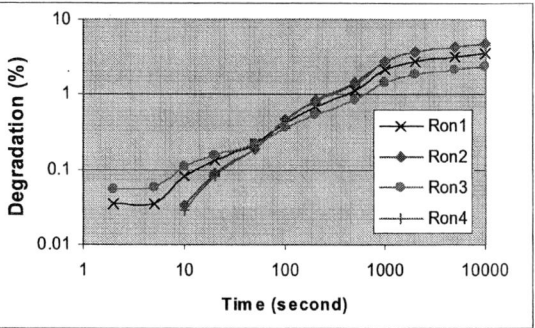

Figure 9: Degradation of Ron vs. stress time.

Fig. 10 shows the degradation of Idsat, GMmax and Vt vs stress times. Due to the increased on-resistance, the Idsat decreases. During the device stress, interface traps increase and charges are injected. They always come together. Such traps can cause the mobility to decrease, resulting in a decrease of the Gm parameter. For Gm degradation, the interface state density in the drift region also plays an important role. Typically for DMOS devices, the Vt does not shift much, as the degradation in the drift region is dominant over the channel degradation by far. The slight decrease seen here may be partly due to the extraction method.

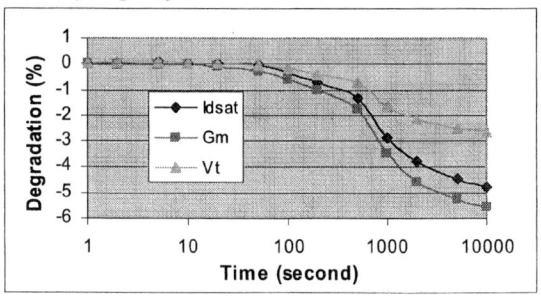

Fig. 10: Degradation of Idat, GMmax and Vt vs. stress time.

In Fig. 11, the degradation of Cgg vs stress time is shown, from which one can see that the Cgg value increases in the depletion region, and remains similar in the accumulation and inversion regions. The injected charge due to stress, e.g. 1e10 to 1e11 charges/cm2, can cause the gate capacitance to increase in depletion. The injected charges are completely mastered by the accumulated or inverted charge in accumulation or strong inversion. This is the reason why the capacitance is not significantly different before and after stress in accumulation and strong inversion.

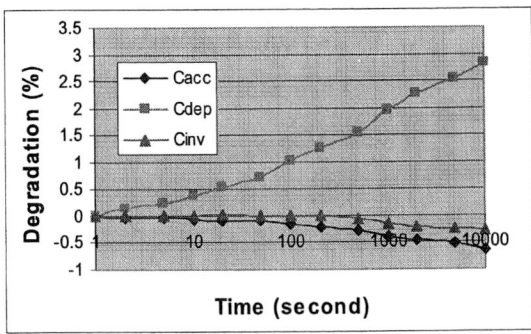

Fig. 11: Degradation of Cgg vs. stress time.

The correlation between the shift of GMmax and Idsat and the change of the gate capacitance in the depletion region is shown in Fig. 12, from which one can see that GMmax and Idsat decrease while the gate capacitance increases. The increase of depletion capacitance (but no shift in function of Vc, as in Fig.8) points towards a negative charge build-up in the drift region only [10], and is consistent with the degradation in on-resistance. The saturating shape of the correlation can be explained by the different dependence of the capacitance (linear) and the on-resistance / GMmax (saturating power law) on the accumulated oxide charge.

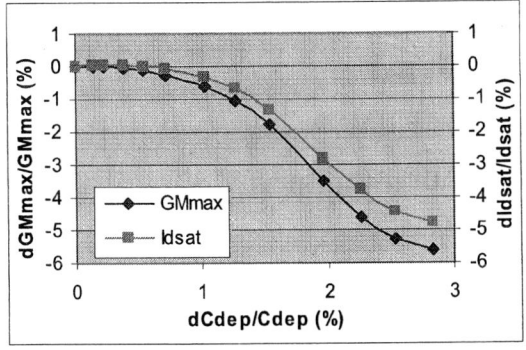

Fig. 12: Correlation between the degradations of GM, Idsat and capacitance.

V. CONCLUSIONS

A measurement technique based on the CBCM principle has been developed for the analogue characterization of current and capacitance degradation. Due to the optimization of switch functionalities and clock signals, the developed technique has a higher resolution, is more flexible and allows extracting a significant amount of statistics. The Id/Vd, Id/Vg, Gm/Vg and CV curves have been successfully measured, from which the key parameters e.g. Vt, Beta, Ron, Idlin, Idsat and the gate capacitance at accumulation and strong inversion have been extracted. Measurement, stress and re-measurement on a logarithmic time sequence are implemented. The degradations of the key parameters are characterized as a measure for the life condition monitoring. The technique is accurate, more flexible and low cost.

ACKNOWLEDGMENT

The author would like to thank Catherine Dekeukeleire and Jean-Francois Cano for their support on the definition of stress conditions.

REFERENCES

[1] C. Hu, S.C. Tam, F-C Hsu P-K Ko, T-Y Chan and K. Terrill, "Hot-Electron-Induced MOSFET Degradation-Model, Monitor, and Improvement," IEEE J. Solid-State Circuits, Vol. SC-20, No. 1, pp 295-305, Feb. 1985.

[2] Specific targeted research project, "Robust mixed-signal design-methodology for Smart-Power ICs (ROBUSPIC)," IST project.

[3] P. Moens, D. Bolognesi, L. Delobel, D. Villanueva, H. Hakim, S.C. Trinh, K. Reynders, D. De Pestel, A. Lowe, E. De Backer, G. Van Herzeele, and M. Tack, "I3T80: A 0.35 um Based System-on-Chip Technology for 42V Batter Automotive Applications," ISPSD 2002, Santa Fe, USA, June 4-7.

[4] P. Heremans, J. Witters, G. Groeseneken and H. E. Maes, " Analysis of the Charge Pumping Technique and Its Application for the Evaluation of MOSFET Degradation," IEEE Trans. Electron Devices, Vol. 36, No. 7, pp13181335, July 1989.

[5] "Hot Carrier Design Considerations for MOS Devices and Circuits," Edited by Cheng T. Wang, Van Nostrand Reinhold, New York, 1992.

[6] J.E. Chung, K.N. Quader, C.G. Sodini, P-K Ko and C. Hu, "The Effects of Hot-Electron Degradation on Analog MOSFET Performance," IEDM, pp. 553-556, 1990.

[7] R. Gregorian and G. Temes," Analog MOS Integrated Circuits for signal Processing," John Wiley & Sons, Inc.

[8] Z. Ning, H.-X. Delecourt, L. De Schepper, R. Gillon, and M. Tack, "Precise Analogue Characterization of MIM Capacitors Using an Improved Charge-Based Capacitance Measurement (CBCM) Technique," *Proc. of ESSDERC 2005*, Grenoble, France, Sept. 12-16, 2005.

[9] E. Takeda and N. Suzuki, "An Empirical Model for Device Degradation Due to Hot-Carrier Injection," IEEE Electron Device Letters, vol. EDL-4, no. 4, pp. 111-113, April 1983.

[10] N. Hefyene, C. Anghel, R. Gillon, 'Hot carrier degradation of lateral DMOS transistor capacitance and reliability issues', Proc. Of IRPS 2005, San Jose, California, Apr. 17-21, pp.551

Mixed Test Structure for Soft and Hard Defect Detection

F. Rigaud[*&†], J.M. Portal[*], H. Aziza[*], D. Nee[†], J. Vast[†], F. Argoud[†] and B. Borot[‡]

[*]IM2NP - Institut Materiaux Microelectronique Nanosciences de Provence, UMR CNRS 6242,
IMT-Technopole de Chateau Gombert, F-13451 Marseille Cedex 20, France
[†]ST Microelectronics, ZI Rousset, BP2, F-13106 Rousset, France
[‡]ST Microelectronics, 850 rue Jean Monnet 38926 Crolles, France
Email: fabrice.rigaud@l2mp.fr

Abstract—The objective of this paper is to present a mixed test structure designed to characterize yield losses due to hard defect and back-end process variation (PV) at die and wafer level. A brief overview of the structure, designed in a ST-Microelectronics' 130nm technology, is given. This structure is based on a SRAM memory array for detecting hard defects. Moreover each memory cell can be configured in the Ring Oscillator (RO) mode for back-end PV's characterization. The structure is tested in both modes (SRAM, RO) using a single test flow. Experimental results are given and confirm the ability of the structure to monitor PV and defect density.

Introduction

The System-on-Chips (SoCs) technology demand grows extremely fast, introducing more and more sophisticated SoCs, especially for advanced data communications and wireless products. This evolution leads to complex problems in terms of design and more specifically in terms of manufacturing test. To achieve SoCs with high performances, advanced process technologies are used. But the actual process technologies are reaching a high degree of sensibility to defect, which may slow down yield ramp-up and reliability. In this background, circuit designers and process engineers have to work in close relations to deal with process variation's (PV) impact on design's performances. Semiconductor industry has adopted a solution based on test chips [1] to evaluate process variation. The SRAM are well known to be a good vehicle to detect hard defect [2], [3], [4]. In the same way ring oscillators [5], [6], [7] are massively used to characterize process variation. Consequently the main objective of this paper is to present a test chip which takes advantages of both structures (SRAM, RO). Section 1 gives a brief overview of the test structure's architecture using a bottom up approach. Section 2 is devoted to the description of the associated test flow and the treatment of the files generated by the tester. The ability of the test chip to detect hard defect and PV is validated (Section 3) with test data results. Finally, the section 4 gives some concluding remarks.

I. Test Structure's Overview

The test structure's overview is given using a bottom up approach from the elementary cell's architecture up to the description of the entire chip. The test structure is designed in a CMOS ST-Microelectronics' 130nm technology. This circuit is based on a classical SRAM memory cell with two pass gates and two inverters, on which a third inverter is added (Fig. 1). When the OSC signal is at zero, the third inverter is

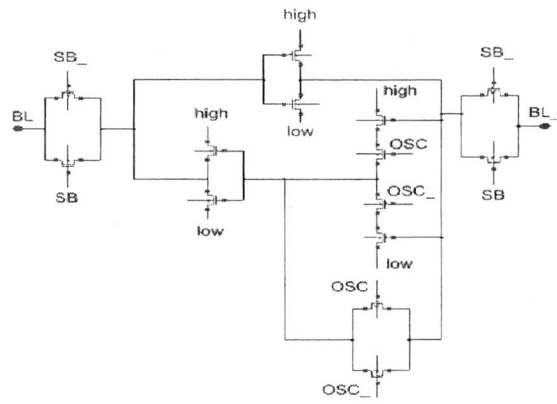

Fig. 1. Schematic view of the basic cell with 2 pass gates 2 inverters and an optional third inverter.

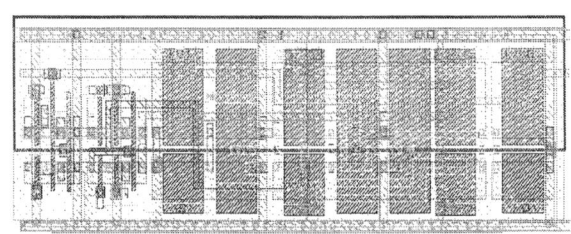

Fig. 2. Memory cell layout (94 μm^2 per bit). Inverters designed with a long channel to reduce the oscillation frequency.

bypassed. By setting up the OSC signal, the three inverters form an oscillation loop. Each individual memory cell is classically selected by setting up the corresponding ROW and COL signals. The layout of the memory cell is given by Fig. 2. This full custom design allows :

- Limiting the oscillation frequency thanks to long channel inverters.
- Interconnecting each inverter with different layers of metal.

This hybrid memory cell is called OSRAM and its area is about 94 μm^2. These OSRAM cells are assembled to make a memory array with a checker board configuration which

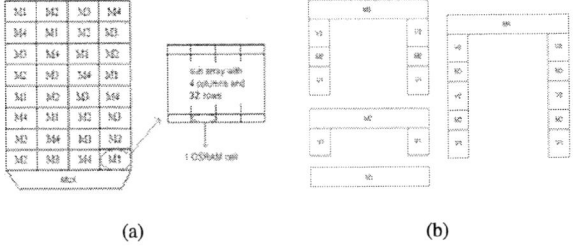

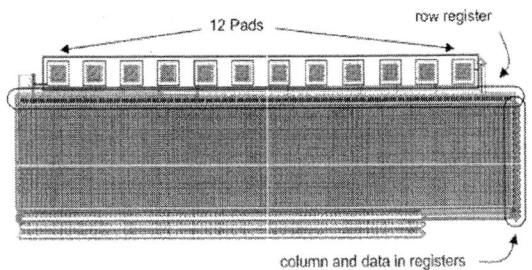

Fig. 3. Memory array topology. Checker board based on sub-arrays with interconnections at different metal layers. M_x routing ends at Metal layer x.

Fig. 4. Layout of the test structure's core(height=1.6mm length=540m).

is given Fig. 3a. The checker board is based on sub-arrays designed using interconnections at different layers of metal. For example a M_x sub-array consists of 4 columns and 32 rows of OSRAM using the metal layer x to interconnect the inverters (Fig. 3b). Indeed the oscillation frequency of a ring oscillator depends on the inverter's front-end but also on the routing between the inverters. Fig. 4 gives the entire layout of the structure with :

- The 4096 bit memory array.
- A column and a row register to address the array.
- A data in register to write the selected cells.
- 12 standard pads.

The total area of the test chip is about 0.86 mm^2.

II. TEST FLOW'S OVERVIEW

The test flow applied to the structure is divided in 3 main steps: standard parametric tests, functional tests, and analogical waveform capture. Those three steps are described below:

- Standard parametric tests: Continuity test, static and dynamic consumptions.
- Functional test: Writing and reading of the register, writing and reading of the memory array.
- Analogical waveform capture: the oscillation waveform of each memory cell is captured and data-logged into a file.

The 3 steps are performed in a single pass (parametric, functional and analog) with a standard production tester (Verigy 93K with a digitizer). After that, the files are transfered on a PC on which a software computes the Fast Fourier Transform (FFT) of each OSRAM. Then, in order to have a higher precision on the measurement, the ninth harmonic is used to draw charts and mappings. The software also links the addresses (row ,column) with the different sub-array types to draw charts by metal layer. In summary, for a 100% yield lot, it takes about 4 hours to test the 150 dies, the output files from the tester occupies 1.4 Go of memory, and the FFT calculation takes about 1 hour and 15 minutes.

III. EXPERIMENTAL RESULTS

In this section, the complete tool set of test data's treatment is presented to evaluate intradie and interdie PV as well as hard defect density. These first results come from the analysis of a 3 wafers lot (4 more wafers are currently processed). The subsection A is devoted to the analysis of a single die to detect intradie variations whereas the subsection B reports the analysis of a single wafer to detect interdie variation of a single wafer. Finally, the sub-section C gives solutions to characterize the variations. All the charts and mappings of this section represent the ninth-harmonic's frequency values of the cell's oscillation.

A. Intradie variation's detection

The test chip is used to collect Pass/Fail and analog data at die level to evaluate defect density and intradie PV. Following the test flow, the hard defects are tracked using the SRAM mode of the array. To achieve this task, a traditional Pass/Fail bitmap approach is applied as presented on Fig. 5a. This first treatment gives the localization and the number of hard defects along the die. Then, a FFT is performed on the analog response of the good dies (RO mode) to evaluate their frequency characteristics. The first representation of the FFT's results is an analog bit mapping of the ninth harmonic as illustrated Fig. 5b. However, because their interconnections are not designed with the same metal layer, the different sub-arrays do not oscillate at the same frequency. Consequently a soft process variation can not be detected with the usual bitmap. So in order to detect these small variation, it is possible to plot an average-relative bitmap. In other words, considering an OSRAM cell of a sub-array M_x with an oscillation frequency F, the *Average-Relative Value* (ARV) is given by F minus the oscillation frequency average of all the M_x OSRAM cells of the die (see (1)). Fig. 6 illustrates this kind of treatment.

$$ARV = F - \frac{\sum F_{OSRAM_{M_x}}}{\#OSRAM_{M_x}} \quad (1)$$

Moreover, to evaluate the intradie process variation, the cumulative distribution and the box-plot by metal layer are computed on Fig. 7. Compared to the bitmap, this chart allows detecting which metal layer suffers a variation. The results obtained are in a good agreement with the ones expected knowing the resistance and capacity introduced by each metal layer and the via stacking condition.

(a)

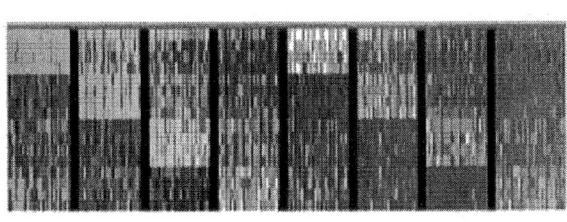

(b)

Fig. 5. Pass/Fail bitmap to localize hard defect in SRAM mode and frequency bitmap to localize process variation in RO mode.

Fig. 6. Bitmap of the average relative frequency.

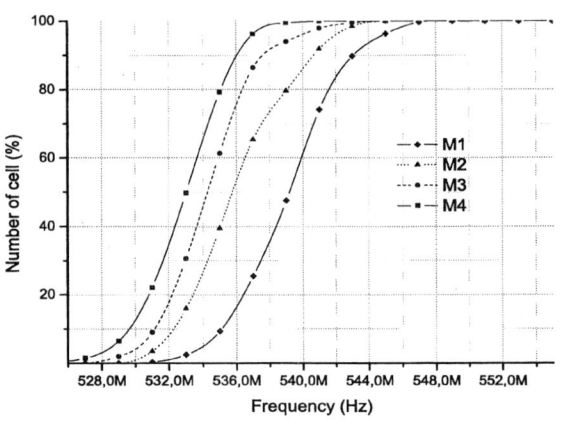

(a)

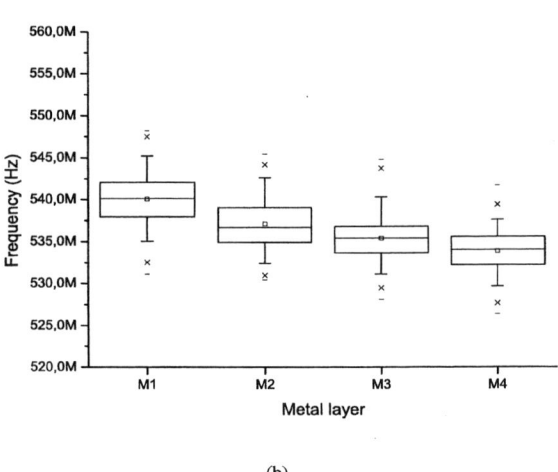

(b)

Fig. 7. Cumulative distribution and box plot at die level.

B. Interdie variation's detection

All the results collected at die level are gathered by wafer to analyze interdie variation or to compare wafer to wafer dispersion. In a first step the wafer map on Fig. 8a represents the number of failing cells of each die along the wafer. In a second step, the wafer map on Fig. 8b represents the mean value of the ninth-harmonic's frequency and allows localizing the interdie variation. The evaluation of this variation is further analyzed using box-plots (Fig. 9) and distributions.

C. Process Variation's characterization

The separated analysis of the different sub-array types (M_1, M_2 ...) can give information on the metal layer variations. For example on Fig. 8b one can observe a variation following the X axis of wafer and the box plot Fig. 9 shows a large distribution on each metal layer. In this example each sub-array type suffers the same variation, consequently the hypothesis is a front-end variation between the left and right edge of the wafer. Even in this case, it is possible to detect and characterize soft back-end variation, using the differences between each sub-array type (M_1, M_2 ...). Following this goal, we build

distributions taking for each $OSRAM_{xy}$ with a frequency F : Frequency($OSRAM_{xy}$)-Frequency($OSRAM_{xy-32}$). Indeed, as described Figure 3a and 3b, a sub-array M_x consists of 32 rows and is followed by a sub-array M_{x+1}. Doing so, one can plot the distributions of the oscillation frequency's differences between each sub-array type. These differences are identified as $Dif(M_1M_2)$, $Dif(M_2M_3)$ and $Dif(M_3M_4)$. Considering that the different sub-array types are designed with the same front-end and routing length, the only disparities in term of back-end resistance are the vias added and the resistivity difference of the top metal layer. For example the oscillation frequency disparity between a $OSRAM_{M_1}$ and a $OSRAM_{M_2}$ is due to the VIA_1 resistance (R_{V_1}) and the resistivity difference between metal layer 1 and 2 ($R_{M_1M_2}$). Then, we can write (2,3,4).

$$Dif(M_1M_2) = a_1.R_{V_1} + b_1.R_{M_1M_2} + c_1 \qquad (2)$$

$$Dif(M_2M_3) = a_2.R_{V_2} + b_2.R_{M_2M_3} + c_2 \qquad (3)$$

$$Dif(M_3M_4) = a_3.R_{V_3} + b_3.R_{M_3M_4} + c_3 \qquad (4)$$

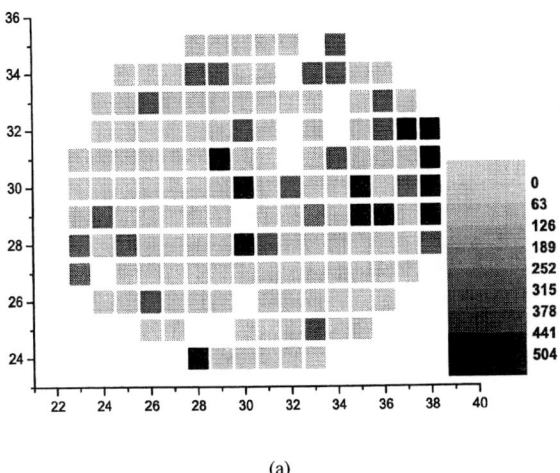

(a)

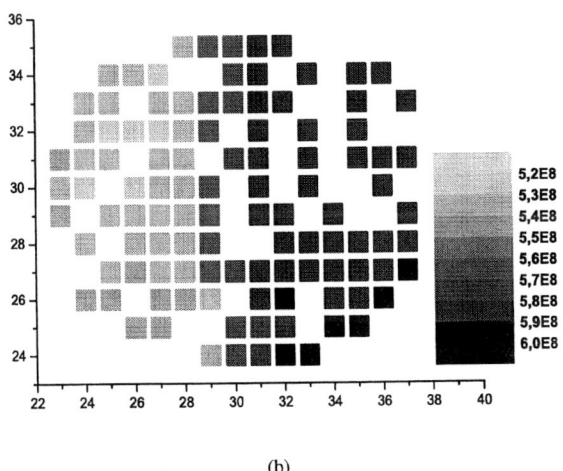

(b)

Fig. 8. Fails' count wafer map to evaluate hard defect density per die in SRAM mode and mean frequency wafer map to detect zone effect.

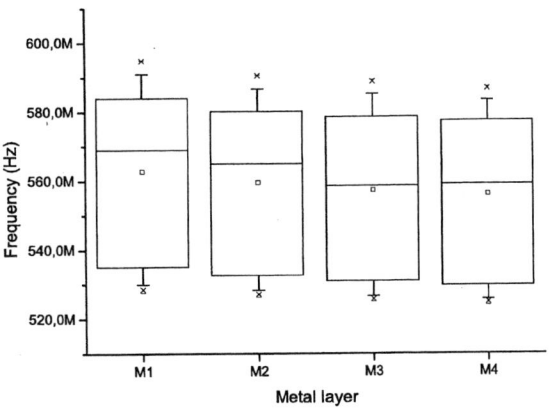

Fig. 9. Box plot of the die's mean frequency in RO mode for an entire wafer to evaluate interdie process variation.

Knowing that the metal resistance is proportional to the line's critical dimension (CD), the CD can be expressed as a function of the frequency differences. In the case of metal 2 this relation is given by (5).

$$CD_{M_2} = a.Dif(M_1M_2) + b.Dif(M_2M_3) + c \qquad (5)$$

Obviously this approach can not be applied to the first and last metal layer.

CONCLUSION

In this paper, a mixed test chip based on a SRAM array which can be configured in a RO array is presented. The associated test flow allowing the extraction of Pass/Fail infor-

mation in SRAM mode and of spectral responses in RO mode, is also described. The test data treatment provides intradie as well as interdie PV's information and defect density's evaluation. Moreover, this test chip keeps a SRAM's functionality and so could be introduced in a product for monitoring of PV and defect density with a low area and routing overhead compared to RO introduction.

ACKNOWLEDGMENT

The authors would like to thank the STMicroelectronics Rousset team for providing wafers, STMicroelectronics Crolles team for designing the structure, and the IM2NP laboratory team for technical support, test material and fruitful discussion.

REFERENCES

[1] Y. Zorian, "What is an infrastructure ip ?" *IEEE Design & Test of Computers*, vol. 23, no. 6, pp. 5–7, May-June 2002.

[2] H. Guo, I. Goh, and J. Zhang, "Integrated yield enhancement strategy for advanced 130 nm beol copper process," in *International Symposium on Semiconductor Manufacturing*, October 2003, pp. 243–246.

[3] T. Zanon, K. Komeyli, M. Ferdman, and W. Maly, "An automated approach using sram bit fail maps," in *International Symposium on Semiconductor Manufacturing*, November 2003, pp. 232–241.

[4] M. Fujii and al, "A large scale, flip-flop ram imitating a logic lsi for fast development of process technology," in *IEEE ICMTS*, March 2007, pp. 131–134.

[5] A. Bassi, L. Croce, A. Bogliolo, and A. Veggetti, "Measuring the effects of process variation on circuit performance by means of digitally-controllable ring oscillators," in *IEEE ICMTS*, March 2003, pp. 214–217.

[6] M. Nourani and A. Radhakrishnan, "Testing on-die process variation in nanometer vlsi," *IEEE Design & Test of Computers*, vol. 23, no. 6, pp. 438–451, November-December 2006.

[7] M. Nekili, Y. Savaria, and G. Bois, "Spacial characterization of process variations via mos transistor time constants in vlsi and wsi," *IEEE Journal of Solid-State Circuits*, vol. 43, no. 1, pp. 80–84, January 1999.

Short-flow Test Chip Utilizing Fast Testing for Defect Density Monitoring in 45nm

Muthu Karthikeyan, William Cote, Louis Medina, Ernesto Shiling, Arthur Gasasira,
Amy Henning, William Ferrante, Mark Craig[†], and Thomas Merbeth[‡]

IBM Systems and Technology Group
2070 Route 52, Hopewell Junction, NY 12533, USA
karthikeyan@us.ibm.com

Advanced Micro Devices[†]
5204 East Ben White Blvd., Austin, TX 78741, USA

AMD Saxony LLC & Co. KG[‡]
Wilschdorfer Landstr. 101, 01109 Dresden, Germany

ABSTRACT

A comprehensive 45nm short-flow test chip was designed and is currently used to improve defect-limited yield. In a novel development to reduce test time, the DC test structures are tested in parallel mode on a functional test platform, resulting in a 5x reduction in test time over conventional parametric testing. The large critical area enables accurate measurement of defect densities down to the ppb-level, while the reduced cycle time of this short-flow test chip makes it an excellent routine defect monitor as well as a test vehicle for evaluating process changes.

Index Terms—**Yield enhancement, Defect-limited yield, Process characterization, Parallel-test.**

INTRODUCTION

Thorough understanding of yield issues is essential for the successful manufacturing ramp of a new semiconductor process technology. The overall yield is the product of two components, the systematic yield that is independent of the chip area, and the random yield that varies with the chip area. In general, random yield-loss is caused by process- or equipment-generated particulates, while systematic yield-loss occurs due to insufficient process window. It is imperative that all yield-loss mechanisms, whether systematic or random, be identified, accurately quantified, and addressed expeditiously. To accurately measure very low defect densities with statistical significance, the test structures must have sufficiently large critical areas. In addition, an ideal defect monitor vehicle must also have a short process time and be amenable to fast testing and data analysis. To this end, short-flow test chips are widely used as defect monitors in semiconductor development and manufacturing lines [1]. They offer both rapid yield learning cycles and large area, thereby alleviating the need for large test structures on product mask-sets.

While a large total testable area is required, a single test structure, such as a serpentine or a via chain, is limited to a certain maximum resistance (and, in turn, size) dictated by tester measurement specifications and normal parasitic leakage currents. Therefore, to achieve the desired defect observability for a given process layer, the required testable area is obtained by tiling many smaller, identical test structures, each individually connected to probe pads. This approach, however, significantly increases the number of measurements and, therefore, the test time on conventional parametric testers that employ serial-mode testing. To reduce the test time, we developed a novel method to test the DC test structures in parallel mode on a functional tester. The use of a digital tester to test yield test structures has been reported previously [2]. In that work, the actual resistances of the individual DUTs were not measured, but only a binary digit for each DUT representing its pass/fail result based on a preset threshold resistance was collected. In contrast, we measure the actual current flow through each test structure using the functional test platform, just as in conventional parametric testing, while still achieving a 5x reduction in overall test time without any trade-off in the required measurement accuracy. The resistance of the DUT is calculated using the measured current and the supply voltage. Based on this calculated resistance value, we are able to classify fails as either hard (catastrophic) or soft (resistive), greatly helping the root cause investigation of yield-loss mechanisms.

TEST CHIP DESIGN

Our test chip measures 20.5 x 32.4 mm and occupies the full reticle field. The test structure contents were partitioned into six chiplets, each measuring 10.1 x 10.65 mm. The test chip was designed for 7 metal levels, and 6 vertical via levels of copper wiring. A cross-sectional schematic of the test chip layer stack is shown in Fig. 1.

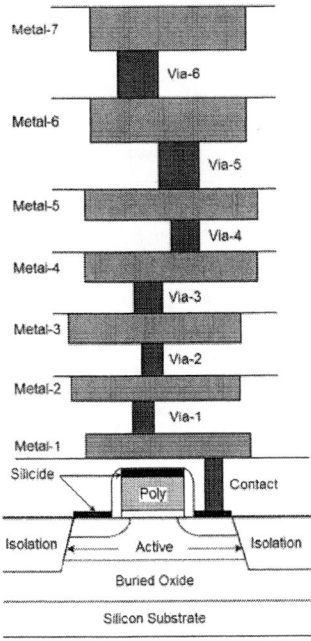

Fig. 1. Cross-sectional schematic of the test chip layer stack (Not to Scale).

The test chip includes the following three types of random defect test structures – Mazes, Multiple Parallel Serpentines (MPS), and Via Chains. Fig. 2 depicts examples of each of the three types. The Mazes and MPS were designed to test the 9 planar conductor levels – Active, Poly, and the 7 metal levels. The Via Chains test the Contact level that connects Active and Poly to Metal-1, and the 6 copper via levels. To maximize area efficiency and reduce the pad count, the pads are not contacted through. Therefore, every test level is completely independent. The Maze design in Fig. 2(a) consists of two combs and an interweaving serpentine for measuring conductor line opens and shorts, while the MPS design in Fig. 2(b) consists of 4 parallel serpentines. By testing each line of the MPS for opens and each pair of adjacent lines for shorts, the defect size distribution for opens and shorts defects is obtained, similar to [3]. Fig. 2(c) shows an example of a standard via chain design for the Via-1 level. In addition to this standard design, several variants with different via densities, via sizes, and via-to-metal enclosures were also included.

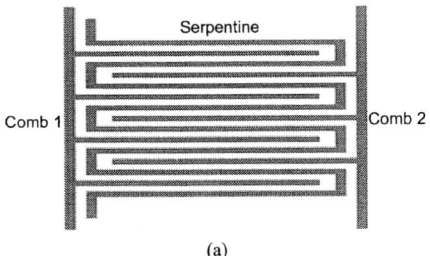

(a)

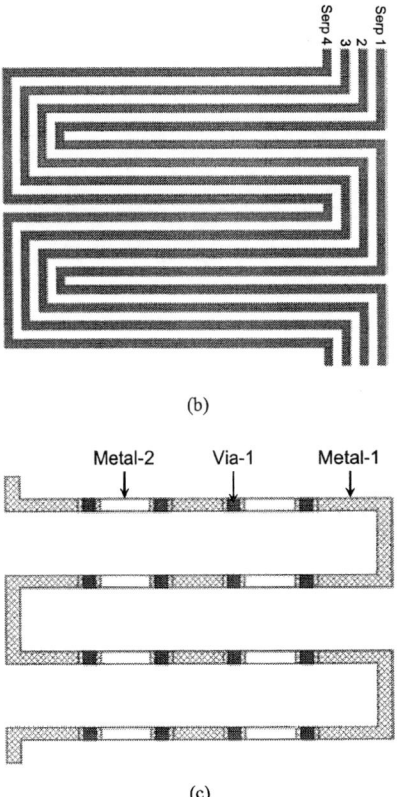

(b)

(c)

Fig. 2. Schematic of the three types of random defect test structures in the short-flow test chip
(a) Maze design consisting of two combs with an interweaving serpentine for measuring line opens and shorts defect density.
(b) MPS design consisting of 4 parallel serpentine wires for measuring the defect size distribution, in addition to the defect density, for line opens and shorts defects.
(c) Example of a Via Chain design for measuring the Via-1 opens fail rate.

The test structure arrays were replicated multiple times across the 6 chiplets as shown in Fig. 3. This allows any systematic defect density variation across the reticle field to be characterized. While the test chip also consists of many test structures designed specifically to measure the impact of various systematic or process window yield-loss mechanisms, over 75% of the area was allocated to random defect monitors. In fact, just the three test structure types discussed here occupy over 42% of the total test chip area. All test structures were wired out to a pad-frame with probe pads arranged in a 2x50 configuration as shown in Fig. 4. Electrical testing was done using a 2x50 probe card. For every probe touchdown, and each measurement type (opens or shorts), all test structures connected to the 100 pads are measured simultaneously.

To illustrate the large critical area available on this test chip, for instance, over 60 km of Metal-1 wiring per wafer and over 22 billion Via-1 vias per wafer are tested for opens/shorts. Other levels have similar extensive coverage, as shown in Table 1, enabling accurate characterization of wafer-to-wafer and intra-wafer defect density variation. An infrastructure was developed to automate standard data analysis and reporting of test chip results.

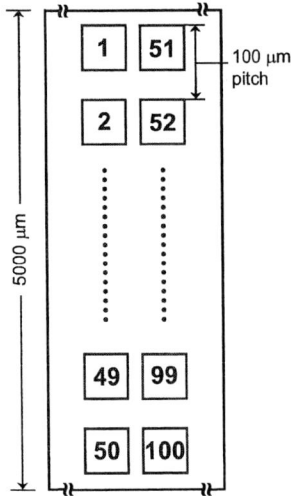

Fig. 3. Floor plan of the test chip with the locations of the Maze, MPS, and Via Chain arrays across the 6 chiplets highlighted. Numbers 1 through 6 represent the six identically-sized chiplets.

Fig. 4. Standard 2x50 pad-frame used for all random defect test structures in the test chip. The width of the array varies depending on the maximum allowable resistance for a single test structure.

Table 1. Coverage of the random defect test structures, by level, as expressed by (a) the length of wiring per wafer for the planar conductor

levels, and (b) the no. of vias per wafer for the vertical via levels. The corresponding test levels are also listed.

Planar Conductor level	Design Rule Pitch	Tested at	Wire length tested for opens/shorts (km per wafer)
Active	Minimum	Silicide	16.1
Poly	Minimum	Silicide	16.0
Metal-1	Minimum	Metal-2	60.3
Metal-2	Minimum	Metal-2	60.4
Metal-3	1.3x Min.	Metal-3	44.0
Metal-4	1.3x Min.	Metal-5	43.8
Metal-5	1.3x Min.	Metal-5	43.8
Metal-6	2x Min.	Metal-7	30.0
Metal-7	2x Min.	Metal-7	30.0

(a)

Via level	Design Rule Size	Tested at	Number of vias tested for opens (billions per wafer)
Contact-Active	Minimum	Metal-1	23.5
Contact-Poly	Minimum	Metal-1	5.4
Via-1	Minimum	Metal-2	22.1
Via-2	Minimum	Metal-3	5.6
	1.3x Min.	Metal-3	16.5
Via-3	1.3x Min.	Metal-5	22.1
Via-4	1.3x Min.	Metal-5	22.1
Via-5	2x Min.	Metal-7	5.5
Via-6	2x Min.	Metal-7	5.5

(b)

TEST METHODOLOGY

Historically, parametric testers have been used to test defect test structures. A traditional parametric tester allows very high precision voltage and current measurements of the order of µV and fA, respectively, but has only a few Source Measure Units (SMU). Therefore, the measurements are made serially using a matrix switching relay, resulting in relatively long test times. While some limited improvement in test time is possible by optimizing the measurement algorithms as described in [4], the steep rise in test structure content with every new technology node demands a more radical approach.

In an innovative effort to significantly reduce test time, we developed a test methodology using a functional test platform to test defect structures in parallel mode. The functional tester consists of 128 pins with a DC Parametric Measurement Unit (PMU) and a test processor embedded in every pin. This allows all test structures contained in a 2x50 pad-frame to be tested simultaneously in a single probe touchdown. This setup offers several selectable ranges and corresponding resolution and accuracy limits for current measurement. The highest level of precision offered by a parametric tester is not required for simple opens/shorts defect testing. We found the measurement range of ± 10 µA with a guaranteed worst-case measurement resolution of 5 nA to be sufficient for testing the defect test structures discussed here. For other test structures, such as active devices, that require a more precise measurement capability, an add-on DC hardware module has been installed. This module offers a measurement resolution down to 100 fA for a current measurement range of ± 2 nA.

Fig. 5 illustrates the procedure for testing the Maze test structure of Fig. 2(a). For testing line opens, a source voltage of 1.5 V is applied to one end of the serpentine with the other end grounded. The current flow through the serpentine is measured at the ground terminal as shown in Fig. 5(a). The combs are disconnected during opens test. For testing line-to-line shorts, a source voltage of 1.5 V is applied to both ends of the serpentine and the leakage current between the serpentine and the combs is measured at the comb ends which are tied together to ground as shown in Fig. 5(b). The Via Chains are tested for opens in the same manner as the Maze, while the algorithm for testing the MPS structure is somewhat more involved.

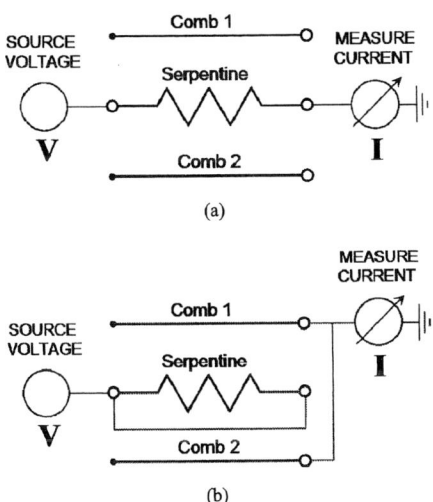

(a)

(b)

Fig. 5. Test configuration for testing the Maze structure of Fig. 2(a)
(a) Opens test to detect breaks in the Serpentine wire.
(b) Shorts test to detect bridging between the Serpentine and the Combs.

With our parallel-test approach, we achieved a 5x test time reduction over conventional parametric testing without any trade-off in the required measurement accuracy. For instance, for a total of 10,496 opens and shorts current measurements, the test time on the functional tester in parallel-mode was only 270 seconds, while the identical set of measurements on a parametric tester in serial-mode took 1,326 seconds. Even on the functional tester, we found the wafer prober stepping time to be a significant contributor to total test time, suggesting an opportunity for further test time reduction. Fig. 6 shows a scatter plot of the opens current of 1,640 individual Metal-2 mazes measured by a conventional parametric tester on the X-axis, plotted against the corresponding parallel-mode functional tester measurement on the Y-axis. The excellent correlation and linearity between the two measurements confirms that our parallel-test methodology is sound.

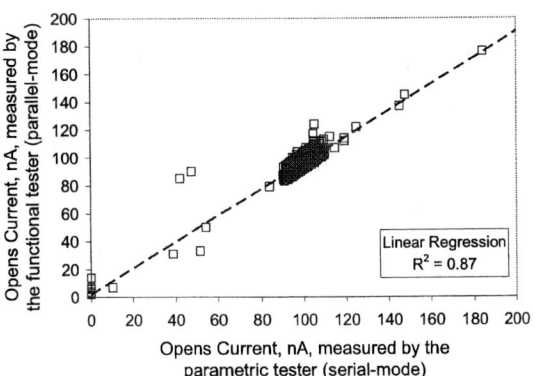

Fig. 6. Scatter plot of opens current measured on 1,640 Metal-2 mazes by a conventional parametric tester vs. corresponding measurements on the functional tester using the parallel-test method described in this section.

TEST CHIP RESULTS AND DISCUSSION

A regular stream of short-flow test chip wafer lots are processed through the line and tested, yielding valuable data. To illustrate the important role of this test chip in 45nm yield learning, Fig. 7 shows the total number of short-flow test chip wafers tested at the Silicide and Metal-2 test levels, by month, for the second-half of 2007.

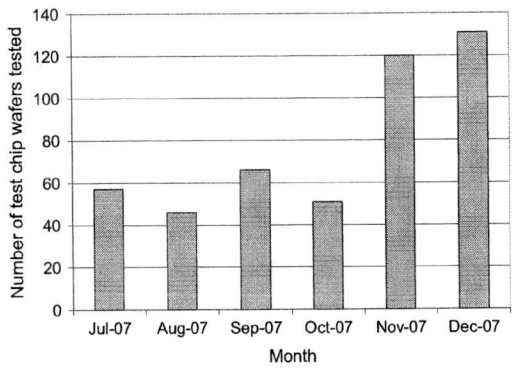

Fig. 7. Total number of short-flow test chip wafers tested at the Silicide and Metal-2 test levels each month during the second-half of 2007.

The test results are processed using automated data analysis routines to generate defect densities (for metal levels) and failure rates (for via levels), by failure mechanism. Summary defect density reports, in the format shown in Table 2, are routinely generated. Yield improvement actions are then prioritized based on actual/target defect densities. In addition, many process split experiments are also conducted on the short-flow test chip lots to speedily evaluate proposed process changes.

Table 2. Standard format of the Defect Density report generated by processing the test results from the short-flow test chip.

Failure Mechanism	For the week ending ...				
	Number of lots	Number of wafers	Actual Defect Density or Failure rate	Target Defect Density or Failure Rate	Ratio: Actual/ Target
Active Opens					
Active Shorts					
Poly Opens					
Poly Shorts					
Contact-Active Opens					
Contact-Poly Opens					
Metal-1 Opens					
Metal-1 Shorts					
Via-1 Opens					
...					
Metal-7 Opens					
Metal-7 Shorts					

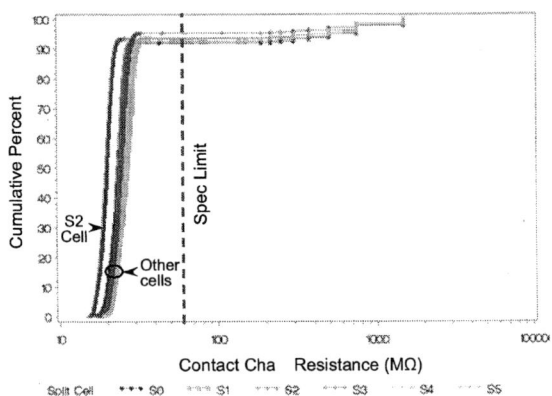

Fig. 9. Cumulative distribution plot of Contact chain resistance, by process split, shows a 20% drop in median resistance for the S2 cell as well as a long tail on the high-end for all cells.

As mentioned earlier, since we measure and store the actual current flow through the test structures, it is possible to distinguish between hard vs. soft fails by varying the resistance spec limit appropriately. To illustrate this capability of our parallel-test approach, Fig. 8 shows the distribution of the measured resistance values of the Via-1 chains on one wafer. The resistance spec limit for Via-1 opens is set at 43 MΩ. It is clear that some chains, though nominally passing, are still abnormal at around 3 to 4x the median chain resistance of 10 MΩ. To understand the root cause of this tail in the distribution, we are able to identify the specific locations of these maverick chains for failure analysis. This would not have been possible if only the digital pass/fail bit information was collected.

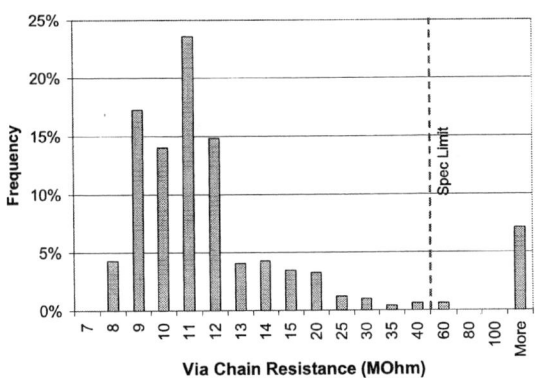

Fig. 8. Histogram of the measured resistance values of the Via-1 chain on the functional tester using parallel, per-pin PMU. The resistance spec limit can be adjusted to identify soft fails.

As a further example, Fig. 9 shows the distribution of the contact chain resistance for a process split lot, grouped by split cell. It is clear that the S2 split cell has 20% lower resistance compared to the other split cells, and secondly, all split cells exhibit a long tail on the high-resistance end. This type of detailed analysis is possible because our parallel-test approach allows large critical areas to be accurately tested in a reasonable time.

Although the test structures described here were primarily intended to measure overall defect densities, they also offer valuable clues on many systematic yield detractors. We provide four different examples. Fig. 10 shows the Via-1 open fail rates, by design type, for 5 different via chain designs. The pure Via-1 open fail rate component was calculated by subtracting the fault density components for Metal-1 and Metal-2 opens from the overall chain yield-loss. If the Via-1 opens were truly random, the fail rates for the different designs should be identical or nearly so. The fact that the Via-1 fail rate strongly depends on specific via chain design parameters suggests that there are several additional systematic yield-limiting factors at play.

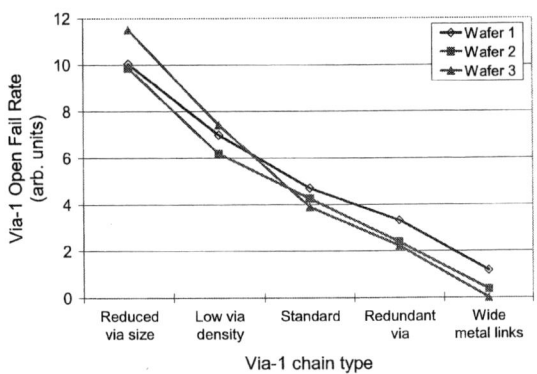

Fig. 10. Normalized Via-1 open fail rate is a strong function of the via chain design parameters indicating that the Via-1 opens are not purely random, but also caused by several systematic yield-limiting factors.

Fig. 11 shows the wafer-average Metal-2 opens defect density for 250 wafers for two different maze designs, plotted against each other. One maze was designed at the minimum design rule pitch, while the other was designed at a pitch 9% above minimum. We see that the average defect

60

density for the minimum pitch maze is generally higher than that for the relaxed pitch maze, especially so for wafers with high defect densities. A reasonable explanation for this difference is a systematic yield-loss factor related to the patterning of the minimum-pitch maze.

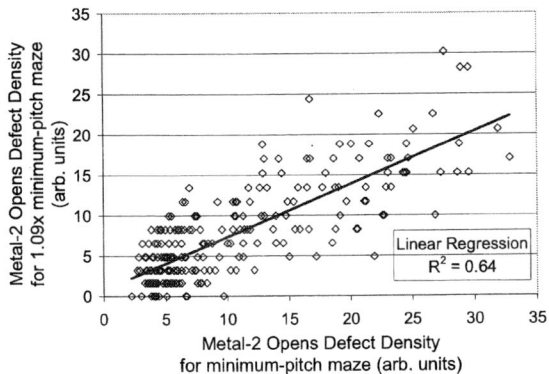

Fig. 11. Scatter plot comparing the wafer-average Metal-2 Opens Defect Density for two maze designs - Minimum-pitch vs. 1.09x minimum pitch.

The Via-1 chains in the test chip are first tested at the Metal-2 test level, and again at the Metal-3 level. The wafer-average Via-1 opens yield obtained at Metal-2 test is plotted in Fig. 12 against the average Via-1 opens yield for the same wafer measured at Metal-3 for 178 wafers. That the Metal-3 test yield is lower than the corresponding Metal-2 test yield for nearly all wafers, confirms that Via-1 opens yield is degraded by subsequent processing. On further investigation, we identified a systematic factor as the root cause for this yield degradation.

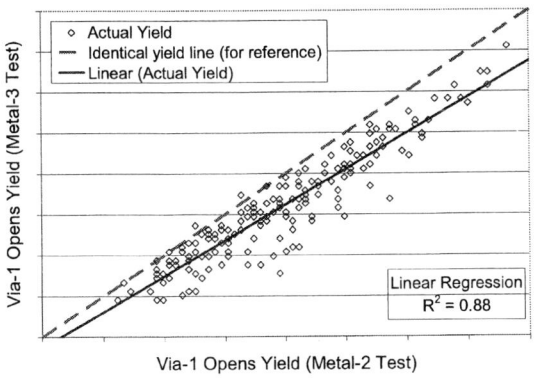

Fig. 12. Wafer-average Via-1 Opens yield at Metal-2 vs. Metal-3 test. The dashed diagonal line represents identical Via-1 open yields at Metal-2 and Metal-3 tests. The solid line is the linear regression fit of actual data.

Finally, the relative frequency (ratio of actual/target defect density) of Metal-2 open defects for different defect size ranges is shown in Fig. 13. This data was obtained from the MPS structure of Fig. 2 (b). It is obvious that both 1-line and

4-line opens are more common than predicted by the critical area model. This observation can be translated into defect sizes. Assuming a perfect circular defect of diameter s, the range of defect sizes that cause n adjacent lines to be open can be expressed as a function of MPS line width (= space), w. For a 1-line open: $w \leq s < 7w$, 2-line open: $3w \leq s < 11w$, 3-line open: $5w \leq s < 15w$, and lastly for a 4-line open: $s \geq 7w$. The large overlap in these ranges is partly because Serps 1 and 4 turn around on themselves. Nevertheless, it is clear that there are two separate systematic yield detractors – one causing small defects (size $< 3w$) and another causing much larger defects (size $> 15w$). This has also been confirmed by inline defect inspection.

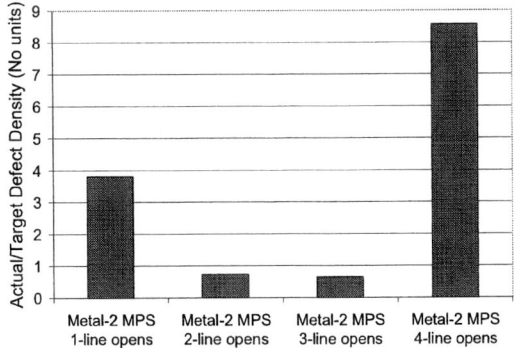

Fig. 13. Relative frequency of Metal-2 opens defects of different defect size ranges obtained from the MPS test structure. The plot shows the median values for 379 wafers.

CONCLUSION

The 45nm short-loop test chip described here accelerates yield improvement by enabling accurate defect density measurements and by providing valuable insights on the top yield detractors. An innovative method for efficiently testing the thousands of defect test structures on this test chip has been demonstrated. This parallel-test approach improves test throughput by 5x over conventional testing.

REFERENCES

[1] H. Nagaishi, M. Fukui, H. Asakura, A. Sugimoto, "Defect Reduction in Cu Dual Damascene Process Using Short-Loop Test Structures," *IEEE Trans. Semiconductor Manufacturing*, vol. 16, no. 3, Aug. 2003, pp. 446-451.

[2] C. Hess, L. Weiland, "Modeling of test structures for efficient online defect monitoring using a digital tester," *Proc. 1994 Int'l Conf. on Microelectronic Test Structures*, IEEE Press, 1994, pp. 108-113.

[3] R. Glang, "Defect Size Distribution in VLSI Chips," *IEEE Trans. Semiconductor Manufacturing*, vol. 4, no. 4, Nov. 1991, pp. 265-269.

[4] C. Hess, et al., "Test Time Reduction Methods for Yield Test Structures," *Proc. 2003 Int'l Conf. Microelectronic Test Structures*, IEEE Press, 2003, pp. 64-69.

SESSION 4
Poster Session

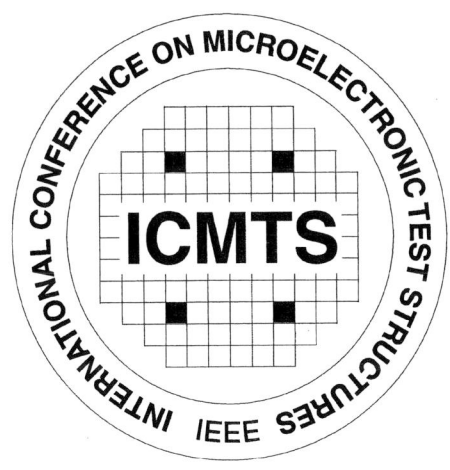

March 25, 15:20–15:55

Co-Chairs: Akis Doganis, *TSMC, Taiwan*

Yoichi Tamaki, *Hitachi Ltd., Japan*

2008 IEEE Conference on Microelectronic Test Structures, March 24-27, Edinburgh, UK

Conduit Diffusion of Dopants in Tungsten Silicide Layers

S Liao, M Bain, P. Baine, D W McNeill, B M Armstrong and H S Gamble

Northern Ireland Semiconductor Research Centre
School of Electronics, Electrical Engineering and Computer Science
Queen's University of Belfast, Northern Ireland, UK

ABSTRACT

Novel test diode structures have been manufactured to characterise dopant diffusion in tungsten silicide layers. Bipolar diode action is demonstrated experimentally for anneal schedules of 30 minutes at 900°C, indicating long-range diffusion of phosphorus (~ 38 μm). The work function of the silicide was found to be 4.8 eV. SIMS analysis shows dopant redistribution is effected by the segregation to the silicide/oxide interface. The concept of conduit diffusion has been demonstrated experimentally for application in advanced bipolar transistor technology.

INTRODUCTION

SOI technology is emerging for application in advanced digital CMOS circuits, advanced analogue bipolar transistor circuits and in novel power device structures. Silicon on silicide on insulator (SSOI) [1] is a novel variant which can provide a low resistance buried collector contact for bipolar transistors. Tungsten silicide (WSi_2) is an ideal buried metallic layer as it is a refractory material that offers low sheet resistance, typically 2 ohms/square for a 200 nm layer and excellent thermal stability. The basic structure of a bipolar transistor manufactured with this technology is shown in Fig. 1.

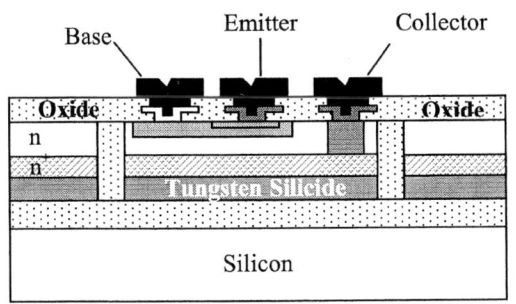

Fig. 1   Bipolar Transistor fabricated on SOI incorporating a buried tungsten silicide conduction layer (SSOI).

The silicide requires a shallow diffused layer to make it ohmic to the collector. Collector parasitic resistance is reduced by a factor of 10 – 100 compared to standard buried n^+ and p^+ implant technology and the overall thickness of silicon can be greatly reduced.

There have been a number of publications concerning the diffusion of dopant species such as boron and phosphorus in WSi_2. This has largely been to investigate the influence of lateral diffusion of these species on threshold voltage stability in MOS transistors. The diffusivity of these dopants in WSi_2 was reported to be approximately 100,000 times greater than that in polycrystalline silicon as shown in Fig. 2 [2][3]. A potential application of this is advocated for SSOI where the dopant species would be implanted into the buried silicide at a remote site as a back end process. A low thermal budget process would diffuse the dopants through the silicide and up into the overlying active silicon layer. This approach greatly simplifies the production of the bipolar transistors and minimises disturbance of the emitter/base doping profiles caused by thermal annealing. Production of complementary bipolar devices with tightly matched properties, low loss microwave diodes in MMICs and smart power ICs with vertical DMOS can also be achieved.

There is limited experimental data on the diffusion of the dopant species in WSi_2 and this paper seeks to establish the viability of the proposed technology through the use of novel test structures and to compare the results obtained with those predicted from simulation.

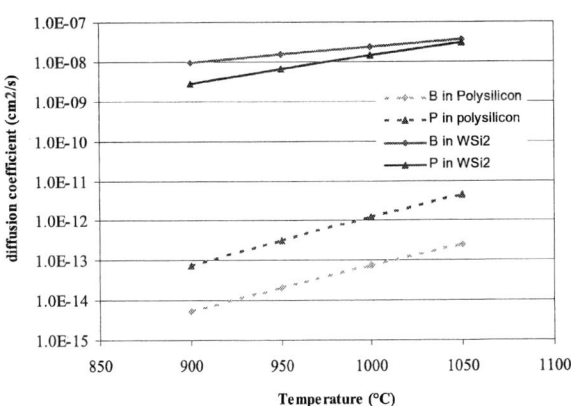

Fig. 2 Diffusivity of phosphorus and boron in WSi_2 and polycrystalline silicon.

978-1-4244-1801-5/08/$25.00 ©2008 IEEE

TEST STRUCTURE DESIGN

In order to evaluate dopant diffusivity in WSi_2 it is not necessary to create buried silicide structures. Instead, surface test structures, similar in design to that shown in Figure 3, have been employed. The dopant implant window is connected to a substrate contact window by a WSi_2 conduit of varying length. The WSi_2 contact to the silicon forms a Schottky barrier diode. Dopant (of opposite type to the substrate) is diffused along the conduit by a thermal anneal in order to form a diffused diode in the substrate. The same structure can then be probed electrically to determine whether a diffused diode has been formed or whether there has been insufficient dopant diffusion and the contact remains a Schottky barrier diode.

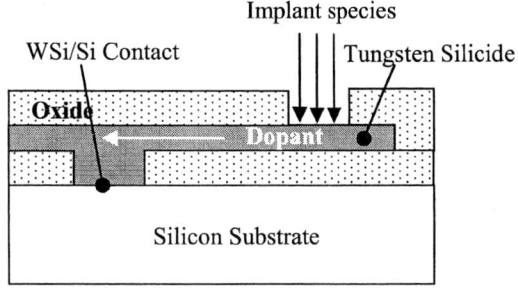

Fig. 3   Conduit diffusion and diode test structure

The test structures consist of 3 short conduits (lengths of 8 µm, 16 µm and 32 µm) which are doped from an implant window at one end, and 3 longer conduits (38 µm, 68 µm and 98 µm) which are doped from twin implant windows at opposite ends. For reference purposes, identical conduit structures are also fabricated without any dopant implant.

Additional van der Pauw structures were used to monitor silicide resistivity. WSi_2 metal gate MOS capacitors were also included to help verify the work function of the silicide and to monitor the field oxide electrical properties. Additional large area diodes were included to permit SIMS analysis.

COMPUTER SIMULATIONS

Silvaco Athena 2D process modelling software has been employed in preliminary design and analysis of phosphorus conduit diffusion. This is not a standard well-documented process, so suitable parameters had to be chosen in the simulation. Diffusivity of phosphorus in tungsten silicide was set to the values given in Fig. 2. Segregation coefficients from silicide to silicon and silicide to silicon dioxide were both set at a value of unity. The transport velocity across these two interfaces was optimised so that this was not limiting diffusion from silicide into silicon or

silicon dioxide. A value of 1×10^{-5} cm/sec was employed for both these interfaces.

Simulations showed that long-range diffusion in silicide is possible if the diffusivity values shown in Fig. 2 are employed. Fig. 4 shows formation of a shallow (0.08µm) diffused diode in the silicon substrate by phosphorus diffusion in the shortest silicide conduit (8 µm).

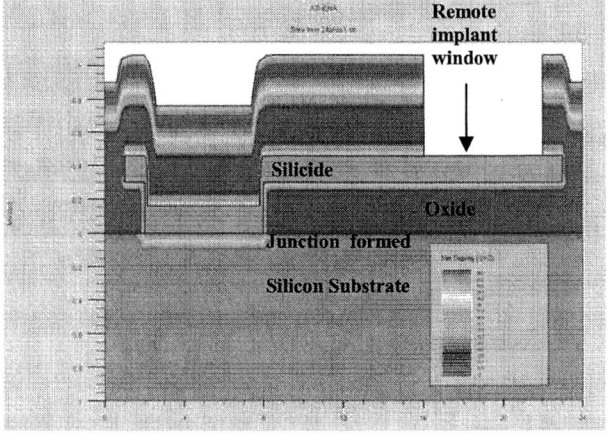

Fig. 4   2D modelling of phosphorus diffusion in an 8 µm conduit following anneal at 900°C for 30 minutes

The phosphorus implant dose was 5×10^{15} cm^{-2} and the anneal was at 900°C for 30 minutes. The rapid diffusion of phosphorus and boron in WSi_2 has been simulated at 900°C and 1000°C. This has shown that diffusion along the entire conduit length is possible even at 900°C in all but the longest conduits. In order to give some quantification of this finding, the resultant n or p junction depth at the silicide-substrate contact has been plotted in Fig. 5 as a function of conduit length for a 30 minute anneal at both 900°C and 1000°C. It is clearly observed that junction depth decreases as conduit length is increased. Shallow junctions are achieved for a 900°C anneal with very limited dopant reaching the contact for the long conduits.

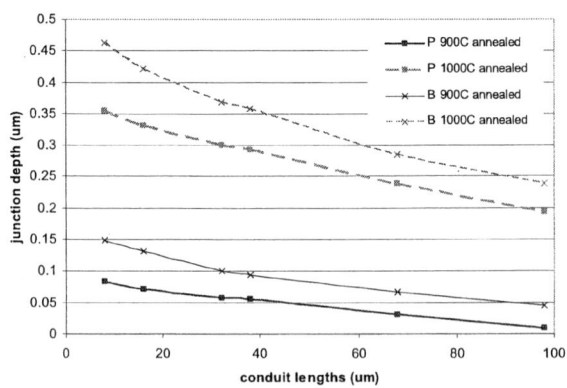

Fig. 5   Effect of conduit length on junction depth for phosphorus and boron annealed at 900°C and 1000°C for 30mins

CONDUIT TEST STRUCTURE FABRICATION

The tungsten silicide material that was used to form the conduits was deposited at 370°C in a Varian 5101 cold-wall LPCVD reactor using WF_6/SiH_4 chemistry. Previous characterisation [4] has shown the as deposited tungsten silicide to be amorphous and silicon-rich. However, subsequent annealing at temperatures of at least 600°C converts the tungsten silicide to the highly conductive tetrahedral WSi_2 phase. The excess silicon stabilizes the silicide and prevents its delamination from dielectric surfaces during the high temperature anneal.

The manufacturing process flow was as follows. 5-12 ohm-cm p type substrates were employed for the phosphorus implanted structures. Substrate contact windows were patterned in a thermally grown 0.45 µm oxide layer. A 0.15 µm tungsten silicide layer was then deposited by LPCVD and patterned. The substrate was then coated with a 0.6 µm LPCVD TEOS oxide deposited at 720°C. The WSi_2 phase was formed by annealing at 1050°C for 2 hours. This anneal also simulated the bond strengthening anneal normally employed when the silicide is incorporated in an SOI structure. Implant windows were patterned and the implantation energy was 55 and 75 keV for boron and phosphorus respectively. A dose of $5\times10^{15}cm^{-2}$ was employed. Dopant was subsequently diffused along the conduits by annealing for 30 minutes at 900°C or at 1000 °C. Aluminium electrodes were formed to allow electrical contact to the diodes. The diode device area was $9 \times 10^6 cm^{-2}$ for all the test structures presented in this work.

RESULTS AND ANALYSIS

Forward current/voltage characteristics for the 3 short conduits are shown in Fig. 6 along with a non-implanted reference diode characteristic. The high current of the reference diode is indicative of a Schottky contact with a barrier height of about 0.5 V for the tungsten silicide to p-type silicon. All 3 implanted conduits show a very significant reduction in forward bias current. This can only be attributed to conduit diffusion of the phosphorus dopant over the full distance of the conduit and the formation of an n-p junction under the silicide contact.

There is a trend towards higher currents as the conduit gets longer. Simulations, as shown in Fig. 5, indicate these junctions will all be less then 90 nm deep. This increase in forward bias current is consistent with shallow emitter behaviour i.e. as the diffused junction becomes shallower

in the longer conduit devices, the hole current injected from the p-type substrate becomes more significant, resulting in an increase in forward bias current.

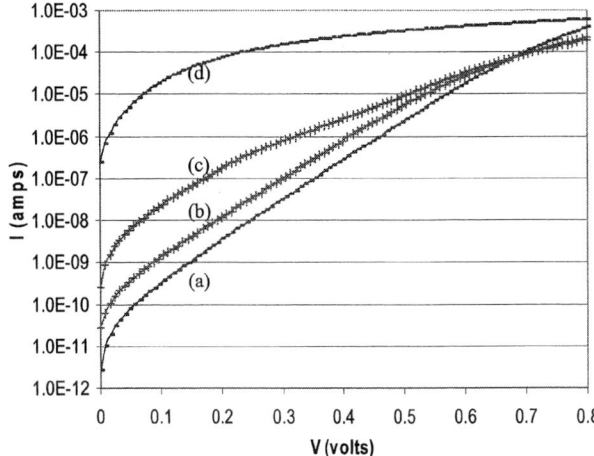

Fig. 6   Forward I-V characteristics of (a) 8µm, (b) 16µm, (c) 32µm conduit lengths implanted remotely with phosphorus and annealed at 900°C for 30mins. (d) Reference device which received no implant.

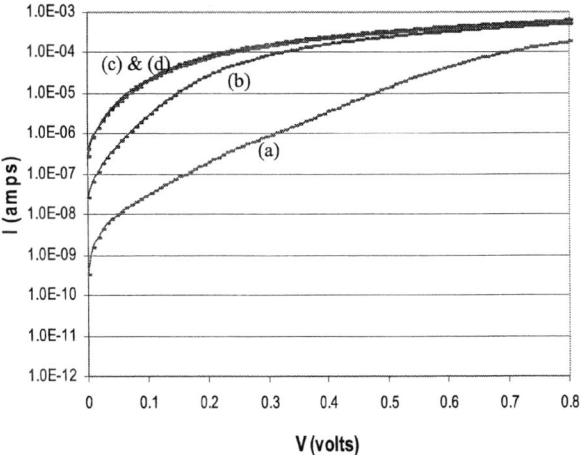

Fig 7   Forward I-V characteristics of (a) 38µm, (b) 68µm, (c) 98µm conduit lengths implanted remotely with phosphorus and annealed at 900°C for 30mins. (d) reference device which received no implant.

Fig. 7 shows the forward current/voltage characteristics of the 3 longer conduits. As the conduit lengths increase, the forward bias current increases. The longest conduit exhibits Schottky barrier characteristics indicating that no phosphorus has reached the diode contact. The 68µm conduit exhibits a small reduction in current indicating that phosphorus has diffused to this diode in small quantity, and has not formed a bipolar n-p diode. The current has been reduced because the diffused n-type layer formed under the silicide is very thin and fully depleted, resulting only in a small enhancement of the effective Schottky barrier height and a resultant fall in current. The 38 µm conduit again exhibits short emitter characteristics in line with the results shown in Fig. 6.

The electrical results correspond to the computer simulations, in which all but the longest conduits were expected to have sufficient phosphorus out-diffusion for bipolar diode formation. The results show great promise with relatively long-range diffusion achieved with a low thermal budget process. The diodes show non-ideality with large n factors of the order 1.5 – 1.6. This will be discussed later.

The forward bias current/voltage characteristic of unimplanted devices yields the Schottky barrier height and therefore the work function of the tungsten silicide. SILVACO Atlas simulation was used to model this structure. A range of different work functions was employed from 4.7 eV to 4.9 eV. The simulated and experimental forward bias characteristics are shown in Figure 8. A tungsten silicide work function of 4.8 eV was found to give the best fit to the experimental results. This result agrees well with the published data [5].

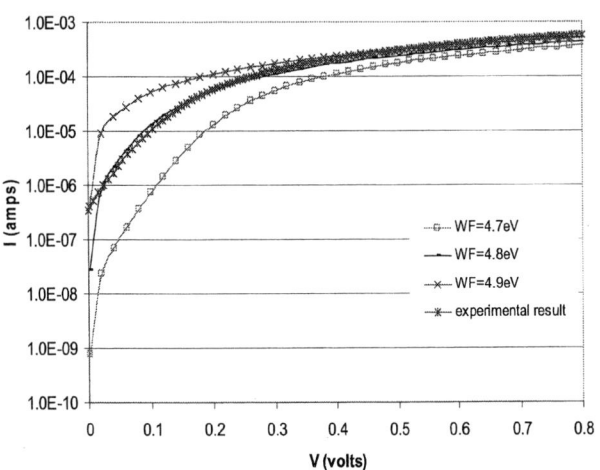

Fig. 8   Experimental forward bias I-V characteristics of unimplanted diode compared to simulation of various workfunctions (WF).

The forward bias current/voltage characteristics of the phosphorus-implanted structures, after annealing at 1000°C for 30 minutes, are shown in Figs. 9 and 10. Fig. 9 shows the short conduit forward bias characteristics (8 μm, 16 μm and 32 μm) and Fig. 10 shows the long conduit forward bias characteristics (38 μm, 68 μm and 98 μm). These results shown similar trends to the devices annealed at 900°C for 30 minutes. The forward bias current increases as the conduit lengths get longer. However, for the same anneal time, higher temperature improves the forward bias characteristics in all diodes. The longest conduit device (98 μm) shows lower current flow than the control Schottky diode. This indicates that phosphorus diffusion

has taken place along the full length of the silicide and into the silicon substrate. Again the amount of phosphorus diffusion into the silicon substrate is relatively small, resulting only in enhancement of the Schottky barrier height. Diffusion at 1000°C appears not to result in the 0.2 μm junction depth as predicted by simulation. The diodes with 38 μm and 68 μm conduits showed a more significant decrease in forward bias current.

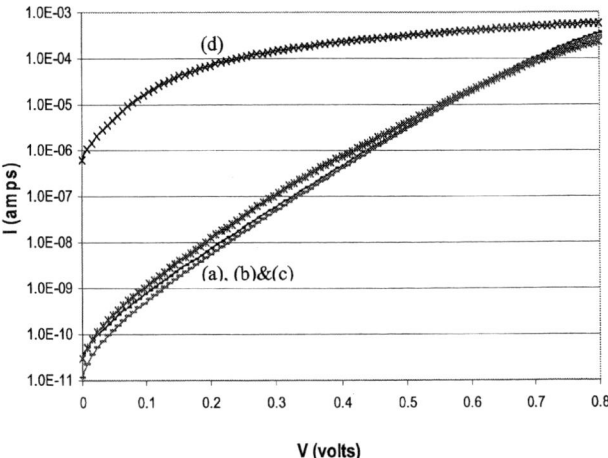

Fig. 9   Forward I-V characteristics of (a) 8μm, (b) 16μm, (c) 32μm conduit lengths implanted remotely with phosphorus and annealed at 1000°C for 30mins. (d) reference device which received no implant.

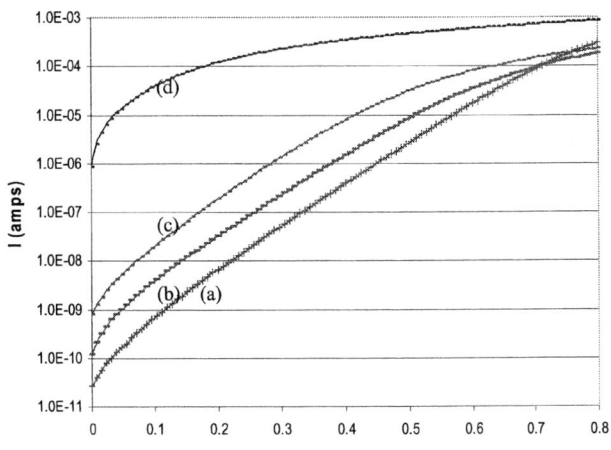

Fig 10   Forward I-V characteristics of (a) 38μm, (b) 68μm, (c) 98μm conduit lengths implanted remotely with phosphorus and annealed at 1000°C for 30mins. (d) reference device which received no implant.

Again, these diode characteristics exhibit non-ideality with n factors of the order of 1.6. MOS C-V characteristics are shown in Fig. 11 for a tungsten silicide gate manufactured on the diode field oxide. This is equivalent to the silicide conduit layer overlying the diode isolation oxide. C-V measurements were performed on the capacitors with a signal frequency of 1 MHz. The MOS devices fabricated with tungsten silicide as a gate material exhibit a threshold voltage of approximately 1.3 V and a flatband voltage of about – 0.8 V. The diode forward bias voltage lies in the range 0 to – 0.8 V. It is clear that throughout this measurement the silicide conduit biases the silicon surrounding the diode into depletion as indicated by the flatband voltage. This enlarges the depletion region area of the n-p diode and exposes a large area of the oxide-silicon interface, which provides additional recombination sites, leading to increased recombination current.

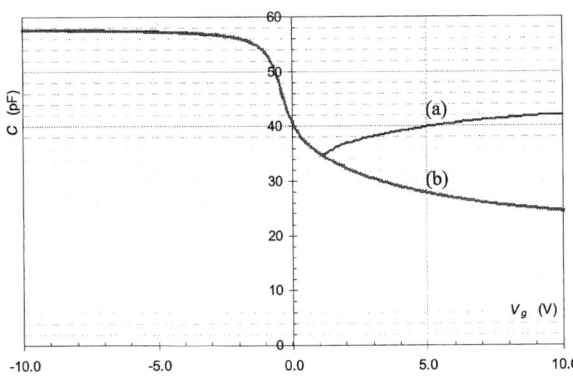

Fig. 11   C-V characteristics of WSi$_2$ MOS capacitor (diameter 1mm). (a) Swept from inversion to accumulation (b) swept from accumulation to inversion.

Recombination current effects are normally characterized by an ideality factor of 2. It is concluded that this gated diode effect with increased surface recombination velocity at the oxide-silicon interface is dominating the I-V characteristics and is responsible for the non-ideality observed. In future test devices, lower resistivity substrates or channel stop technology will be employed to avoid the field region depletion under the conduit.

Reverse bias characteristics show significant leakage current and spread in results. This is thought to be due to gated diode effects. When the reverse bias exceeds 1.3 V, strong inversion creates an n-channel in the silicon under the conduit, surrounding and extending beyond the n-p junction. Application of increased reverse bias voltage will result in a reverse current dominated by edge leakage effects around the periphery of the channel region. Localised breakdown will lead to high leakage and the variable results observed.

SILVACO Atlas 2D modelling was employed to model a simplified diode structure. A silicide workfunction of 4.8 eV and a substrate doping of 4.5×10^{14} cm^{-3} were employed. Previous process simulation showed junction depths of about 100 nm were achieved for 900°C anneals.

This simplified model therefore assumed an n-junction depth of 100 nm with a Gaussian profile and several values

of surface concentration. The impact of the additional recombination sites due to gated diode action was approximated by adjustment of the bulk lifetime. Agreement between experimental I-V characteristics and simulations was achieved using a bulk lifetime of 0.01 μs and a surface concentration of the order of 7×10^{17} cm^{-3}, as shown in Fig. 12.

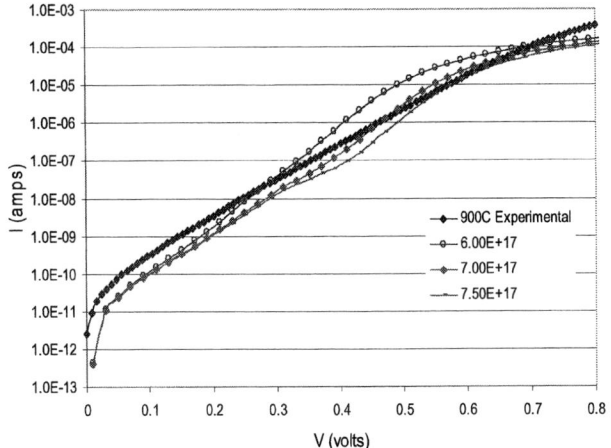

Fig.12   Experimental forward bias I-V characteristics of 8μm conduit diode with P and annealed at 1000°C 30mins compared to simulation of various surface concentrations with a constant junction depth of 100nm and a lifetime of 0.1μs.

This doping concentration was less than anticipated, so SIMS analysis was used to profile the doping concentration in a large area diode. The profile obtained is shown in Fig. 13. It can be seen that the peak phosphorus concentration under the silicide is of the order of 1×10^{18} cm^{-3}. There is also significant segregation of the dopant to the upper oxide-silicon interface. The profile therefore confirms that phosphorus has diffused over long range but that significant phosphorus is retained at the silicide interface with underlying and overlying oxide layers [6]. This remains to be overcome if conduit diffusion of phosphorus is to be fully exploited. The introduction of alternative materials at this interface such as polysilicon or silicon nitride may reduce the segregation and this will be investigated with new test structures.

Test structures were manufactured with boron implants, and anneals at 1000°C for 30 minutes. The forward characteristics for the three shortest conduits are shown in Fig. 14. Similar results were achieved with the longer conduits. All the diodes exhibit excellent ideality with n factors of the order of 1.02. The field regions under the silicide conduits will be in accumulation during diode forward bias measurement and hence the oxide-silicon interface will play no role in diode operation. These preliminary results appear very promising for long range diffusion of boron in tungsten silicide.

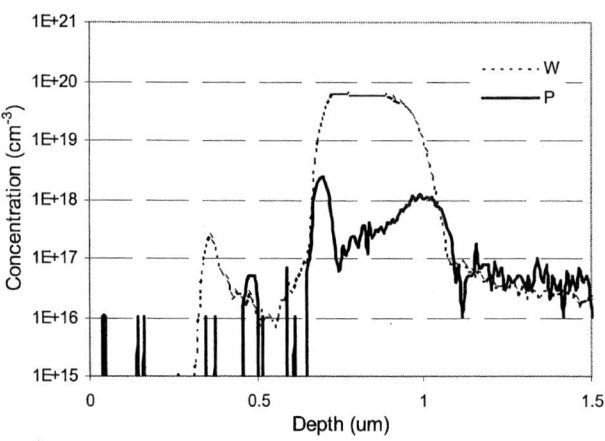

Fig. 13   SIMS depth profile showing phosphorus (P) and tungsten (W) for contact window to silicon region after annealing at 900°C for 30 mins.

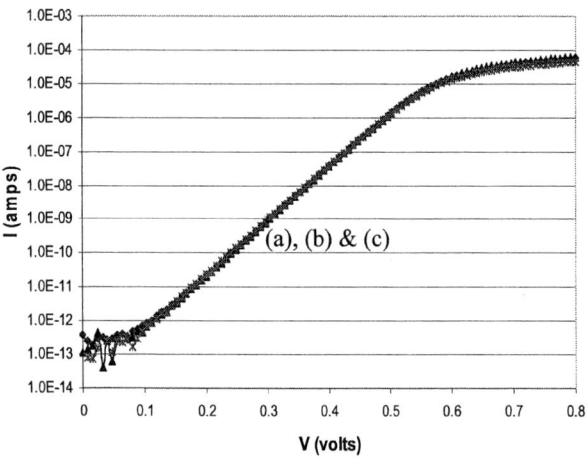

Fig. 14   Forward I-V characteristics of (a) 38μm, (b) 68μm, (c) 98μm conduit lengths implanted remotely with boron and annealed at 1000°C for 30mins.

CONCLUSIONS

A novel test structure has been developed to characterize dopant diffusion in tungsten silicide. Experimental results indicate that rapid diffusion of phosphorus has occurred over long range (~ 38 μm), for a low thermal budget heat treatment at 900°C for 30 minutes. This is extended to over 90 μm for 1000°C anneals for 30 minutes. The test structure has exhibited diode characteristics ranging from Schottky diode to short emitter diode characteristics. Non-ideal characteristics were achieved for n-p diodes. This was attributed to MOS gated diode effects resulting in increased recombination current. Whilst long-range diffusion has been achieved, preferential segregation of phosphorus to silicide-oxide interfaces has been observed, resulting in reduced doping concentration in the diffused diodes. Initial results with p-n diodes made by conduit diffusion of boron show nearly ideal characteristics. The concept of using rapid diffusion of dopants in silicide conduits has been established

REFERENCES

[1] H.S. Gamble, Proc 10th Intl Symp ECS vol.2001-3, 2001, pp.1-12

[2] C.L. Chu, K.C. Saraswat and S. S. Wong IEEE Transactions on electron devices, Vol. 39, No. 10, Oct 1992, pp 2333 - 2340

[3] SILVACO Athena user menus

[4] M.F. Bain, B.M. Armstrong and H.S. Gamble. Vacuum 64 (2002) pp 227-232

[5] K.C. Saraswat, F. Mohammadi "Work function of WSi2 ", IEEE electron device letter, VOL.EDL-1, NO.2, Feb 1980 pp 18-19

[6] F.C. Shone, S.E. Gabseb, D.B. Kao, K.C. Saraswat. and J.D. Plummer, IEDM Tech.Dig. (1986) pp 534-537

A Novel High Speed Automatic Layout System to Place and Route Test Structures for Parametric Test Capability

Andrew J. West[*], Samrat Mondal[#], Devjyoti Patra[#], Kalyan Goswami[#], and Shamik Sural[#]

[*]National Semiconductor Corporation, Santa Clara, CA, U.S.A.

[#]Indian Institute of Technology, Kharagpur, India

ABSTRACT

In this paper, we created a generalized framework for the automated placement and routing of analog test structures. We exploited the concept of terminal properties when placing and routing the test structures and generated a library of place and routing strategies for different architectures. This new approach significantly reduces layout time, maximizes the re-use of place and route routines, and facilitates the introduction of a holistic parametric test design flow.

INTRODUCTION

The generation of a test chip layout for process development can be an expensive and time consuming task. It often consumes valuable time that could be better spent on other, more challenging process development tasks. With the faster turn-around times between test chip tape-outs and ever decreasing process development timelines that are forced on process engineers, the need for automation within the entire process development cycle is a critical factor. Some of the test chip workload can be handled by contract layout engineers but often this still requires the process engineer to devote time to define the test structure and parameter variations, visually inspect the final layout, and generate the test chip documentation. We have applied a parameterized layout system to automatically generate, place, and route the parameterized device cells utilizing an object-ware JavaScript based language [1]. The main benefit of the automatic placement and routing of parameterized cells is the reduction in time to generate the layout. The tool also maintains the terminal-to-pad connections with the device parameters, from which the test chip documentation can be generated in electronic and hardcopy formats.

As new processes are introduced, process engineers have to redesign the layouts to match the design rules of the new technology [2]. To reduce the turnaround time of test chip design cycle and minimize the redundancy of layouts when a new technology is adapted, a parameterized approach for layout development is required.

A parameterized layout framework supports the representation of the devices and the placement and route routines in the form of templates, which when provided with the design rules as values and the various design parameters, produces a physical layout [3]. The layouts of analog circuits and devices involve many more complexities than the digital ones that make digital layout automation approaches infeasible for analog circuits [4] [5] [6]. Most analog CAD tools today are targeted for circuit generation using a macro cell-based approach. When used for the layout automation for a large set of discrete devices, these approaches result in huge computational overhead. In reference [7], the layout synthesis problem and technology mapping are addressed only for array-type analog test structures. In reference [8], the proposed tool can handle different types of test structures, but the terminal-to-pad connections are predefined for a specific layout configuration and require manual intervention to alter the configuration.

In the framework selected for this work, the layout generation tool utilizes a library of JavaScript modules. These modules are split into four module types:

- Device: Generates the actual device layout and defines the terminal position and attributes.
- Pad Ring: Defines the layout for the probe pads to which the device terminals are connected.
- Placement: Places and orientates the device relative to the Pad Ring and determines the terminal-to-pad connections.
- Routing: Generates the terminal-to-pad interconnect utilizing the placement terminal locations and pad connections.

In the first of four phases of layout generation, the definition phase, the user selects the Device, Pad Ring, Placement, and Routing modules [1]. The device parameters are supplied in a spreadsheet format. The Pad Ring, Placement, and Routing modules can also be supplied with user variables to control their behavior during later phases. After the user is satisfied with the device parameters the tool can assemble the layout. The next phase generates the device layout, at which point the Device and Pad Ring templates are executed using the supplied variables generating the layout for every device type variation. On completion of this phase we enter the placement phase where every device is passed through the Placement module for quad (pad ring), position, orientation, and terminal assignment. During the Routing phase the terminal assignments are used to generate the inter-connect layout between the device terminals and pads, completing the automatic layout. The user can then review the layout and accept the generated output. The output can be exported in GDS format and transferred to a traditional layout tool for verification checks prior to tapeout.

MOTIVATION

The drive behind this work is to build a library of Place and Route routines that can support a wide variety of device designs with similar placement requirements but with different terminal characteristics. The routines must generate an optimized layout within a reasonable amount of time. We called this new technique Terminal Matching Placement And Routing (TEMPLAR). Pattern matching is traditionally used in the verification and analysis of VLSI layouts [9]. In this work we used a Pattern Matching approach to identify the devices with similar placement and routing characteristics and placed and routed them together. This paper focuses on the efficiency of the approach in reducing the computational overhead of the test chip placement and routing problem. Most of the layout intricacies are handled in the device construction and placement phases of layout synthesis so that the routing phase deals with simple point-to-point routing.

DEFINITION AND PROBLEM FORMULATION

Consider a set of devices D of cardinality d as an input. We assume that various layout characteristics are set as terminal attributes in the device template and the numbers of pads in the test chip are fixed.

Any arbitrary device d_i from the set D can be represented by its height and width of the device $\{h_i, w_i\}$ and the set of terminals $\{T_1, T_2, ..., T_n\}$, where each terminal T_j is represented by a set of terminal attributes, $T_j = \{c_1, c_2, ..., c_m\}$, where c_k is a terminal constraint specified in terms of a terminal attribute. In general, c_j can be any attribute such as terminal type, terminal angle, width of the terminal, HiMetal layers, LowMetal layers, etc. These attributes together form a terminal pattern for terminal T_j of device d_i.

Each test structure has a set of terminals. A block diagram of a typical MOSFET test structure is shown in Fig 1. A test structure is a generic object that has n number of terminals, and each terminal has m attributes. A Place and Route strategy must be created

for each combination of n and m. Each terminal has a set of terminal properties known as terminal attributes. For example, a terminal T_j may have the following attributes:

- Name: Specifies each terminal must have a name and a device can have more than one terminal with the same name.
- Angle: Indicates the direction the terminal connection should exit the device.
- Type: Indicates whether the terminal should be connected to a unique pad, a common bus, or a shared pad.
- Width: Specifies the width of the terminal connection.
- HiMetal and LoMetal: Indicates if multilayer routing is required. When HiMetal and LoMetal are the same value the connection is drawn with a single layer. When these values are different, the connection is drawn with multiple layers beginning with LoMetal and continuing to HiMetal.

The terminal attributes are specified in the device template and their values are analyzed by the TEMPLAR module defined in the placement algorithm.

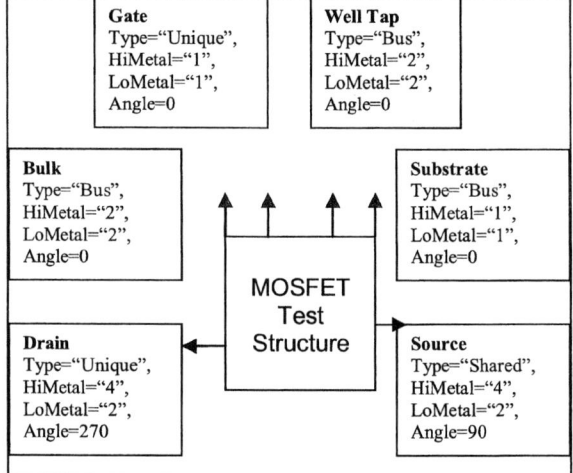

Fig 1. A typical MOSFET Test Structure and its terminal pattern

METHODOLOGY

The main objective of the TEMPLAR module is to identify the set of devices having similar placement and routing criteria. The devices having different placement and routing constraints are placed separately from one another. To identify the set of similar devices the terminal properties of the devices are exploited. A pattern is generated from a subset of the terminal properties such as Type, HiMetal, LoMetal, Angle, etc. This pattern is compared with the patterns of the other devices. If the patterns match, the devices are placed together. The TEMPLAR module loops until all of the devices are placed. The main objective of the TEMPLAR module is to make routing simpler, which in turn allows the algorithm to run faster, reducing the total turn around time.

In Fig 2, the total flow of the TEMPLAR module is shown. At the beginning of the placement phase, the first device from the master list is selected and its pattern is generated. The first device is removed from the master list and put into a new array. The next device from the master list is selected and its pattern is generated. This pattern is compared with the pattern of the first device. If the patterns match, the selected device is placed in the same array as that of the first device and is removed from the master list. If the patterns do not matched, the selected device is retained in the master list. The process is repeated for every device in the master list. On completion of the process iteration, the matched devices are placed and routed.

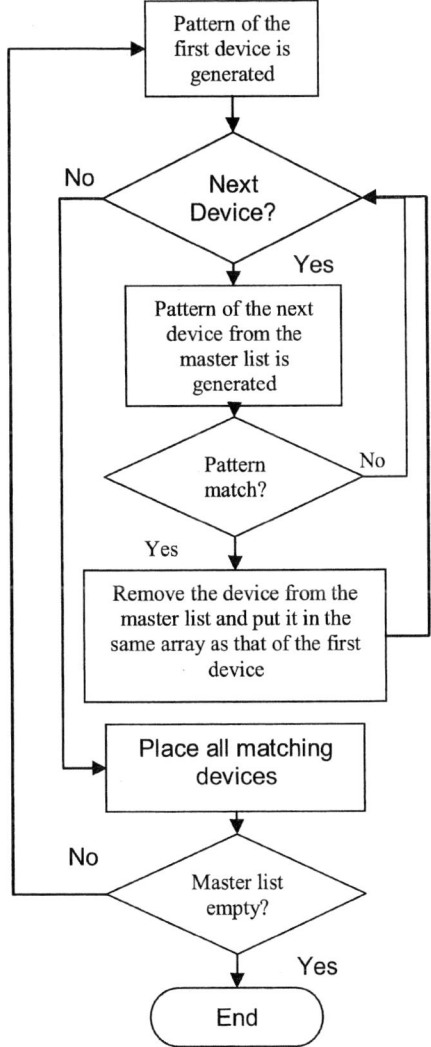

Fig 2. Flow of work of TEMPLAR module

The Placement module works on the premise of minimizing the complexity of the Routing modules. For example, depending on the placement orientation, clockwise or counterclockwise ordering of the terminals is performed. The terminals are ordered to minimize the opportunity for overlapping of connections during the routing phase of layout generation. The rotation and reflection of the device is another important factor. A device placed with a non-optimal rotation angle may further complicate the terminal-to-pad routing. Adjusting the orientation of the device ensures that the bus terminals are closer to the center of the quad with unique terminals closer to the pads, and shared terminals closer to the matching shared terminal of the adjacent device. All of these orientation and location decisions are controlled by the attributes assigned to the terminal during the device layout phase.

The placement routines do not use the terminal names to determine the placement and terminal-to-pad connections. This enables the placement routine to be reused by a similar set of devices with different terminal names but similar terminal constraints or a different number of bused connections. The different device types could be placed and routed within the same layout generation session. Prior to the TEMPLAR technique, a new placement routine would be required to handle every terminal name change or terminal addition/deletion.

RESULT AND PERFORMANCE

To illustrate the power of this approach, the TEMPLAR technique was used to efficiently layout a set of test structures using one placement and one routing module, with three example MOSFET devices and a capacitor device. An example of the layout generated is shown in figs 3, 4, 5 and 6.

In Fig 3, the pattern matched test structures (see fig 1 for an outline of the device and the terminals) are placed along the top and bottom inside edges of the quad in pairs except for a large structure at the left end and one small structure in the top right inside corner of the quad. Structures placed along the top inside edge are either rotated by 180 degrees or reflected in the x axis. For structures placed along the bottom inside edge, they are either unchanged or reflected in the y axis. These rotations and reflections allow simple connections to the bus lines running along the middle of the quad, and to simplify the shared Source terminal pad connection with corresponding terminal of its paired structure. The Gate and Drain terminals for each structure are routed directly to a unique pad. The three bus terminals Bulk, Well Tap, and Substrate are connected to the corresponding bus line running along the middle of the quad. The Bulk and Well Tap buses are connected to pads on the middle bottom of the quad. The Bulk routing for structures placed along the bottom inside edge of the quad use a MET3 bridge to avoid shorting to the Well Tap bus line and the Well

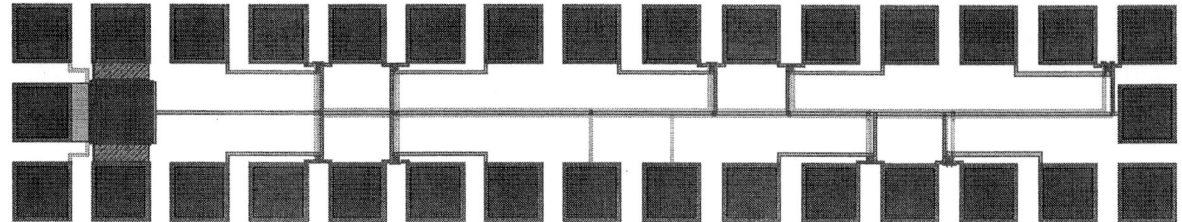

Fig 3. Example placement of a Six terminal structure with three buses, two unique and one shared terminal.

Tap routing uses a MET3 bridge for structures placed along the top inside edge of the quad. The Substrate bus uses MET1 for its routing and does not conflict with the other buses. The large MOSFET structure at the left end of the quad is rotated by 90 degrees counterclockwise. All the terminals are connected to unique pads with the exception of the substrate terminal which is connected to the corresponding bus.

In Fig 4 the pattern matched test structures are as in fig 3 except the Well Tap bus has been omitted for this device type. The Well Tap bus line was not added to the quad and therefore the Bulk routing does not use MET3 bridging. The bottom left pair of devices are shifted to the right, utilizing the pad freed by the omission of the Well Tap bus.

In Fig 5 the Bulk and Well Tap terminals have been omitted from the device and therefore the Bulk and Well Tap buses has been omitted from the layout. The Substrate terminal is not connected to the bus but is connected directly to a pad. Again, the bottom left pair of structures are moved to the right utilizing the pads freed by the omission of the Bulk and Well Tap buses.

Fig 6 is a quad of two terminal structures with the unique connection attribute placed along the middle of the quad, and each terminal is connected to a separate pad.

These four device types were placed with a single Placement and Routing module utilizing a different combination of devices types or number of devices for each test run. The execution times for different combinations are shown in Table 1.

Table 1: Placement and Routing Module Timing Results

No. of structures	Types of devices placed	Approximate time for the Placement module	Approximate time for the Routing module
43	2, 4, 5 and 6 term	6 sec	5 sec
75	2, 4 and 6 term	9 sec	19 sec
100	6 term	5 sec	58 sec
100	4 and 6 term	5 sec	49 sec
100	2, 4, 5 and 6 term	17 sec	55 sec

This test was conducted on a standard machine with 2 GB RAM and 1.9 GHz P4 processor. The time taken by Placement and Routing modules largely depends on the types and number of devices used. The fast turnaround of layout generation enables the device engineer to alter the mix of devices and device parameters to maximize quad usage and still out-perform the manual layout of the test structures.

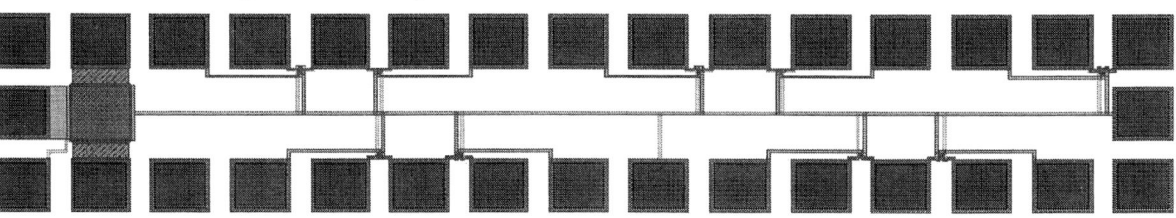

Fig 4, Example placement of a five terminal structure with two buses, two unique and one shared terminal.

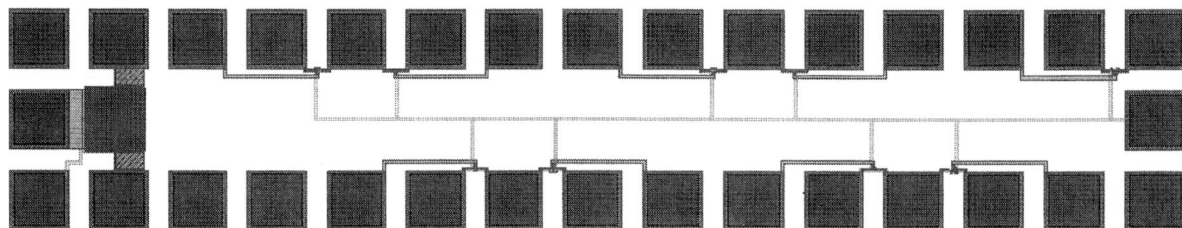

Fig 5, Example placement of a four terminal structure with one buses, two unique and one shared terminal.

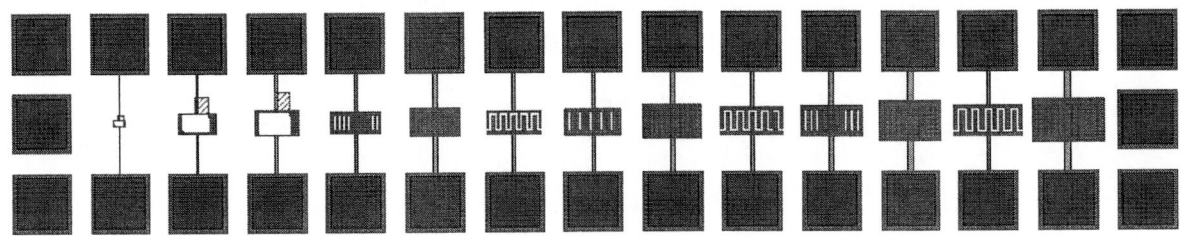

Fig 6, Example placement of a two terminal test structure with two unique terminals.

CONCLUSION

In utilizing the TEMPLAR technique we were able to compress four separate place and route strategies into a single strategy that works for any test structure with a similar placement requirement irrespective of the structure terminal names. This advanced automatic place and route system enables fast and accurate generation of GDS layout for any device architecture (such as MOSFET, Capacitor, LDMOS, EEPROM, and GOI devices). The library of place and route strategies developed provides the guidance to the system based on the device placement style rather than the device type and variation. This technique significantly reduces the number and complexity of the placement and routing scripts required. As a result, we have taped-out MOSFET, LDMOS, EEPROM, and GOI devices using the system.

In addition, the ability to regenerate an identical layout has been invaluable. When performing last minute modifications to the structure template for a layer re-spin, we can regenerate the GDS layout and import the replacement GDS into the original test chip while maintaining the appearance of unaltered layers.

REFERENCES

[1] Stone Pillar Users Guide Test Chip Builder, Volume 1 2007

[2] C.H. Diaz et al., "Process and circuit design interlock for application-dependent scaling tradeoffs and optimization in the SoC era", *IEEE Journal. of Solid-State Circuits*, Vol. 38, No. 3, Pages: 444-449, 2003.

[3] N. Jangkrajarng, S. Bhattacharya, R Hartono, C.-J. Richard Shi, "IPRAIL - Intellectual property reuse-based analog IC layout automation". *Integration, the VLSI Journal* Vol. 36, Issue. 4, Pages: 237-262, 2003.

[4] M. C. Sechen and A. L. Sangiovanni Vincentelli. Timberwolf 3.2: "A new standard cell placement and global routing package". *23rd Conference on Design Automation*, Pages: 432-439, 1986.

[5] R. A. Rutenbar and J. M. Cohn, "Layout tools for analog ICs and mixed-signal SoCs: a survey," *Proceedings of the 2000 International Conference on Computer Aided Design*, Pages: 76-83, 2001.

[6] Gielen and L. Carley, "Computer-aided design of analog and mixed signal integrated circuits," *Proceedings of IEEE*, Vol. 88, No. 12, Pages. 1825–1854, 2000.

[7] G. Van der. Plas, J. Vandenbussche, G.G..E. Gielen, W. Sansen, "A layout synthesis methodology for array-type analog blocks", *IEEE Transactions on Computer-Aided Design of Integrated Circuits and Systems*, Vol. 21, No. 6, 2002.

[8] L. Kasel, C. C.McAndrew, P. Drennan, W. F. Davis, R. Ida, "Automated Generation of SPICE Characterization Test masks and Databases", *Proceedings of. IEEE 1999 International. Conference. on Microelectronic Test Structures*, Vol 12, Pages: 74-79, 1999.

[9] M. Niewczas, W. Maly and A. Strojwas. "A pattern matching algorithm for verification and analysis of very large IC layouts". *Proceedings of the 1998 International Symposium on Physical Design*, Pages: 129- 134, 1998.

A Test Structure for Channel Length Engineering of NAND Gates in Standard Cell Library

T. Matsuda[1], *Member, IEEE*, Y. Sugiyama[1], J. Takakuwa[1], H. Iwata[1] and T. Ohzone[2]
[1]Department of Information Systems Engineering, Toyama Prefectural University.
matsuda@pu-toyama.ac.jp
[2] Dawn Enterprise, Nagoya, Japan.

ABSTRACT

A channel length engineering technique for optimization of primitive cells in standard cell libraries is proposed and a test structure to analyze the operation performance and leakage current of 3-input NAND is presented. Since the topmost transistor (N1) in the three series connected n-MOSFETs of 3-input NAND has the largest V_{DS}, subthreshold leakage current can be reduced by optimizing a channel length L of N1. The leakage current of NANDs for input vector of (0, 1, 1) decreases by about 22 ~ 40 % with the change of L (N1) from 0.1 to 0.11 μm. The channel length engineering of series connected MOSFETs provides a leakage reduction method for standard cells without significant increase of delay time, maintaining the same cell size.

INTRODUCTION

The increase of leakage currents in sub-100 nm CMOS VLSI has become a serious issue. Many technologies have been introduced for the leakage reduction in process, device, circuit and system levels [1]-[5]. The tradeoff between cost and performance is a difficult challenge in VLSI design. Among major leakage mechanisms, subthreshold leakage is sensitive to threshold voltage (V_{th}), drain-source voltage (V_{DS}) and channel length. Series connected MOSFETs have the effect of reducing subthreshold leakage due to the decrease of V_{DS} of one or more MOSFETs in the line [1]. In standard cell libraries used in digital LSI design, leakage reduction techniques for primitive cells such as NAND/NOR gates, which have series connected MOSFETs inherently, are necessary while maintaining the high performance. In this paper, a test structure to analyze the operation performance and leakage current of 3-input NAND is presented. And a channel length engineering technique for primitive cells in standard cell libraries is proposed to optimize the performance and leakage.

CHANNEL LENGTH ENGINEERING OF NAND GATES

Fig. 1 shows a schematic of 3-input NAND with a table of maximum V_{DS} of each n-MOSFET simulated by 90 nm CMOS device parameters. The topmost N1 in the three series connected n-MOSFETs has the largest V_{DS} of 1.0 V

at supply voltage V_{DD} = 1.0 V and input vector of (A, B, C) = (0, 1, 1), while the maximum V_{DS} of N2 and N3 are less than that of N1 by 0.17 ~ 0.19 V. Subthreshold leakage current can be reduced by optimizing channel lengths of series connected n-MOSFETs in NAND gates, because it depends on V_{DS} and channel length [1]. Although primitive cells are usually designed with the same minimum design rule, channel length engineering in standard cells can allow the leakage current reduction without significant increase of delay time.

TEST STRUCTURES

Fig. 2 illustrates a ring oscillator/NAND chain in our test structure fabricated with 90 nm standard CMOS process.

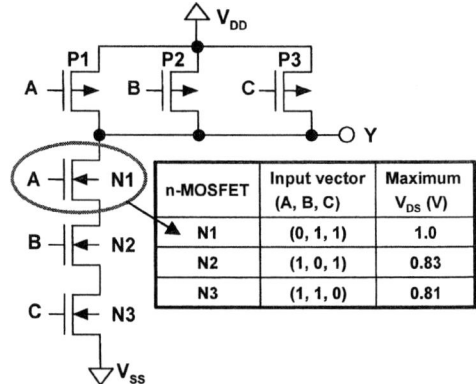

Fig. 1. Schematic of 3-input NAND with a table of maximum V_{DS} of each n-MOSFET simulated by 90 nm CMOS device parameters. N1 has the highest V_{DS} of 1.0V at V_{DD} = 1.0 V and input vector of (A, B, C) = (0, 1, 1).

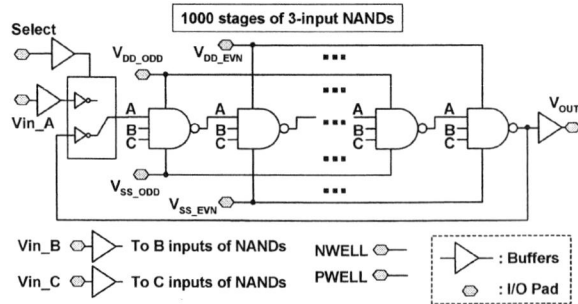

Fig. 2. Ring oscillator/NAND chain block in the test structure, which can be operated as either of two configurations by a multiplexer switch. Odd- and even-numbered stages have independent V_{DD} lines to measure the current of two different static states.

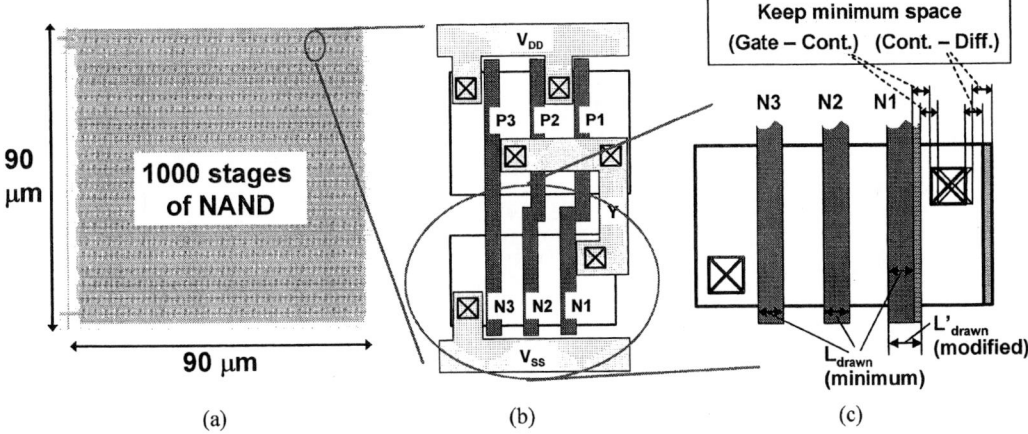

(a)   (b)   (c)

Fig. 3. (a) Layout of a ring oscillator/NAND chain block, (b) simplified layout of NAND gate and (c) layout of n-MOSFETs. L of N1 is modified by 0.10, 0.11, 0.15, 0.2 μm (drawn value in CAD). Spaces of "Gate - Contact" and "Contact - Diffusion edge" maintain minimum design rules.

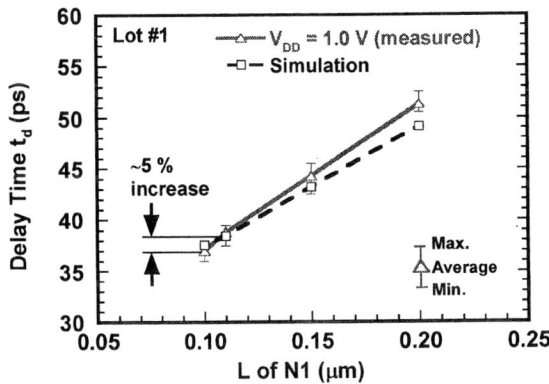

Fig. 4. Dependence of delay time t_d per NAND on L (N1), which were simulated and measured with the ring oscillator configuration. 5 dies were measured.

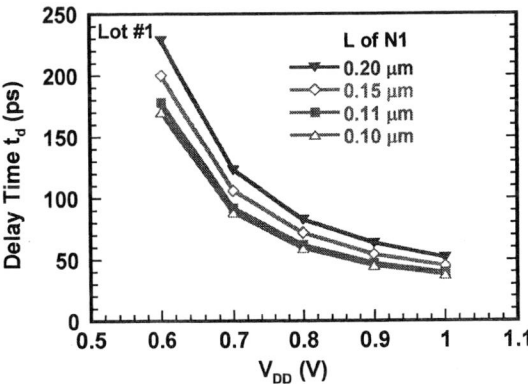

Fig. 5. t_d - V_{DD} characteristics of the ring oscillators with the different L (N1). Average values of 5 dies are presented, since variations are small.

It can be operated as either of two configurations by a multiplexer switch. Since odd- and even-numbered stages have two different static states of "1" and "0" alternately, power lines are separated to measure the current for each state independently. Four blocks of the ring oscillator/NAND chain have the different channel length L of N1 respectively, where the values of L (N1) correspond to 0.10, 0.11, 0.15, 0.2 μm (drawn value in CAD). L of the other MOSFETs is designed with the minimum design rule of 0.10 μm, which corresponds to 90 nm on wafer. The spaces of "Gate - Contact" and "Contact - Diffusion edge" in mask layers maintain minimum design rules to lower the change of parasitic effects for the different L (N1), as shown in Fig. 3. There is no area overhead by changing L (N1) as a standard cell. The number of stages of NANDs is n = 1000 to measure delay and leakage easily. Fan out of NANDs is 1 in the test structure. Typical values of p-MOSFET V_{th} in lot #1 and #2 were about -0.28 and -0.20 V, while those of n-MOSFET were about 0.23 and 0.21 V, respectively.

EXPERIMENTAL RESULTS

Fig. 4 shows a dependence of delay time t_d per NAND on channel length L of N1, which were simulated and measured with the ring oscillator configuration at $V_{DD} = 1.0$ V. Output of every NAND was connected to input A of the next stage and output of the last stage was fed back to the first stage through the multiplexer switch. The other inputs B and C were fixed to high level to keep NANDs in active state. The delay time t_d per NAND is calculated from $t_d = 1/(2nf)$, where f is an oscillation frequency of the ring oscillator. The measurement data are in good agreement with the simulation. The delay time t_d is nearly proportional to L (N1), and increases by about 5 % with L (N1) from 0.10 to 0.11 μm.

Fig. 5 shows t_d - V_{DD} characteristics of the ring oscillators with the different L (N1). Average values of 5 dies are presented, because the variations are small. t_d of

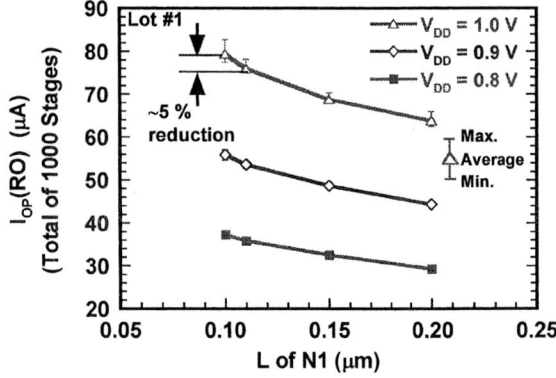

Fig. 6. L (N1) dependence of operation currents I_{OP}(RO) (total of 1000 stages) of the ring oscillators at different V_{DD}.

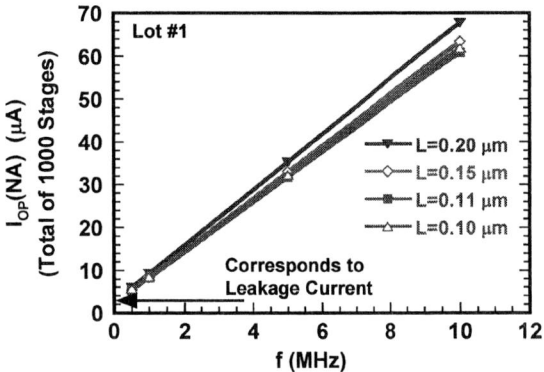

Fig. 7. Operation current I_{OP}(NA) (total of 1000 stages) of NAND chains versus frequency of input pulses.

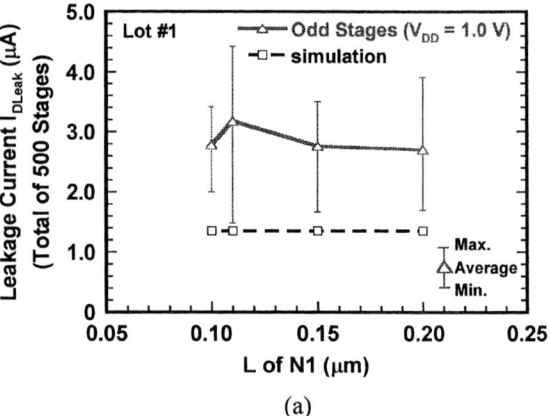

(a)

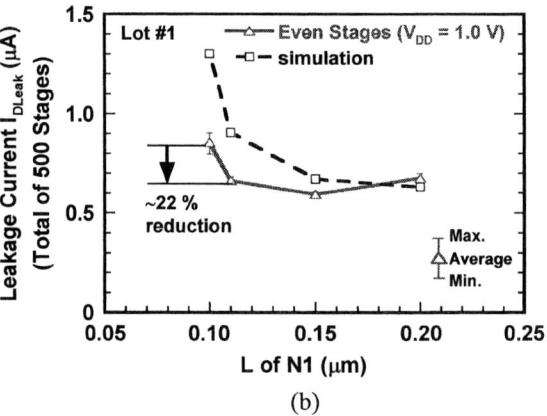

(b)

Fig. 8. Leakage currents of (a) odd-numbered stages (input vector = (1, 1, 1)) and (b) even-numbered stages (input vector = (0, 1, 1)) with different L of N at V_{DD} = 1.0 V. 5 dies of Lot#1 were measured. I_{DLeak} decreases by about 22 % with L (N1) from 0.10 to 0.11 μm.

NAND increases with the decrease of V_{DD} regardless of L (N1). Since the tendency of t_d of NAND with L (N1) of 0.10 μm is quite similar to that of 0.11 μm, slight change of L (N1) does not have much effect on t_d - V_{DD} characteristics.

Fig. 6 shows L (N1) dependence of operation currents I_{OP}(RO) of the ring oscillators at different V_{DD}. The amount of reduction of I_{OP}(RO) is about 5 % with the change of L (N1) from 0.1 to 0.11 μm at V_{DD} = 1.0 V. Since the decrease of oscillation frequency f for the change of L (N1) is also 5 % and I_{OP}(RO) is in proportion to f, the substantial I_{OP} of 0.11 μm is nearly the same as that of 0.1 μm.

The test structure can be operated as NAND chain by switching the multiplexer. Operation currents I_{OP}(NA) of NAND chains with various frequencies of input pulses as shown in Fig. 7. In general, the operation currents I_{OP}(NA) can be expressed as follows:

$$I_{OP}(NA) = I_{stand\text{-}by} + C_L V_{DD} f$$

where $I_{stand\text{-}by}$ and C_L are leakage current of the stand-by state and load capacitance, respectively. The leakage current $I_{stand\text{-}by}$, which is calculated by I_{OP}(NA) values at f = 0 in Fig. 7, are between 2 to 5 μA.

The odd- and even-numbered stages of the NAND chain have two different static states alternately. When input B and C of NANDs are fixed to high level and terminal Vin_A is set to 0 V in the NAND chain configuration, input vectors for the odd- and even-numbered stages become (A, B, C) = (1, 1, 1) and (0, 1, 1), respectively. The odd-numbered stages exhibit higher leakage current independently of L (N1) as shown in Fig. 8(a). In contrast, the leakage current I_{DLeak} of the even-numbered stages, of which input vector of (0, 1, 1), decreases by about 22 % with the change of L (N1) from 0.10 to 0.11 μm at V_{DD} = 1.0 V as shown in Fig. 8(b).

Since all of p-MOSFETs and n-MOSFETs become off- and on-state, respectively, p-MOSFETs form dominant

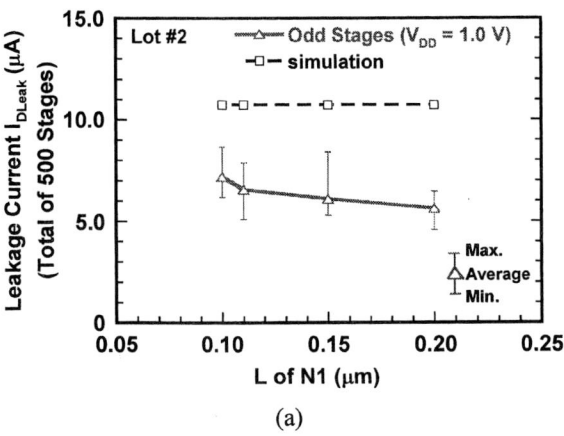

(a)

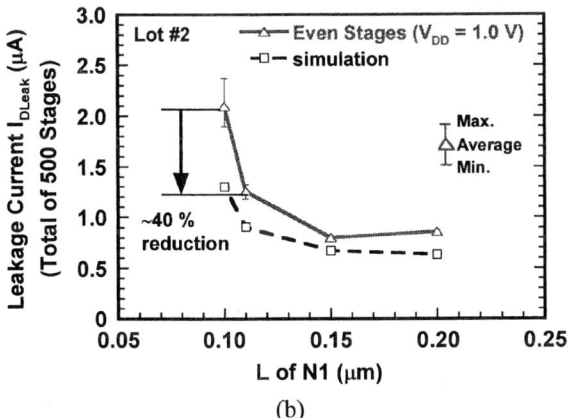

(b)

Fig. 9. Leakage currents of (a) odd-numbered stages and (b) even-numbered of Lot #2, which has lower V_{th} than lot #1. I_{DLeak} decreases by about 40 % with L (N1) from 0.1 to 0.11 μm.

leakage paths at the state of (1, 1, 1) input vector in the odd-numbered stages. The leakage current of odd-numbered stages is thus independent of L (N1) and the variations are large. When the input vector is (0, 1, 1), P1 of parallel connected p-MOSFETs turns on and N1 of series connected n-MOSFETs goes off state. Since N1 rules the leakage current of even-numbered stages in this state, the channel length of N1 has an effect on leakage current reduction and the variations are very small due to the stacked effect of MOSFETs [2]. The reduction of the leakage current becomes saturated for the larger L (N1), because the other components of leakage, such as gate oxide leak, get dominant.

The NAND chains in another wafer (lot #2) with lower p-MOSFET's V_{th} have higher leakage current for (1, 1, 1) case as shown in Fig. 9. Consequently, the leakage of

p-MOSFETs is considered to be mainly subthreshold current. As a result, a channel length engineering, which optimizes L (N1), can reduce the leakage current of NAND for the output state of "1", while the "0" output state requires the optimization of p-MOSFETs.

CONCLUSION

A test structure of ring oscillator/NAND chain was presented and the operation performance and leakage current were analyzed. Since the topmost transistor N1 in the three series connected n-MOSFETs of 3-input NAND has the largest V_{DS}, subthreshold leakage current can be reduced by optimizing L (N1). The leakage current of NANDs for input vector of (0, 1, 1) decreases by about 22 ~ 40 % with the change of L (N1) from 0.10 to 0.11 μm. In the standard cell based design, a combination of device size optimization within a standard cell and input vector control technique will be effective in leakage reduction. As a method of device size optimization, a channel length engineering of series connected MOSFETs such as a slightly larger L for N1 provides leakage reduction without great increase of delay time.

ACKNOWLEDGMENT

The VLSI chip in this study has been fabricated through the chip fabrication program of VLSI Design and Education Center(VDEC), the University of Tokyo, with the collaboration by STARC, Fujitsu Limited, Matsushita Electric Industrial Company Limited., NEC Electronics Corporation, Renesas Technology Corporation, and Toshiba Corporation.

REFERENCES

[1] S. Narendra, V. De, S. Borka, D. A. Antoniadis, A. P. Chandrakasan,"Full-chip subthreshold leakage power prediction and reduction techniques for sub-0.18 μm CMOS," IEEE J. of Solid-State Circuits, vol. 39, No. 2, pp. 501-510, 2004.

[2] R. Rao, A. Srivastava, D. Blaauw, D. Sylvester, "Statistical estimation of leakage current considering inter- and intra-die process variation," IEEE/ACM Intl. Symp. Low Power Electronics and Design (ISLPED), pp. 84-89, 2003.

[3] D. Lee, D. Blaauw, D. Sylvester, "Gate oxide leakage current analysis and reduction for VLSI circuits," IEEE Trans. Very Large Scale Integration (VLSI) Systems, vol. 12, No. 2, pp. 155 - 166, 2004.

[4] M. Nomura, Y. Ikenaga, K. Takeda, Y. Nakazawa, Y. Aimoto, Y. Hagihara, " Delay and Power Monitoring Schemes for Minimizing Power Consumption by Means of Supply and Threshold Voltage Control in Active and Standby Modes," IEEE J. of Solid-State Circuits, vol. 41, No. 4, pp. 805 - 814, 2006.

[5] A. Agarwal, C. H. Kim, S. Mukhopadhyay, K. Roy, K, "Leakage in nano-scale technologies: mechanisms, impact and design considerations," Design Automation Conference, pp. 6 - 11, 2004.

Test Structure for Characterising Low Voltage Coplanar EWOD System

Yifan Li, Mita Yoshio∗, Les Haworth, William Parkes, Masanori Kubota∗, Anthony Walton

School of Engineering and Electronics, University of Edinburgh, Edinburgh, UK, EH9 3JF

Email: y.li@ed.ac.uk

∗University of Tokyo, Tokyo, Japan

Email: mita@ee.t.u-tokyo.ac.jp

Abstract—This paper presents test structures designed for studying the relationship between the operation voltage and the configuration of electrode area for coplanar EWOD (ElectroWetting On Dielectrics) devices. Robust anodic Ta_2O_5 dielectric and thin aFP (amorphous Fluoropolymer) have been used to fabricate the structures. Test structures have been used to characterise the significant contact angle change on asymmetric configurations, 114° to 81° on CYTOP (amorphous fluoropolymer from Asahi Glass Co. Ltd.) with an applied voltage of less than 20V. This demonstrates that by modifying the design, the operating voltage can be reduced by a factor of two, compared to the existing symmetric coplanar EWOD structures. Droplet manipulation on a coplanar EWOD system with this new design has been successfully demonstrated, with a driving voltage of 15V.

I. INTRODUCTION

A. Electro-Wetting on Dielectrics (EWOD) and driving voltage

Lab-on-a-chip (LOC) and bio-MEMS systems, which can manipulate and analyse biological fluidic samples in micro- and nano-litre scales, have emerged as a solution for automating repetitive laboratory tasks [1], [2]. Digital microfluidic, involving manipulating liquid flows in droplet form, provide a potentially reconfigurable method of obtaining a bio-MEMS system [2], [3]. EWOD technology is an attractive option that has a low power consumption making it well suited for the design and manufacture of microfluidic systems [2]. EWOD uses solid-liquid surface tension as a driving force, which can be controlled by applying a suitable voltage across a droplet sandwiched between a top ground electrode and an array of driving electrodes covered by a two layer dielectric underneath [1]. By doing so, the selected area on the bottom dielectric surface will be charged and turned from hydrophobic to hydrophilic, and the droplet will be attracted toward the hydrophilic region.

The contact angle of a droplet is the measure of the surface hydrophobicity. The amount of contact angle change due to the EWOD operation is used to judge whether droplet manipulation is achievable. The voltage applied giving this contact angle change can be named the driving voltage V_D. Obviously, a lower driving voltage V_D is preferred in controlling microsystems. Earlier work on electrowetting arrays required driving voltages in the range 80 - 100V [4]. More recently and with a more judicious choice of materials, processes and dielectric

thickness, the voltage required to manipulate droplets has been reduced below 15V [4].

B. Coplanar EWOD systems

Recently there has been increasing interest in coplanar microfluidic EWOD devices where a top plate electrode is not required to drive droplets of liquid. In practice, the top cover plate provides many advantages (e.g. reliable droplet volume control, gravity insensitivity, reduced evaporation loss) [5]. However, one drawback of using a two-plate system is the restrictions imposed by the top-plate, which needs to provide a conductive ground electrode. If this requirement can be removed, then other functions such as contact sensing, and integration of other microfluidic actuation methods such as dielectrophoresis (DEP) and surface acoustic waves (SAW) can be incorporated. Moreover, the packaging processes is simplified when a cover plate can be assembled without a specific gap and no electrical connections.

This has led to the need for test structures to characterise this technology and the first test structure to address this requirement was reported in [5]. This structure has been used to help characterise the relationship between electrode gap ratio and the contact angle change for a coplanar device and has provided the data required to modify the Young-Lippmann equation to model this relationship [5]. The technology and architecture reported in [5] required nearly 100V to modify the contact angle from 110° to 80° and 65Vac to move droplets. Clearly, a lower operating voltage is desirable [4], [6] and this paper presents an improved test structure modified for characterising such low voltage coplanar EWOD device architectures. In addition the test structure reported in this paper can be used to characterise not only the electrode gap dimensions but also the electrode area ratio.

II. BACKGROUND THEORY

A. EWOD driving voltage and test device

Figure 1(a) shows an experimental configuration for evaluating an EWOD system in terms of driving voltage in a contact angle measurement system [1], [4]. The metal wire inserted into the droplet represent the top ground electrode in the two-plate EWOD device provides a voltage reference for the bottom driving electrode where the hydrophobic dielectric

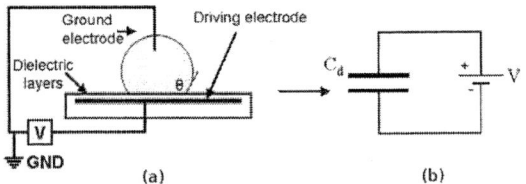

Fig. 1. (a) Side view of an EWOD driving voltage evaluation experimental configuration for two-plate EWOD devices. (b) Equivalent device circuit of (a).

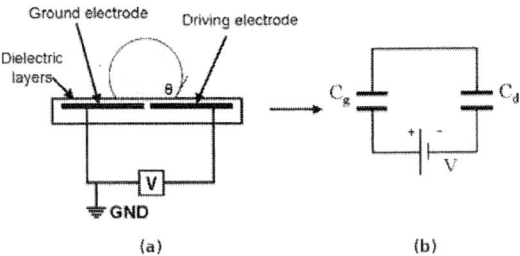

Fig. 2. (a) Side view of an EWOD driving voltage evaluation experimental configuration for coplanar EWOD devices. (b) Equivalent device circuit of (a).

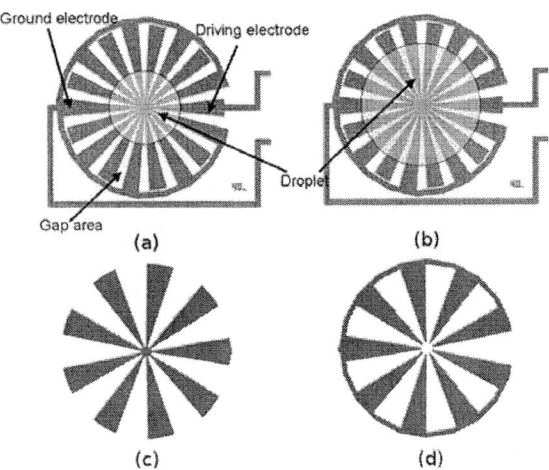

Fig. 3. Top view of a droplet sitting on a general coplanar configuration test structure with two radially wedge shaped electrodes. (a) Without voltage applied. (b) With voltage applied, showing the covered electrode area ratio and ratios of the three phase contact line length on each electrode remain the same when the droplet spread out. (c) Driving electrode. (d) Ground electrode.

surface is switched to hydrophilic by applying voltage. The device consists of dielectric layers coating electrodes and helps to evaluate the driving voltages of EWOD devices fabricated using various of processes by performing contact angle - voltage experiments.

The relationship between the average surface charge density (the dielectric property), the applied voltage, the solid surface initial hydrophobicity and the contact angle of the droplet is given by the Young-Lippmann equation

$$\cos\theta(V) - \cos\theta(0) = \frac{\varepsilon_r \varepsilon_0}{2\gamma_{lg} t} V^2 \qquad (1)$$

where γ_{lg} is the liquid-gas surface tension (the value of water-air surface tension is 0.072N m^{-1}), t is the thickness of the dielectric layer [1], [7], V is the voltage applied, ε_r is the dielectric relative permittivity (κ), $\theta(0)$ is the initial droplet contact angle and $\theta(V)$ is the contact angle when voltage V is applied. Figure 1(b) shows the equivalent circuit of the test device. By changing the dielectric layer properties, and hence the driving electrode capacitance C_d, the driving voltage can be modified [4].

B. Coplanar EWOD electrode arrangements

The major design difference between the two-plate and coplanar EWOD systems is the arrangement of electrodes. The ground electrode is placed coplanar with the driving electrode on the bottom plate in coplanar EWOD systems [5], [8]–[10]. Figure 2(a) shows the coplanar EWOD configuration reported in [5], [11].

Since the voltage is applied between two electrodes placed coplanar on the surface, the equivalent device circuit is now modified, as shown in fig. 2(b). The average charge density

and the voltage across the dielectric layers underneath the liquid now relate to not just the dielectric properties, but the area ratios of each electrode as well. The capacitances of dielectric layers over them [C_d (driving electrode), C_g (ground electrode)], and the gap area between the electrodes determine the voltage required to drive the droplets. These all depend on the area ratios design and the Young-Lippmann equation (1) needs to be modified, to include the electrode area and gap area ratios [5]. Therefore, the test device for EWOD driving voltage evaluation shown in fig. 1 cannot be used.

C. Test structures for coplanar EWOD evaluation

Figure 3 shows a test structure reported for evaluating both the layout and the dielectric layers of a general coplanar configuration [5]. The radially wedge shaped electrodes are the driving and ground electrodes and the droplet is positioned in the centre. As shown in fig. 3(a) and (b), this ensures a constant area ratio for each electrode and the gap area between the electrodes follows the wetting of the droplet (fig. 4) . The driving electrodes in the test structure are all electrically connected at the centre (fig. 3(c)) with the ground electrodes connected around the circumference of the structure (fig. 3(d)).

On this test structure, the dependence of the solid-liquid surface tension γ_{sl} at applied voltage V on the entire area covered by the droplet can be modified from the original Lippmann's equation [1], [5]:

$$\Delta\gamma = \Sigma\gamma_{sl}(V) - \Sigma\gamma_{sl}(0) \qquad (2)$$

in which,

$$
\begin{aligned}
\Delta\gamma &= \gamma_{sl}(V) - \gamma_{sl}(0), \\
\Sigma\gamma_{sl}(V) &= \gamma_{sl}^d(V) + \gamma_{sl}^g(V) + \gamma_{sl}^{gap}(V), \\
\Sigma\gamma_{sl}(0) &= \gamma_{sl}^d(0) - \gamma_{sl}^g(0) - \gamma_{sl}^{gap}(0)
\end{aligned}
$$

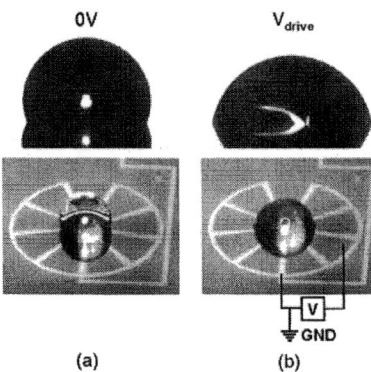

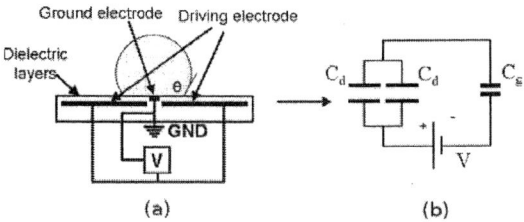

Fig. 4. Side view (top) and 3-D view (bottom) of a droplet sitting on a EWOD coplanar test structure with wedge shaped electrodes. (a) Without voltage applied. (b) With voltage applied.

Fig. 5. (a) Side view of an EWOD driving voltage evaluation experimental configuration for coplanar EWOD devices with ground lines. (b) Equivalent device circuit of (a).

where $\gamma_{sl}^d(V)$, $\gamma_{sl}^g(V)$, $\gamma_{sl}^{gap}(V)$ and $\gamma_{sl}^d(0)$, $\gamma_{sl}^g(0)$, $\gamma_{sl}^{gap}(0)$ are the solid-liquid surface tensions with or without applied voltages (V or 0) for driving electrode, ground electrode and gap area respectively.

The area of each wedge covered by the droplet is much smaller than the total area covered by the droplet and thus the changes in surface tension above the electrodes and the gap area is averaged. Based upon this assumption, equation (2) can be re-written as [5]:

$$\Delta\gamma = -\frac{\varepsilon_r\varepsilon_0}{2t}\left(\frac{A_d}{A_t}V_d^2 + \frac{A_g}{A_t}V_g^2 + \frac{A_{gap}}{A_t}V_{gap}^2\right) \quad (3)$$

Combining equation (3) with the Young-Lippmann equation (1), the modified Young's-Lippmann equation for general coplanar structures becomes:

$$\Delta\cos\theta = -\frac{\varepsilon_r\varepsilon_0}{2\gamma_{lg}t}\left(\frac{A_d}{A_t}\left(\frac{A_g}{A_d+A_g}\right)^2\right.$$
$$\left. +\frac{A_g}{A_t}\left(\frac{A_d}{A_d+A_g}\right)^2\right)V^2 \quad (4)$$

where $\Delta\cos\theta = \cos\theta(V) - \cos\theta(0)\gamma_{sl}(0)$ is the interfacial tension with no voltage applied, A_d, A_g and A_{gap} are the areas over the driving electrodes, the ground electrodes and the gap areas between the electrodes, respectively, A_t is the total combined area, and V_d, V_g and V_{gap} are voltages across the dielectric layers above each electrode areas [5].

When the area of the ground electrode is equal to the area of driving electrode, the voltage required for a given contact angle change in this design is minimised. Hence, the electrode sizes in the general coplanar systems were kept identical, and the test structure presented was only designed to study the gap ratio effect [5].

From equation (4), it can be determined that when the gap area is minimised, the minimum critical driving voltage can be achieved. Compared with the two plate configuration, the minimum value is doubled in the general coplanar system if the dielectric layers have the same thicknesses.

The experimental results reported in [5] confirmed the theoretical calculation. With 300nm PECVD SiO$_2$ and 200nm CYTOP® and a 2% gap ratio, the critical driving voltage was between 100 and 120V and for droplet manipulation 65 to 80V AC driving voltages were required. Moreover, the high voltage required resulted in dielectric breakdown before contact angle saturation was reached [5].

III. NEW TEST STRUCTURE DESIGN

Figure 5(a) shows that a ground line was deposited over the insulating dielectric layer in the gap between the driving electrodes to build the coplanar EWOD systems as described in [8], [9]. The insulating and the aFP (amorphous Fluoropolymer, such as CYTOP® and Teflon-AF®) layers on the driving electrode are modeled as a capacitor C_d (fig 5b). The ground line has only the aFP layer covering it, and is modelled by the capacitor C_g (fig 5b). Since the aFP layers are typically much thinner than the insulating layer (and cannot withstand large voltages), the majority of the potential will be dropped across C_d [4], [9]. This suggested that coplanar devices including a ground line could have a lower driving voltage than the general coplanar devices due to the larger potential drop across the dielectrics on the driving electrodes. Ideally, if the ground line and the gap area size can be reduced to very close to zero, the critical driving voltage can approach that in the two plate systems. Such a system can facilitate a low voltage coplanar EWOD system, potentially having a critical driving voltage close to the 15V presented in [4] for a two-plate system.

A. Modified test structure for low voltage coplanar configuration

To enable full characterisation, a new test structure architecture (shown in fig. 6(a) and (b)) was designed to study the electrode area ratio effect and also quantify the performance improvement when the dielectric is removed from the ground electrode.

In this new design, anodic Ta$_2$O$_5$ was used as the insulating layer. The structure shown in fig. 6(a), is anodised by grounding the "ground" electrodes during the anodisation process. The creation of the dielectric using anodisation for coplanar devices including a ground line results in Ta$_2$O$_5$ being only grown on the driving electrodes and so no lithography or etching is required to remove unwanted dielectric on the ground electrodes.

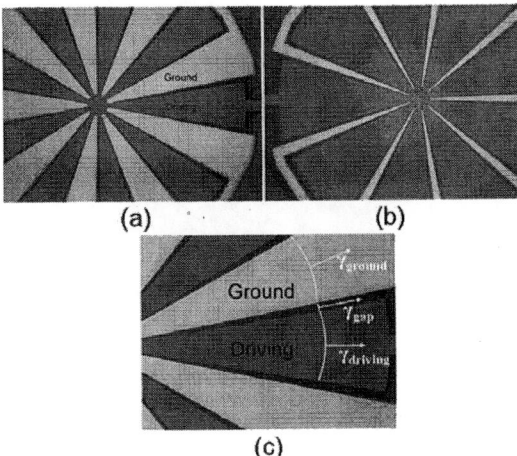

Fig. 6. EWOD coplanar test structure with: (a) Wedge shaped electrodes with identical sizes but no ground insulating coatings (driving electrode with Ta_2O_5 and CYTOP® - ground electrode has CYTOP® only); (b) Same type of dielectric coating as in (a), but with a different electrode area ratio ($A_{driving}$: A_{ground} = 8 : 1); (c) Surface tension distribution along the contact line (white curves show part of the droplet outline) when a droplet is wetting the surface.

Fig. 6(b) shows an example of a coplanar device including ground line with Ta_2O_5 coated driving electrodes that are larger than those shown in fig. 6(a).

Changing the area ratio of both electrodes will result in a different surface tension distribution along the three phase contact line during wetting (fig. 6(c)). By looking for a maximum average surface tension change using the same voltage in equation (2), the optimum electrode ratio for low voltage coplanar EWOD can be determined. The shape of the droplet, when viewed in the horizontal plane, enabled the contact angle to be measured in the traditional manner.

As described by Yi et al. [5], the changes in interfacial energy above each electrodes and the gap area were assumed to be averaged along the three-phase contact line. As discussed in [4], the voltage drop across the aFP layers on the ground electrodes was neglected and so for the coplanar device configuration with a ground line, the relationship between the contact angle and applied voltage can then be determined as:

$$\cos\theta(V) - \cos\theta(0) = -\frac{\varepsilon_r\varepsilon_0}{2\gamma_{lg}t}\frac{A_d}{A_t}V^2 \qquad (5)$$

In this case A_d is the area of the driving electrode, and A_t is the total area (driving + ground electrode areas + the gap area). When A_d approaches A_t, the required driving voltage reduces to the levels reported for two-plate EWOD system.

For comparative studies, both general coplanar test structures with gap ratios from 10% to 60%, and coplanar with ground line test structures and electrode area ratios (A_g : A_d) from 1:1 to 1:8 (10% gap ratio) were designed.

IV. SAMPLE FABRICATION AND EXPERIMENT RESULTS

A. Sample fabrication

The test structures were fabricated on Si substrates with a $0.5\mu m$ layer of SiO_2. A layer of 500nm tantalum was then sputtered and patterned to create the wedge shaped test electrodes. To prevent dielectric breakdown failure before contact angle saturation occurred [5], a thick Ta_2O_5 layer (180nm) was anodised at 100V. A layer such as this can theoretically sustain 100V DC before breaking down (this voltage should have the same polarity as the anodisation).

The ground electrodes of the test structures with coplanar ground lines were not anodised. For the general coplanar devices, the ground electrodes were also anodised at the same voltage.

After the anodisation, 35nm of CYTOP® was spin coated on the entire area to make the surface hydrophobic.

B. Experiments and results

The characterisation system used to study the contact angle as a function of applied voltage has been shown in [4]. Additionally, the shape of the wetting droplet (shown previously in fig. 4) was observed from the top and side, using a camera systems similar to [4], [12].

1) General coplanar configuration characterisation: Performing measurements of contact angle as a function of voltage on the general coplanar configuration test structures has demonstrated the advantages of anodic Ta_2O_5 over PECVD SiO_2. The thicker film, which can theoretically sustain 100V, enables the system to achieve a critical voltage of 35V DC (voltage to drive the contact angle into saturation) with a gap ratio of 10%. This compares to the value of over 100V measured on samples with 300nm PECVD SiO_2 and 200nm CYTOP®. Moreover, there was no dielectric breakdown when voltages up to to 60V DC were applied, enabling the study of the saturation phenomenon on EWOD coplanar structures.

Figure 7 shows that the contact angle changes follow the theoretical values given by equation (4) before saturation sets in. For structures with a 20% gap ratio, the critical driving voltage was around 40V DC. Due to early contact angle saturation, the contact angle of droplets operated on the 40% and 60% gap ratio test structures did not reach 81°. The reason for this will be discussed in the following section.

2) Characterisation of devices with coplanar ground line: Following the lower driving voltage obtained in the general coplanar EWOD characterisation, measurements were performed on devices including a coplanar ground line. With the same dielectric coatings, the test structures with no Ta_2O_5 dielectric on the reference electrode and a 10% gap, A_g:A_d = 1:1 (A_d:A_t = ~1:2), exhibited a reduced driving voltage of around 25V (fig. 8). Changing the A_g:A_d ratio to 1:8 (A_d:A_t = ~8:9), reduced this voltage further to between 17 and 20V (figure 8). These results clearly indicate that the driving voltage is reduced by half for the new coplanar architecture.

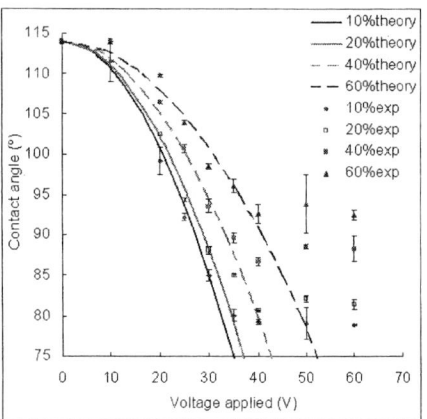

Fig. 7. Contact angle as a function of applied voltage for general coplanar configuration test structures as a function of gap ratio (10%, 20%, 40% and 60%), when A_d:A_t = 1:2 (A_d:A_g = 1:1);

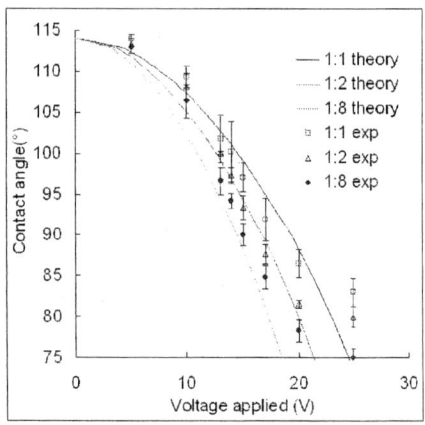

Fig. 8. Contact angle as a function of applied voltage for structures with a coplanar ground line as a function of A_g:A_d ($\frac{1}{1+A_d:A_t}$) ratio (from 1:1 to 1:8) with a 10% gap. (Since the results of electrode ratio of 1:5 were very close to those of ratio 1:8 , it was omitted for clairity.)

3) Results comparison: Figure 9 shows the relationship between contact angle change and applied voltage for EWOD coplanar electrodes. In terms of reducing the driving voltage, the coplanar design with ground line obviously has an advantage. By reducing the ground electrode size while increasing the driving electrode size, a lower critical voltage can be achieved. The lowest value achieved, which is roughly the same as that obtained using a two-plate configuration was around 15V.

V. DISCUSSION

A. Early contact angle saturation tests in the general coplanar configuration

Before reaching 81°, early saturation was observed for the 40% and 60% gap ratio samples. This is due to the saturation of γ_{sl}^d and γ_{sl}^g. Despite the different gap ratios, the voltage across the insulating and aFP dielectric layers

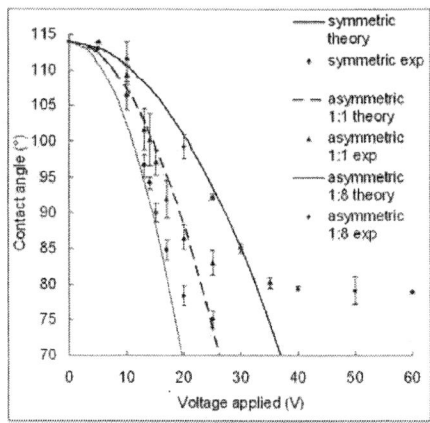

Fig. 9. Contact angle as a function of applied voltage for coplanar test structures with the general coplanar configuration with 1:1 A_g:A_d ratio, and with a coplanar ground line with 1:8 coating A_g:A_d ratio (all gap ratios = 10%) .

on all the samples were the same. Hence, referring to the modified Lippmann equation (2), γ_{sl}^d and γ_{sl}^g on all the test structures consisting of the same dielectric layers should reach the saturation value at the same applied voltage (in this case 35V to 40V). Since the wettability of the gap area did not change during the test, a larger area of unchanged γ_{sl}^{gap} will result in an earlier overall contact angle change.

B. Surface tension distribution on coplanar EWOD structures

During the design evaluation test, the shapes of the droplets were observed not only from the side using the contact angle measurement system, but also from the top using the droplet manipulation observation system.

Figure 10 shows the top view of wetting droplets on the general coplanar test structure with different gap ratios. The liquid-solid contact line roughly followed a circular shape. Although there was no electrowetting force in the gap area, the Laplace pressure created by γ_{sl}^d and γ_{sl}^g inside the droplet formed a γ_{sl}^{gap}. When the gap ratio is relatively small, the assumption made previously that the contact angle measured can represent the average surface tension is still applicable. This is confirmed in figure 7, where results of 10% to 40% gap ratio met the theoretical curve better than those of 60%. These results agreed with Wang et al.'s observation which showed the imbalance of electrowetting between identical dielectric surfaces with different electric field polarity [13].

In fig. 11, the top view of wetting droplets on the coplanar test structure including a ground line with different electrode ratios is shown. From this figure, the area with highest wettability was on the driving electrodes on both samples. Wetting on the ground electrodes and gap area could hardly be seen, due to the fact that most of the voltage applied was dropped across the dielectric layers deposited on the driving electrodes.

Hence, γ_{sl}^{gap} and γ_{sl}^g were both formed mainly by the Laplace pressure. When the electrode ratio A_g:A_d was larger

84

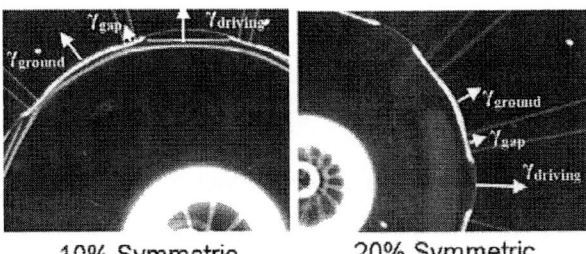

10% Symmetric   20% Symmetric

Fig. 10. Top view of wetting droplets on the general coplanar test structure with different gap ratios. (The nautilus shaped reflection is the droplet centre, while radial lines are the edge of the electrodes.)

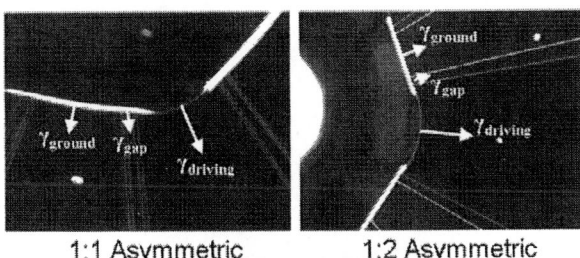

1:1 Asymmetric   1:2 Asymmetric

1:5 Asymmetric

Fig. 11. Top view of wetting droplets on the general coplanar test structure with different gap ratios. (The nautilus shaped reflection is the droplet centre, while radial lines are the edge of the electrodes.)

than 1:5, the liquid-solid contact line during wetting was not close to circular (clearly shown in figure 11). As a result, the contact angle measured can not accurately represent the average surface tension. This may explain the slight mismatch of the experimental contact angle results in figure 8.

VI. DROPLET MANIPULATION

Based on the results from the coplanar test structures, coplanar EWOD electrode arrays were designed using parameters derived from results obtained in the previous sections. The ground electrodes were routed in the gaps between the electrodes and interconnects.

For fabricating this structure, tantalum was sputtered and patterned as $5\mu m$ wide metal tracks for the ground electrodes, and 1mm × 1mm squares for the driving electrodes. The gap between the driving electrodes and ground electrodes were $3\mu m$ (meaning the gap and ground electrode area ratios were below 1%). The driving electrodes were then anodised at 50V to form a 95nm anodic Ta_2O_5 insulating layer. After 16nm

Teflon-AF® was spin coated, the droplet could be manipulated at a speed of ~16mmsec^{-1} with applied voltage as low as 15V.

VII. CONCLUSIONS

In this paper, an improved coplanar EWOD test structure has been presented. It has been modified for characterising low voltage coplanar EWOD device architectures and can be used to characterise electrode gap dimensions and the electrode area ratio. This characterisation was an important component in the development of a new EWOD architecture that reduced the droplet driving voltage.

Characterising the electrode gap dimensions in the general coplanar configuration test structures using anodic Ta_2O_5 for comparison with the work in [5], the significantly reduced critical driving voltage (reduced from around 100V to around 40V) shows the advantages of using Ta_2O_5 as a dielectric insulation in the coplanar EWOD structures. Changing the electrode ratio in the coplanar configuration including a ground line helped to optimise the design and reduce the critical voltage to less than 20V, while having the same dielectric layer composition.

Finally, successful low voltage coplanar EWOD droplet manipulation has been demonstrated as a result of the design and process optimisation.

ACKNOWLEDGMENT

The authors would like to acknowledge the financial support of EPSRC, the IIS (part of the Edinburgh Research Partnership in Engineering and Mathmatics (ERPEM)),the RASOR project (EPSRC and BBSRC) and GOLEM project (EC).

REFERENCES

[1] F. Mugele and J.-C. Baret, "Electrowetting: from basics to applications," *J. of Phys.: condensed matter*, vol. 17, pp. R705–R774, 2005.
[2] K. Chakrabarty and F. Su, *Digital Microfluidic Biochips*. CRC, 2006.
[3] A. Wixforth, "Acoustically driven programmable microfluidics for biological and chemical applications," *JALA*, pp. 399–405, 2006.
[4] H. Moon, S. Cho, R. Garrrell, and C. Kim, "Low voltage electrowetting-on-dielectric," *J. of Appl. Phys.*, vol. 92, pp. 4080–4087, 2002.
[5] U.-C. Yi and C.-J. Kim, "Characterization of electrowetting actuation on addressable single-side coplanar electrodes," *J. of micromechanics and microengineering*, vol. 16, p. 20532059, 2006.
[6] Y. Li, W. Parkes, L. Haworth, A. Stokes, K. Muir, P. Li, A. Collin, N. Hutcheon, R. Henderson, B. Rae, and A. Walton, "Anodic Ta_2O_5 for CMOS compatible low voltage Electrowetting-On-Dielectric device fabrication," in *ESSDERC 2007*, 2007, pp. 446–449.
[7] G. Lippmann, "Relations entre les phenomenes electriques et capillaires," *Ann. Chim. Phys.*, vol. 5, p. 494, 1875.
[8] R. B. Fair, "Digital microfluidics: is a true lab-on-a-chip possible?" *Microfluid Nanofluid*, vol. 3, pp. 245–281, 2007.
[9] M. G. Pollack, A. D. Shenderov, and R. B. Fair, "Electrowetting-based actuation of droplets for integrated microfluidics," *Lab on a chip*, vol. 2, pp. 96–101, 2002.
[10] S. K. Cho, H. Moon, and C.-J. Kim, "Creating, Transporting, Cutting, and Merging Liquid Droplets by Electrowetting-Based Actuation for Digital Microfluidic Circuits," *J. of Microelectromechanical systems*, vol. 12, pp. 70–80, 2003.
[11] U. Yi and C. Kim, "EWOD actuation with electrode-free cover plate," in *Transducer 2005: 13th Int. Conf. on Solid-State Sensors, Actuators and Microsystems*, vol. 1, 2005, pp. 89–91.
[12] U. Herberth, "Fluid manipulation by means of Electrowetting-On-Dielectrics," Ph.D. dissertation, Albert Ludwigs University, 2006.
[13] S.-K. Fan, H. Yang, T.-T. Wang, and W. Hsu, "Asymmetric electrowettingmoving droplets by a square wave," *Lab on a Chip*, vol. 7, 2007.

2008 IEEE Conference on Microelectronic Test Structures, March 24-27, Edinburgh, UK
4.5

Measurement of the MOSFET Drain Current Variation Under High Gate Voltage

Tetsuo Chagawa[1], Kazuo Terada[1], Jianyu Xiang[1], Katsuhiro Tsuji[1],
Takaaki Tsunomura[2] and Akio Nishida[2]

Faculty of Information Sciences, Hiroshima City University[1]
3-4-1, Ozuka-Higashi, Asa-Minami-Ku, Hiroshima, 731-3194, JAPAN
MIRAI-Selete[2]
16-1, Onogawa, Tsukuba, Ibaraki 305-8569, JAPAN

ABSTRACT

The method for accurately measuring the drain current of the MOSFETs, which are integrated in an array and are biased at high gate voltage, is studied. Feedback loop in Kelvin connection is made by software to obtain both accurate and stable measurement. The experimental data show that this Kelvin method is accurate and it is applicable to evaluate the accuracy of the conventional Kelvin method using the hardware feedback loop.

INTRODUCTION

To estimate the MOSFET parameter variation, many test circuits have been proposed, in which a large number of MOSFETs are integrated and the method to save the area for large probing pads is adopted. One method uses the decoder circuit for selecting an MOSFET and the transfer gate for connecting it to the common probing pads [1-3]. The test circuits using this method are, for example, called DMA (Device Matrix Array)[2], TSA (Test Structure Array)[3] and so on. It enables the measurement flexible. It, however, requires the transfer gate which is inserted into the current path for the MOSFET under test (DUT), and therefore, obstructs accurate MOSFET drain current measurement. To remove its influence, Kelvin connection is used in those test circuits. However, the Kelvin connection measurement using the semiconductor parameter analyzer is not perfect, but makes significant error when the MOSFET is measured under high gate voltage. This error is caused by the protection circuits in Kelvin force-sense feedback circuit. In this paper, we study the influence of the protection circuits and propose accurate method to measure the drain current variation under high gate voltage.

MEASUREMENT USING KELVIN CONNECTION

A   Measured Data

Figure 1 shows the measured channel resistances for 16 MOSFETs, whose design channel width/length (*W/L*) are 6.0/0.4 μm, in the DMA-like test circuit fabricated by 0.35-μm technology. The measurements are done both without and with the Kelvin connection of Agilent 4156A. They are indicated as "No Kelvin" and "Hard Kelvin", respectively The MOSFETs are placed on a line and the lengths of the metal wiring to the probing pads decreases as the address increases. It is found that the channel resistance monotonically increases with the wiring length, even when the Kelvin connection measurement is done.

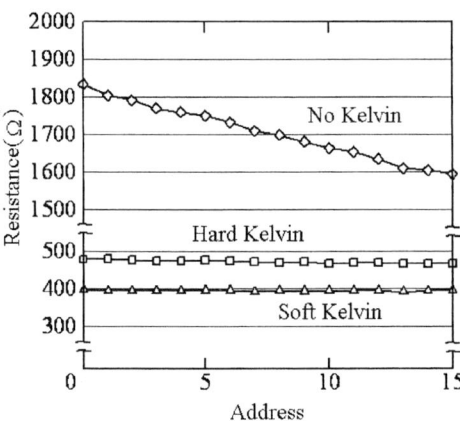

Fig. 1. Measured channel resistances for 16 MOSFETs in the DMA-like test circuit (*W/L*= 6.0/0.4 μm).

Figure 2 shows the cell structure in the test circuit. When this cell is selected, nodes V_{DF}, V_{DS}, V_{SF} and V_{SS} are connected to common probing pads. In this measurement, V_{DF} and V_{DS} are connected to Kelvin force and sense terminals of one SMU (Source Monitor Unit). V_{SF} and V_{SS} are connected to those of another SMU. The distance from the DUT to the joining point of force and sense lines is invariant for all the cells. It, therefore, is found from Fig. 1 that the influence of the wiring resistance outside the cell cannot be removed by this Kelvin connection measurement. As explained in the next section, this is

978-1-4244-1801-5/08/$25.00 ©2008 IEEE
86

because of the protection circuit in the Kelvin force-sense feedback circuit.

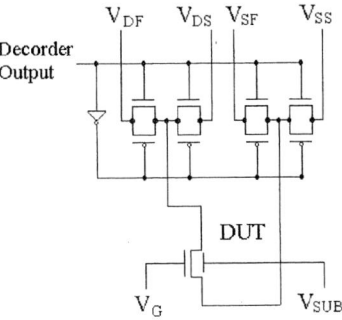

Fig. 2. Cell structure in the test circuit

B. Kelvin Connection Measurement

SMU in Agilent 4156A has a protection circuit between the force and sense terminals in order to avoid abnormal situation. Since the force terminal bias is increased to set the sense terminal at the desired voltage, if the sense terminal is not contacted, namely electrically floating, the force terminal bias is being increased to the limit of the SMU's capability. It is very dangerous. After the experiments to examine the protection circuit, we found that we can consider it as a resistor of 10 kΩ in the voltage range used here. So, the equivalent circuit of the measurement setup for the test circuit can be considered as shown in Fig. 3, where R_1, R_2, x (=10kΩ) and y denote the resistance for the force terminal, the MOSFET under test, the resistor in the protection circuit and the resistance for the sense terminal, respectively.

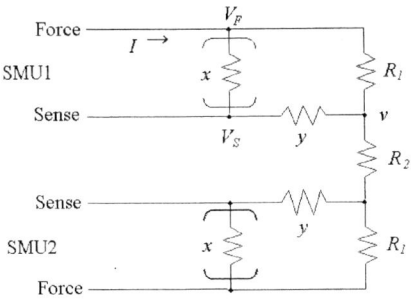

Fig. 3. Equivalent circuit of the measurement setup

Assuming this equivalent circuit being symmetrical at the center of R_2, voltage v applied to R_2 is expressed by voltage V_S monitored by the Kelvin sense terminal as:

$$(1) \quad v = \frac{V_S}{1 + \dfrac{2R_1 y}{R_2(R_1 + x + y)}}.$$

The difference between v and V_S is caused by the current flowing to the sense node and becomes the measurement error. When the gate voltage for the MOSFET under test is high, its channel resistance R_2 is about 400Ω. Since the channel resistance for the transfer gate is about 300Ω, R_1 and y are 300Ω plus the wiring resistance. Therefore, error of a few percent may be caused in the measured value by the protection circuit, depending on the wiring resistance. This is significant for measuring the current variation.

SOFTWARE KELVIN METHOD

To solve the above-mentioned problem, we tried to feed the sense node voltage back to the force node using software. The sense nodes are connected to VMU (Voltage Monitor Unit) having high input impedance and then, no current flows to the sense node. The sense node voltage is fed back to the force node through the computer and the SMU. In the computer, the feedback voltage is checked and the sense node voltage is converged to the given value. We call it Software-Kelvin method and the conventional one Hardware-Kelvin method, hereafter.

Figure 4 shows σ_{IDS}/I_{DS}-V_G relations, where σ_{IDS}, I_{DS} and V_G are the standard deviation, the average for the drain current and the gate voltage, respectively. It is found that the σ_{IDS}/I_{DS} values measured by Hardware-Kelvin method deviate from them by Software-Kelvin method in high V_G region. It is considered that this difference is caused by the metal wiring resistance in the sense terminal (part of y) and the current flowing through it. This difference is a few percent, but significant compared with accurate σ_{IDS}/I_{DS} value. Figure 1 also shows the channel resistance measured by Software-Kelvin method, which is seen to be independent of the address.

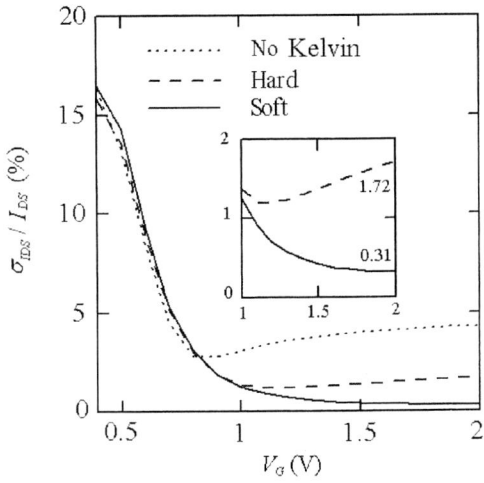

Fig. 4. σ_{IDS}/I_{DS}-V_G relations (W/L=6.0/0.4 μm).

NEW TEST CIRCUIT

The measurement time by Software-Kelvin method is several hundred times longer than that by Hardware-Kelvin method. It, therefore, is not suitable to measure many samples. Based on the above-mentioned experience, we have recently developed new DMA-like test circuit using 65-nm technology.

A   Design

This test circuit contains 16K cells. All MOSFETs used in the circuits other than DUTs, which are decoders, transfer gates and so on, are designed with 0.6-μm channel and 3-V supply technology. The gate, source and drain terminals of a DUT are connected to common probing pads through the transfer gates. Two probing pads and wirings are provided to one DUT terminal, which are for Kelvin force and sense terminals of an SMU. These wirings are joined near the DUT. There are two transfer gates between the probing pad and the DUT terminal. One is for selecting one row/column line and the other is for selecting one cell on the row/column line. Total channel resistance for those transfer gates are designed to be less than 4-kΩ. The DUTs are designed with 65-nm channel and 1.2-V supply technology. Their channel resistance is estimated to be more than 5-kΩ, when $W/L=1$. Figure 5 shows the test circuit layout.

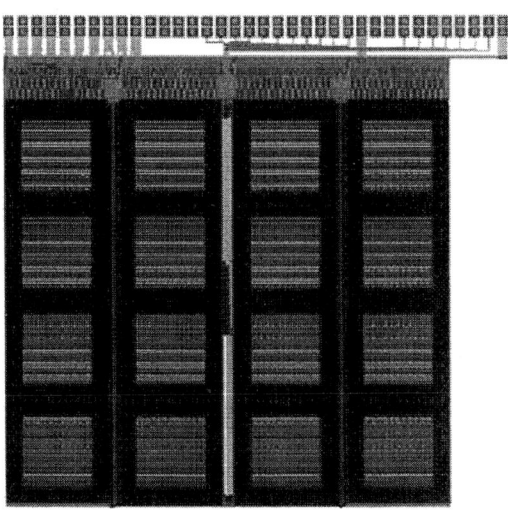

Fig. 5.   Test circuit layout

We measure this test circuit with Agilent N9201A, whose SMU has 50-kΩ protection resistor. It is found from Eq. (1) that the error caused by the protection circuit is less than about 0.1% for the DUT of $W/L=1$. Therefore, if the drain current variation is larger than a few percent, the errors caused by the Hardware-Kelvin

method become negligible except for special DUTs having large W/L.

B   Measurement Results

Samples are made by the fabrication process including both 65-nm and 0.6-μm CMOS technology. Figure 6 shows I_{DS} vs. V_G relation measured by Hardware-Kelvin method, where I_{DS} is average value for 64 MOSFETs of $W/L = 500/50$ nm. Figure 7 compares the I_{DS} data between measured by Hardware- and Software-Kelvin methods. It is found that the relative difference is less than 1 % even for high V_G, which coincides with the estimation with Eq. (1).

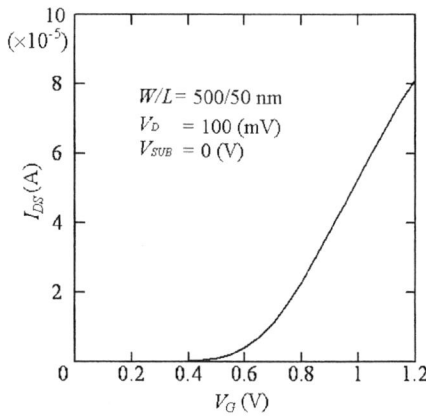

Fig. 6. I_{DS} vs. V_G relation measured by Hardware-Kelvin method. Average value for 64 MOSFETs of $W/L = 500/50$ nm

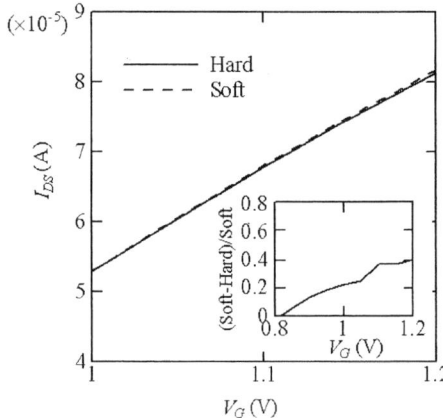

Fig. 7 Difference of I_{DS} between measured by Hard- and Software-Kelvin methods as a function of V_G. Average value for 64MOSFETs of $W/L = 500/50$ nm

Figure 8 shows I_{DS} and σ_{IDS}/I_{DS} measured at V_G=1.2 V as a function of L^{-1}. Those are calculated from the measured I_{DS} data for 32 MOSFETs of W=2 μm. It is found that I_{DS} measured by Software-Kelvin method is a slightly larger than that by Hardware-Kelvin method for large conductance (short channel) devices. It is less than 2 % for L=50 nm MOSFET. The σ_{IDS}/I_{DS}

values in Fig. 8 are measured by Hardware-Kelvin method and is more than 3 % for L=50 nm MOSFET. It, therefore, is considered that these σ_{IDS}/I_{DS} values are larger than the error caused by the protection resistor and are meaningful. Since σ_{IDS}/I_{DS} @ V_G=1.2 V increases as L^{-1} increases, it is considered that σ_{IDS}/I_{DS} expresses the variation of the channel length.

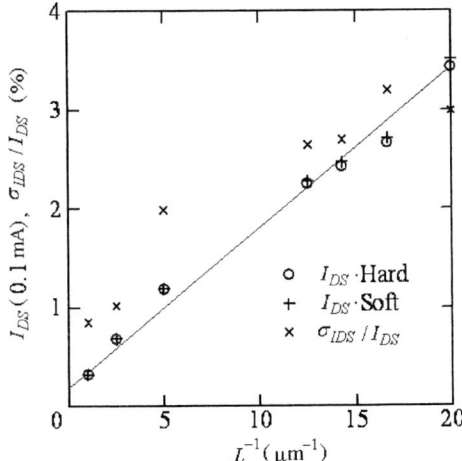

Fig. 8. I_{DS} and σ_{IDS}/I_{DS} measured at V_G=1.2 V as a function of L^{-1}.

DISCUSSION

As mentioned above, the current value measured using the Kelvin terminals of the present SMU depends on the conductance of both the transfer gate and the DUT. On the other hand, if we make the protection resistor sufficiently large, namely x in Eq. (1) infinity, the current value becomes independent of those conductance. To confirm this fact, we have measured the MOSFET drain current when changing the supply voltage to the peripheral circuits. It changes the conductance of the transfer gate. Figure 9 shows the measured drain current as a function of the supply voltage for the transfer gate V_{DD}. This is measured with Agilent-4156A. It is confirmed from Fig. 9 that the current measured by Software-Kelvin method is not affected by the protection circuit and then is probably accurate.

Figure 10 shows the convergence of the sense node voltage for drain, V_{DS}, and that for source, V_{SS}. To set V_{DS}=0.1 V, the force node voltage for drain, V_{DF}, is increased by $(0.1-V_{DS})/2$, till $|0.1-V_{DS}|$ becomes less than 0.1 mV. V_{SS} is also set at 0 V by decreasing the force node voltage for source, V_{SF}, in the same method. It is found that V_{DS} and V_{SS} converge well but it takes many iterations.

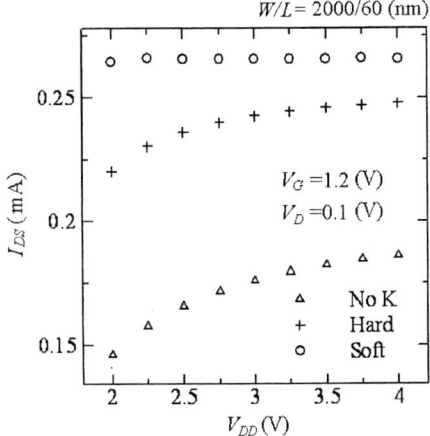

Fig. 9. I_{DS} as a function of the supply voltage for the transfer gate V_{DD}

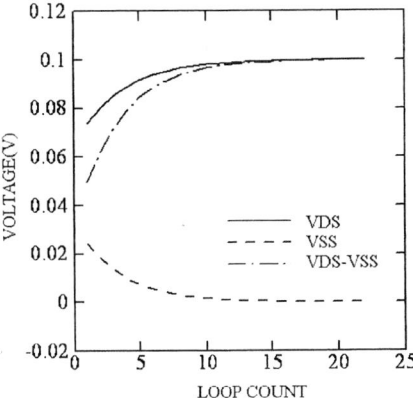

Fig.10. Convergence of the sense node voltage for drain, V_{DS}, and that for source, V_{SS}. MOSFET in the new test circuit is used as DUT, whose W/L=2/0.06 μm. V_G=1.2 V

CONCLUSION

The problem caused by the Kelvin measurement is discussed. To solve this problem, Software-Kelvin method is proposed and it is shown that this method is accurate but takes long measurement time. To measure many samples, we use Hardware-Kelvin method and monitor its accuracy using Software-Kelvin method. It is confirmed that the measurement error caused by the Kelvin measurement can be suppressed less than the MOSFET drain current variation under high gate voltage by the appropriate test circuit design.

ACKNOWLEDGEMENT
This work is supported by NEDO.

REFERENCES
[1] Y. Shimizu, et.al., Proc. ICMTS, p.49-54, (2002).
[2] S. Ohkawa, et.al., Proc. ICMTS, p.70-75, (2003).
[3] K. Doong, et.al., Proc. ICMTS, p.98-103, (2006)

2008 IEEE Conference on Microelectronic Test Structures, March 24-27, Edinburgh, UK

4.6

Spacing Impact on MOSFET Mismatch

A. Cathignol[1,3], S. Mennillo[2], S. Bordez[1], L. Vendrame[2], G. Ghibaudo[3]

[1] STMicroelectronics - Crolles 2 Alliance, 850 rue Jean Monnet 38926 Crolles Cedex, France
[2] STMicroelectronics, Advanced R&D, NVMTD-FMG, via Olivetti 2, 20041 Agrate Brianza, Italy
[3] IMEP, Minatec, INPG, 3 Parvis Louis Néel, BP 257, 38016 Grenoble, France

ABSTRACT

Many test structures embedded in various technologies were measured to study the spacing impact on MOSFET mismatch. This impact is showed to highly depend on technology, device family, device type and bias conditions. The study of spatial correlation allows to properly model spacing impact on mismatch: this analysis -in this paper focused on mosfets- may be extended to any device. Finally, a worst case model that only requires standard matched pairs at minimum spacing is proposed to provide designers the maximum matching degradation that may affect spaced devices.

Index Terms—**mismatch, matching, fluctuations, variability, spacing, correlation**

INTRODUCTION

System On Chip (SOC) applications often require matching properties between spaced devices, e.g. current mirroring or sensing circuits in Non Volatile Memory (NVM) blocks. Sources of mismatch are usually separated into local and global fluctuations [1, 2, 3]; the former are characterized by a short correlation distance and depend on device area, while the latter are characterized by a wafer or die level correlation distance and mainly depend on spacing. Thanks to many MOSFET test structures embedded in various technologies, it will be shown that spacing impact on mismatch highly depends on technology, device family, device type and bias conditions. As a consequence modeling spacing impact on mismatch would be too silicon area consuming. However, by using standard matched pairs designed at minimum spacing, we will suggest a model that helps designers providing for each area the maximum spacing induced degradation that can occur.

TEST STRUCTURES AND EXPERIMENTAL EVIDENCE

We have investigated the impact of global fluctuations on several technologies, in particular a 45nm and a 65nm stand-alone NVM flash technology (NVM45, NVM65), a logic 45nm and 65nm technology (CMOS45, CMOS65), by means of matched pairs of different MOSFET families (thickness oxide ranging from 20Å to 150Å), geometries and spacings. Threshold voltage (Vt), current gain (K) and drain current (Id) at several bias conditions were measured showing different behaviors. Fig. 1 and Fig. 2

respectively depict, for each measured parameter, a clear degradation and a constant trend of matching as device spacing gets larger, highlighting the role of device area, as observed in [4]. Indeed Table I shows that large devices suffer from a higher degradation of matching: such degradation is computed as the relative increase of mismatch with respect to the minimum distance d_{MIN}. The extent of the degradation is also highly dependent on technology, transistor family and device type. For instance Fig. 3 shows a difference in the matching degradation of P-MOS and N-MOS. Finally, the dependence on gate bias (Vg) of current mismatch at increasing spacing is given in Fig. 4 and Fig. 5, where matching degradation at three different Vg is plotted versus spacing: an increase in Vg corresponds to a higher or constant degradation of current mismatch. To model the complex behavior of matching with spacing it is convenient to focus on the correlation between matched devices [5,6].

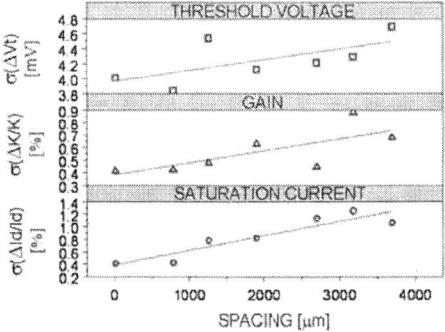

Fig. 1: Threshold voltage, gain and current mismatches versus spacing for a thick oxide transistor from NVM65 technology. Device size: W=20µm, L=0.6µm.

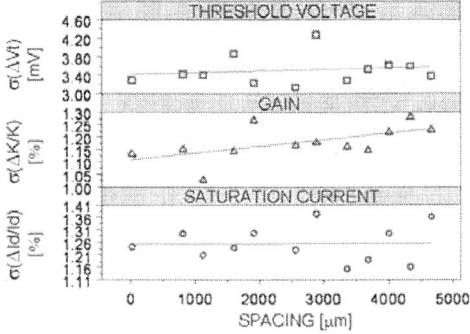

Fig. 2: Threshold voltage, gain and current mismatches versus spacing for a thin oxide transistor from NVM65 technology. Device size: W=1µm, L=1.4µm.

978-1-4244-1801-5/08/$25.00 ©2008 IEEE

90

Table I: Matching degradation of *saturation current at maximum overdrive* for thick oxide transistors from NVM65 technology.

Device type	Area [μm²]	Matching degradation at 4000μm [%]
P	3.29	32
P	9.36	167
P	12.00	205
N	1.46	6
N	3.12	47
N	4.80	62

ABOUT THE CORRELATION BETWEEN MATCHED DEVICES

We remind that the link between mismatch $\sigma(\Delta P)$ and device correlation $\rho(P_1,P_2)$ is given by (1) which can be simplified into (2) assuming $\sigma(P_1) \approx \sigma(P_2) = \sigma(P)$. Indeed it was verified experimentally, by placing test structures in different die positions (Fig. 6), that process gradients act on P_1 (or P_2) average (Fig. 7a) but not on its standard deviation (Fig. 7b).

$$\sigma^2(\Delta P) = \sigma^2(P_1) + \sigma^2(P_2) - 2\rho(P_1,P_2)\sigma(P_1)\sigma(P_2) \quad (1)$$

$$\sigma^2(\Delta P) = 2\sigma^2(P)(1 - \rho(P_1,P_2)) \quad (2)$$

When spacing increases, gradients at die level induce a decrease of device correlation and consequently a degradation of matching. An example of the opposite decreasing/increasing trends of correlation/mismatch versus spacing is reported in Fig. 8. The correlation value is also function of device area: Fig. 9 shows that it is possible to distinguish two regions of small/large devices that are little/very correlated and for which mismatch is higher/lower than single device fluctuations. The trend of mismatch versus spacing for devices in these regions is respectively similar to the ones reported in Fig. 1 and Fig. 2. Starting from few assumptions concerning the contribution of local and global fluctuations to mismatch, device correlation can be expressed as a function of device area and spacing.

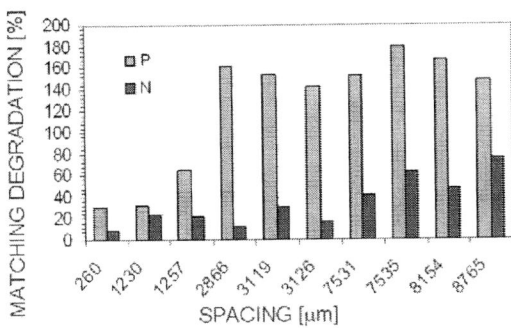

Fig. 3: Difference in matching degradation of *threshold voltage* between thin oxide P-MOS and N-MOS (CMOS45 technology). Device size: W=10μm, L=3μm.

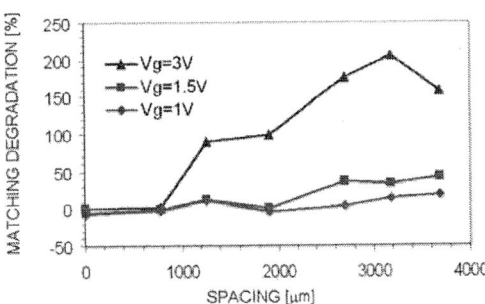

Fig. 4: Matching degradation of *saturation current* at different gate biases for a thick oxide p-channel transistor from NVM65 technology (device geometry: W=20μm, L=0.6μm).

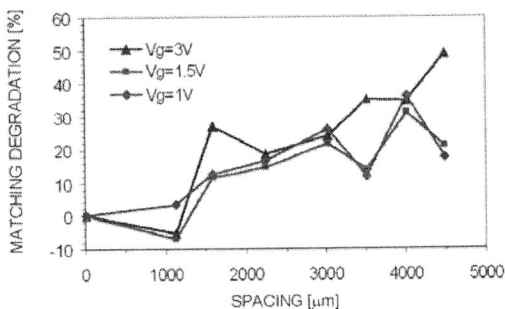

Fig. 5: Matching degradation of *saturation current* at different gate biases for a thick oxide p-channel transistor from NVM65 technology (device geometry: W=5.48μm, L=0.6μm).

A. Correlation as a function of device area and spacing

We consider that P_1 and P_2 are the sums of a local (x) and a position-dependent (z) contribution:

$$P_1 = A/\sqrt{2WL} \cdot x_1(0,1) + z_1(P_{10},G)$$
$$P_2 = A/\sqrt{2WL} \cdot x_2(0,1) + z_2(P_{20},G) \quad (3)$$

where

- $x(\mu,\sigma)$ and $z(\mu,\sigma)$ are normally distributed with mean μ and standard deviation σ,
- A is the mismatch coefficient of Pelgrom model [1], i.e. $\sigma(\Delta P)=A/\sqrt{WL}$,
- W and L are device width and length,
- z_1 and z_2, which are position-dependent, represent for simplicity $z_{r1\theta1}$ and $z_{r2\theta2}$ where r and θ are the coordinates of the two devices on the wafer,
- P_{10} (P_{20}) is the average of P_1 (P_2) on the considered wafer,
- x_1 and x_2 are not correlated, since they describe a random process,
- z and x are not correlated, since they depict fluctuations acting at different scales,
- correlation between z_1 and z_2 depends on transistor geometry and spacing.

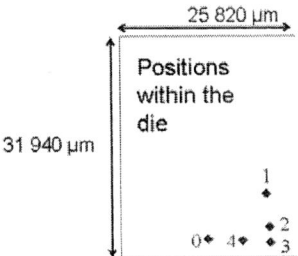

Fig. 6: Sketch of the position of the scribes used to investigate the impact of within-die fluctuations on mean and standard deviation of threshold voltage, gain and current for CMOS45 thin oxide transistors

a)

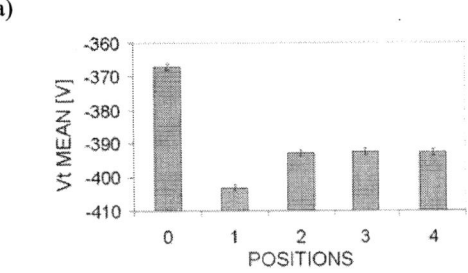

b)

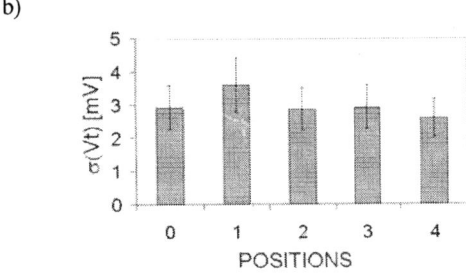

Fig. 7 a,b: Particular case observed on one wafer of mean (a) and standard deviation (b) of *threshold voltage* for CMOS45 thin oxide p-channel transistors placed in the scribes sketched in Fig. 6; 95% confidence bars are superimposed to the plots. Device size: W=10μm, L=3μm.

At minimum spacing z_1 and z_2 are physically very close (at the wafer scale) and correlation between z_1 and z_2 can be considered equal to one. So, the covariance of P_1 and P_2 is given by (4). Finally, using (P_1, P_2) correlation definition (5) and thanks to (3) and (4) (P_1, P_2) correlation is given by (6).

$$cov(P_1, P_2) = G^2 \qquad (4)$$

$$\rho(P_1, P_2) = \frac{cov(P_1, P_2)}{\sigma(P_1)\sigma(P_2)} \qquad (5)$$

$$\rho(P_1, P_2) = \frac{2G^2WL}{A^2 + 2G^2WL} \qquad (6)$$

At this stage we have provided a model for (P_1, P_2) correlation as a function of device area. Practically, G^2 can be determined using (4) and it can be experimentally verified that it is independent from

device area. Indeed Fig. 10 shows an example of G^2 plotted as a function of device area in the case of the threshold voltage of thin oxide transistor pairs at minimum spacing from CMOS65 technology. It must be noticed that G^2 can be meaningful only in the range of geometries for which (P_1, P_2) correlation is itself meaningful. In Fig. 11 (P_1, P_2) correlation is plotted versus device area using the new model (6). A good agreement between the provided model and the experimental correlation for minimum-spaced pairs is observed.

Now the case where spacing is not minimal is considered. As spacing increases, (z_1, z_2) correlation drops and can no longer be considered equal to 1. So, (P_1, P_2) covariance becomes:

$$cov(P_1, P_2) = G^2 \rho(z_1, z_2) \qquad (7)$$

Finally, (P_1, P_2) correlation becomes:

$$\rho(P_1, P_2) = \frac{G^2 \rho(z_1, z_2)WL}{A^2/2 + G^2WL} \qquad (8)$$

By combining (9) -that is deduced from (3)- and Pelgrom model [1] reminded in (10), the behavior of (P_1, P_2) correlation as a function of both device area WL and spacing d can be deduced (11).

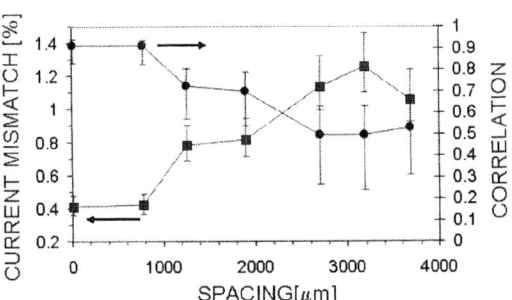

Fig. 8: *Saturation current* mismatch and correlation versus device spacing with 95% confidence bars for matched thick oxide p-channel transistors with W=20μm and L=0.6μm. (NVM65 technology).

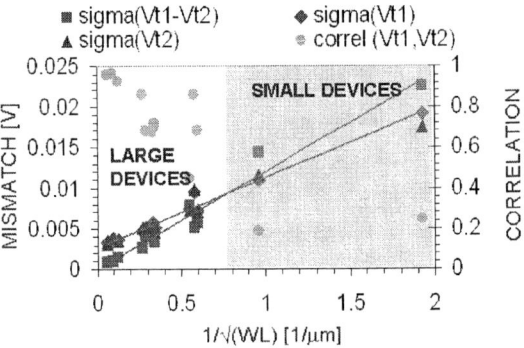

Fig. 9: *Threshold voltage* mismatch, correlation and spread of Vt_1 and Vt_2 versus $1/\sqrt{(WL)}$ for thick oxide p-channel matched pairs. (NVM65 technology).

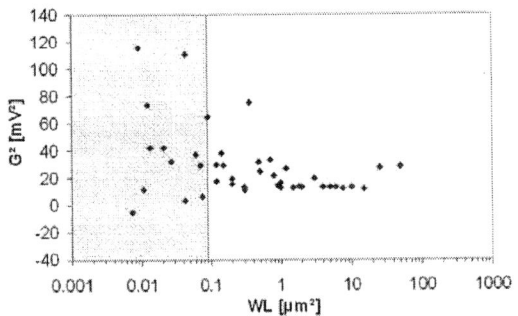

Fig. 10: Experimental G^2 versus device area for the *threshold voltage* of thin oxide p-channel transistor pairs at minimum spacing from CMOS65 technology. The values of G^2 are not meaningful in the range of geometries highlighted in the figure, where (P1, P2) correlation is not significant [7].

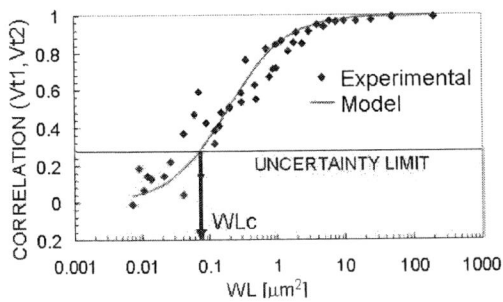

Fig. 11: Experimental *Vt* correlation versus device area for thin oxide p-channel transistor pairs at minimum spacing from CMOS65 technology. The model equation (6), with coefficient A=2.6mV·µm G=3.7mV, is superimposed to the plot. The highlighted region represents the values of correlation which can not be considered significantly different from zero with 99.9% confidence and 80 tested dice [7]. WLc is the critical area.

$$\sigma^2(\Delta P) = \frac{A^2}{WL} + 2G^2\big(1 - \rho(z_1, z_2)\big) \qquad (9)$$

$$\sigma^2(\Delta P) = \frac{A^2}{WL} + S^2 d^2 \qquad (10)$$

$$\rho(P_1, P_2) = \frac{G^2 WL}{A^2/2 + G^2 WL} - \frac{S^2 d^2 WL}{A^2 + 2G^2 WL} \qquad (11)$$

According to (11) device correlation behaves as a quadratic function of spacing. Fig. 12 shows a good agreement between the model provided in (11) and the experimental correlation for spaced devices. It is worth noting that this agreement is valid only for a limited range of spacings: for large values of *d* the trend may no longer be quadratic, since experimental correlation decreases down to 0 and then keeps constant around such value. For some devices we have not observed a fully quadratic trend (Fig. 13), possibly due to second order effects not considered in the quadratic model; however, the numerical approximation of correlation

provided by model (11) is still quite good and within the statistical uncertainty affecting experimental correlation. At this stage we have provided a model for device correlation that accounts for both device area and device spacing. The above formula (11) clearly depicts that the highest values of correlation, and thus the best matching, are reached for minimum-spaced pairs; while, as spacing increases, correlation and matching tend to degrade due to global fluctuations. It also shows that such a degradation is more pronounced on large area pairs.

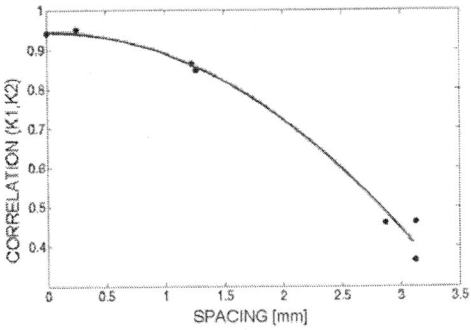

Fig. 12: Experimental K correlation versus spacing for spaced thin oxide p-channel transistor pairs from CMOS45 technology (device area=30µm²). The model equation (11), with coefficients A=1.33%·µm, G=0.7% and S=0.00024%/µm, is superimposed to the plot.

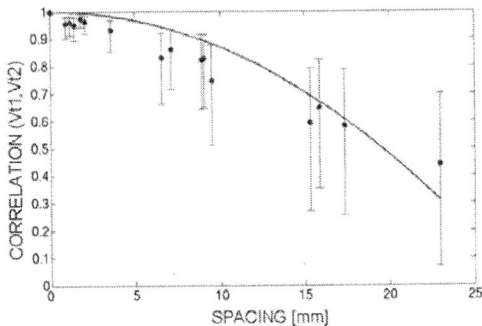

Fig. 13: Experimental *Vt* correlation versus spacing for spaced thin oxide p-channel transistor pairs from NVM45 technology (device area=450µm²) with 99% confidence bars. The model equation (11), with coefficients A=4.8mV·µm, G=2.18mV and S=0.11mV/µm, is superimposed to the plot.

B. The role of gate bias

Using (1) and Pelgrom formula [1] reminded in (12), it is possible to express the mismatch of current as a function of gate bias Vg and of the correlations of global fluctuations impacting gain and threshold voltage, namely $\rho(z_{K1}, z_{K2})$ and $\rho(z_{Vt1}, z_{Vt2})$:

$$\sigma^2\left(\frac{\Delta Id}{Id}\right) = \sigma^2\left(\frac{\Delta K}{K}\right) + \alpha^2 \frac{\sigma^2(\Delta Vt)}{(Vg - Vt)^2} \qquad (12)$$

$$\sigma^2\left(\frac{\Delta Id}{Id}\right) = \frac{2\sigma^2(K)}{K^2} \cdot \left[1 - \frac{G_K^2 \rho(z_{K1}, z_{K2})WL}{A_K^2/2 + G_K^2 WL}\right] +$$
$$+ 2\alpha^2 \frac{\sigma^2(Vt)}{(Vg-Vt)^2} \cdot \left[1 - \frac{G_{Vt}^2 \rho(z_{Vt1}, z_{Vt2})WL}{A_{Vt}^2/2 + G_{Vt}^2 WL}\right] \quad (13)$$

where α is equal to 1 or 2 respectively in linear and saturation conditions. The above formula highlights that at low values of gate bias the trend of current mismatch degradation versus spacing is highly dependent on the behaviour of $\rho(z_{Vt1}, z_{Vt2})$ and G_{Vt}; as Vg increases, the contribution of $\rho(z_{Vt1}, z_{Vt2})$ and G_{Vt} to total current mismatch tends to decrease being multiplied for the term $1/(Vg-Vt)^2$ and, as an effect, $\rho(z_{K1}, z_{K2})$ and G_K impacts get more and more pronounced. This dependence of current mismatch on gate bias and on the correlations of global fluctuations impacting gain and threshold voltage is clearly depicted in Fig. 14, 15 and 16 for a NVM45 thin oxide p-channel transistor. In Fig.14 we have plotted Vt and K correlation versus spacing: Vt correlation shows a consistent decreasing trend, while K correlation is less impacted by device distance. In Fig. 15 and 16 we have reported the behavior of current mismatch degradation at increasing spacing both in saturation and linear conditions: such degradation is significantly higher at low values of gate bias.

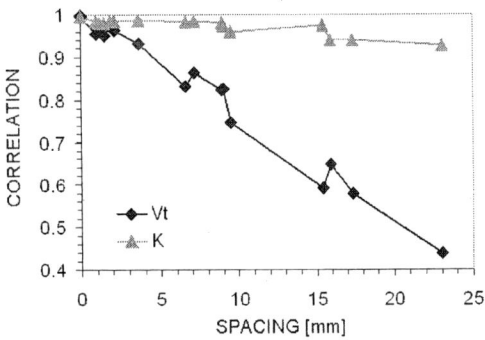

Fig. 14: *Threshold voltage* and *gain* correlation versus spacing for a p-channel thin oxide transistor from NVM45 technology (device geometry: W=30μm, L=15μm).

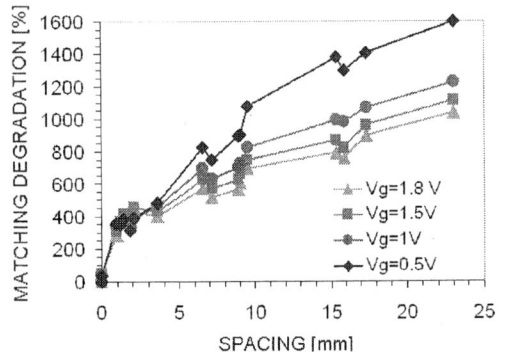

Fig. 15: Matching degradation of *saturation current* at different gate biases for the transistor of Fig. 14.

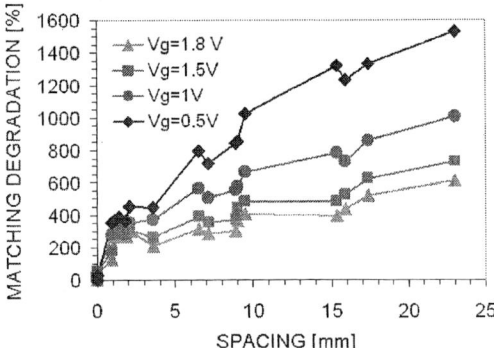

Fig. 16: Matching degradation of *linear current* at different gate biases for the transistor of Fig. 14.

TOWARDS INCLUDING IMPACT OF SPACING ON STOCHASTIC MISMATCH INTO MODELS

A full characterization of mismatch versus spacing - for instance using the spectral model described in [8] or based on the model developed above (with the parameters A_{Vt}, G_{Vt}, S_{Vt}, A_K, G_K, S_K) – would require too many test structures including several geometries and spacings. However standard minimum-spaced test structures can provide some very useful guidelines about the geometries which may be impacted by spacing and, for such geometries, the maximum degradation that may affect their matching performance. Notice that (14) highlights the link between mismatch $\sigma_{\Delta P}(d)$ and correlation ρ_d for pairs at distance d. The worst value of mismatch is clearly reached when ρ_d is null. Therefore, if correlation at minimum spacing is already null ($\rho_d=0$), mismatch cannot be worse and consequently no impact for spacing may be expected. On the contrary, if correlation at minimum spacing is not null, mismatch can degrade with spacing and the upper limit of this degradation is when correlation becomes 0. In other words, we use the simple idea which consists in noting that the worst mismatch case is obtained when correlation drops to a null value and mismatch is given by (15). So, spacing impact can be quantified by Mismatch Maximum Increase Coefficient (MMIC) defined in (16) where ρ_0 is correlation at minimum spacing.

$$\sigma^2_{\Delta P}(d) = 2\sigma_P^2 (1 - \rho_d(P_1, P_2)) \quad (14)$$

$$\sigma^2_{\Delta P}(d)_{MAX} = 2\sigma_P^2 \quad (15)$$

$$MMIC = \frac{1}{\sqrt{1 - \rho_0(P_1, P_2)}} \quad (16)$$

$$\sigma_{\Delta P}(d)_{MAX} = MMIC \cdot \sigma_{\Delta P}(d_{MIN}) \quad (17)$$

The correlation at minimum spacing is obtained thanks to the model that was suggested in Part A. As an example, Fig. 11 reports the modeled correlation versus device area (6) for the threshold voltage of thin oxide transistors from CMOS65 technology and the maximum value of correlation that can be considered not significantly different from zero [7]: the cross between the two lines provides the critical device area WLc over which spacing may have some impact. By the way it is worth noting that WLc is highly dependent on technology, device family and device type. Then MMIC can be plotted as a function of device area (Fig. 17). For instance, it can be read on the graph that for 1 μm^2 matched devices, a threshold voltage matching degradation up to 30% may be expected when a not minimal distance separate matched devices.

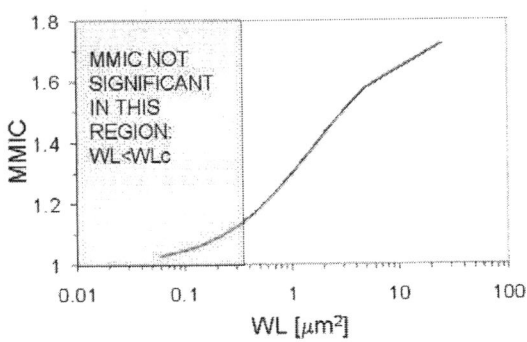

Fig. 17: Mismatch Maximum Increase Coefficient MMIC=$1/\sqrt{(1-\rho_0)}$ versus device area for the *threshold voltage* of thin oxide p-channel transistor pairs from CMOS45 technology. Notice that ρ_0 is the modelled correlation at minimum spacing based on equation (6), with coefficients A=3mV·μm and G=2.6mV.

CONCLUSIONS

Thanks to the experimental results collected on several technologies, we have investigated the impact of spacing on mismatch and the role played by device family, device type, device area and gate bias. The analysis of correlation allowed to properly describe spacing impact on mismatch. Finally, an easy to use model that only requires standard matched pairs at minimum spacing was built up to predict maximum degradation that can affect their matching properties.

ACKNOWLEDGEMENTS

The work was partially supported by MEDEA+ T-201 NEMeSyS project.

REFERENCES

[1] M. J. M. Pelgrom, A. C. J. Duinmaijer, A. P. G. Webers, "Matching properties of MOS transistors", *IEEE J. of Solid-State Circuits*, vol. 24, n. 5, pp. 1433-1440, 1989

[2] P.G. Drennan, "Device mismatch in BiCMOS technologies", in *Proceedings BCTM*, 2002, pp.104-111

[3] J. B. Shyu, G.C. Temes, K. Yao, "Random errors in MOS capacitors", *IEEE Journal of Solid-State Circuits*, v SC-17, n 6, p 1070-6, Dec. 1982

[4] U.Schaper, C. Linnenbank, "Comparison of distance mismatch and pair matching of CMOS devices", in *Proceedings ESSCIRC*, 2003, pp. 703-705

[5] M. Conti, P. Crippa, S. Orcioni, C. Turchetti, "Statistical modeling of MOS transistor mismatch based on parameters'autocorrelation function", in *Proceedings ISCAS*, 1999, vol. 6, pp. 222-225

[6] H. Tuinhout, "Transistor matching", ICMTS'98 Tutorial Short Course

[7] G. Casella, R. Berger, "Statistical Inference", Duxbury Advanced Series, 2000

[8] J. Oehm, U. Grünebaum, K. Schumacher, "Mismatch effects explained by the spectral model", in *Proceedings ICECS*, 1999, vol. 3, pp. 1519-1523

Highly Automated Test Chip Layout and Test Plan Development for Parametric Electrical Test

Ann Gabrys, Wendy Greig, Andrew J. West, Philipp Lindorfer, and William French

National Semiconductor Corporation
Santa Clara, CA, USA

ABSTRACT

This work outlines a fully integrated device development procedure that automates test chip development, including placement and routing algorithms, and electrical test program generation. This procedure improves over classic test chip and electrical test program development by reducing the development timeline and allowing more complete and elegant experimental device design, as well as eliminating many of the opportunities for human error while maximizing reuse between technologies.

INTRODUCTION

Process and device development requires several test chip learning cycles involving numerous variations of multiple device structures. In the past, due to the complexity of the device structures and the number of possible experimental parameters involved, manual development of test chips and electrical tests have limited the practical scope of device design experiments and the rigor of optimization efforts. A number of methods have been developed that address the test chip layout bottleneck [1-5]. Using a commercial software system [6], a novel methodology has been applied to process development, as outlined in Fig. 1. A central database - containing all process layout rules, electrical device parameters and layer/grid information - is accessed by a fully parameterized layout system to create the test chip GDSII files. It also saves the information for each instantiated device back into the database which then enables automatic generation of electrical test plans and all relevant documentation. The sections of this paper discuss the details involved at each stage in the methodology. Since the focus is on parametric layout and test plan generation, for the purposes of this discussion it is assumed that all layout rules, process layers, and GDSII/grid information exist in the database.

DEVICE TEMPLATE DEVELOPMENT

Traditionally, parameterized cells have been written in procedural scripting languages (e.g. SKILL). The code written for each cell is, by necessity, separate and

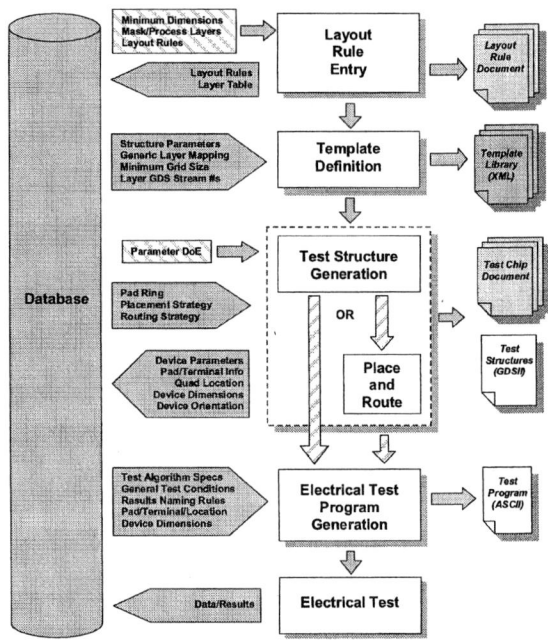

Fig. 1. Overview of the automated device/process development flow.

unique for each device type included in a test chip. By definition, there can be little to no reuse between templates, apart from shared sections of code with multiple locations for edit. In addition, significant modifications are typically required when porting a cell from one process to another due to, among other things, layer name changes, differences in feature sizes, rule scaling, and manufacturing capability.

In this approach the device templates are constructed from hierarchical, reusable, process independent objects written in an object-aware JavaScript based language. These objects, which are stored in a central database, serve as building blocks to construct more and more complex templates, including device, pad ring and interconnect templates. The objects are defined using generic layer names and automatically generated design rule names as variables, which are supplied from the central database for a specific instantiation. The internal device template dimensions can be defined as user-specified parameters, layout rule variables, or a combination of both, which the template also accesses from the database. These features facilitate template reuse across multiple processes with minor customization. Fig. 2 illustrates the device template definition environment, as well as the device parameter table that the user defines for test structure autogeneration. The sample JavaScript code illustrates the use of device variables specified in the table to set internal device dimensions in the template. The use of this environment enables the construction of extremely complex and powerful templates with a multitude of structural and dimensional variables. For example, NLDMOS and PLDMOS templates have been written with ten to fifteen internal sizing parameters, variable designations for body drift and epi implants, and enable the user to choose between: extended drain and extrinsic drain incorporating an oxide drift region; non-RESURF and RESURF (with and without substrate isolation); and butting body tie-downs or isolated body contacts.

TEST CHIP STRUCTURE AUTO-GENERATION, PLACEMENT AND ROUTING

Test structures are generated by creating a spreadsheet of layout parameters that the device template uses to automatically create the final test structures. Placement and Route templates control the placement and routing method used during the generation of test module layout. These templates are organized into a hierarchical library of placement strategies. Since each strategy targets different layout optimization criterion (e.g. maximum pad sharing, minimum parasitic resistance, terminal bussing, etc.), the test chip layout is dependent on the strategy selected "in press" [7]. By selecting the relevant placement and routing modules, the devices are then placed into a designated pad ring and the terminals connected to pads. The completed structures are exported in GDSII format to a Cadence database for mask generation. Fig. 3 shows one of twenty completed laterally diffused metal oxide semiconductor (LDMOS) transistor test modules, a subset of an experiment simultaneously modifying five internal structure parameters, which were automatically placed and routed. The structure information, including

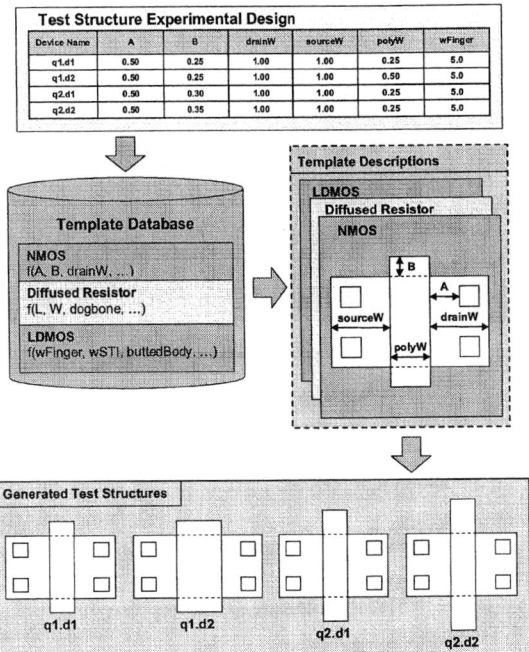

Fig. 2. Illustration of the template development workspace (left) and the table of device structure parameters (right) used to define the device test structures for autogeneration. User-defined device parameters contained in the spreadsheet can be used to define internal dimensions of the device template.

terminal/pad assignments, device orientation, module/row location, and device size, were saved back to the database for use during test plan generation. Fig. 4 illustrates a large NLDMOS array generated by the software that was then manually placed into a standard pad ring, bypassing the place and route portion of the flow. As with the routed module, the device structure parameters and size were automatically saved to the database. The terminal/pad assignments, orientation, and module/row locations were manually updated once layout was completed.

ELECTRICAL TEST PROGRAM AUTOMATION

In parallel with test chip development, generalized test algorithms (e.g. Idsat, Vt, Beta) are created which call out generic terminal names and have test conditions

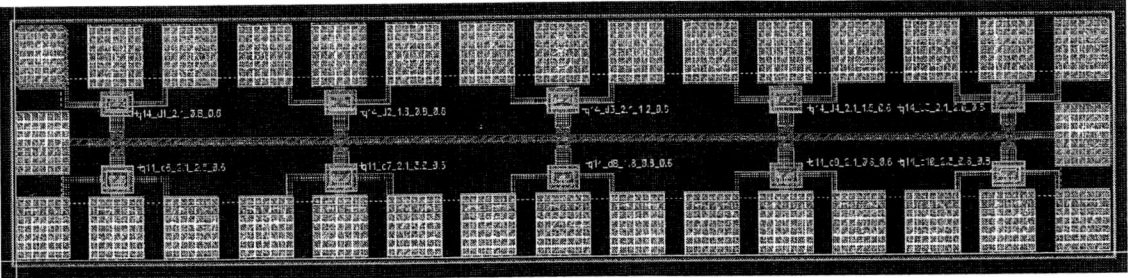

Fig. 3. Screen capture of an automatically generated and routed test structure module.

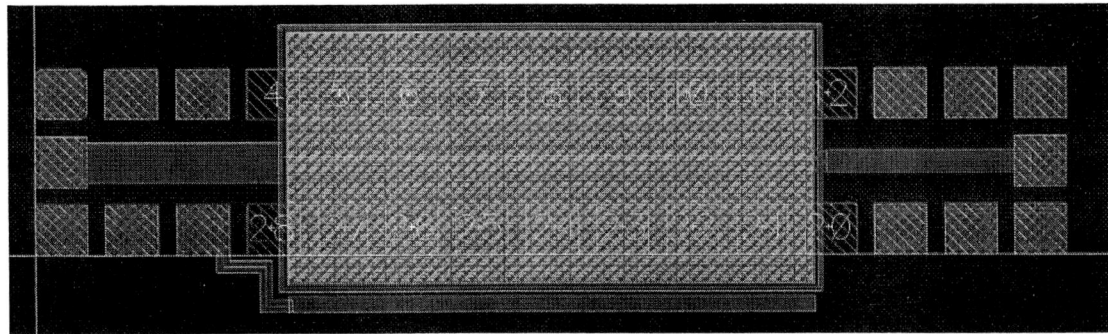

Fig. 4. Screen capture of a generated NLDMOS array manually placed in a standard pad ring.

that scale with device dimensions. After the completion of test chip assembly, the database information stored from the test chip is used to assign the requested test procedures to each specific device.

Since the test algorithms are generically defined objects, the database information is used to scale the test conditions and appropriately assign terminals to pads/pins. Fig. 5 illustrates this procedure, where the information is stored into the database during test chip definition and generation.

Once the test plan is generated and testing is complete, results are saved back into the database for analysis within the software suite. To facilitate experimental

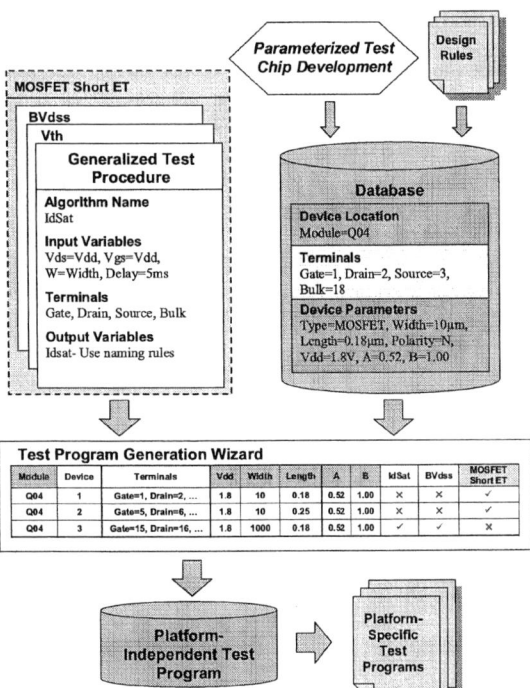

Fig. 5. Test program generation flow. Test algorithm information and test chip information stored in the database port directly into the test plan generation tool. Tests can be specifically assigned to individual device test structures in procedure groups using a check matrix array system.

design analysis, unique test names are automatically generated using a predefined algorithm. The names are then linked back to the full definition of each device structure in the database. After structure testing, results are saved back into the database for analysis. Results are associated with the original experimental design through their unique test names which are associated in the database with all experimental layout design parameters. As a result, any device performance metric can easily be analyzed according to any of the experimental design variables, enhancing the efficiency and effectiveness of the analysis.

The automated procedure not only eliminates human error associated with creating test programs from printed documentation, but allows test engineers to spend more time developing optimal test algorithms and generalizing test conditions. The reduced time and engineer resource requirements also enable testing of longer, more complete device experiments. Typical automated program generation takes about a day, independent of program size. As illustrated in Fig. 6, application of the automated flow tripled the potential test plan size, allowing the device engineer to fully realize the advantages of the parametric layout solution. The integrated flow further facilitates statistical analysis device data since it automatically associates each test result with the experimental design, rather than simply a set of pads and a die location.

DATA ANALYSIS

Implementing such a powerful, automated system enables engineers to design combined process and device structure experiments capable of generating an overwhelming quantity of parametric data. Depending on the size of the test chip reticle, the number of devices, and the parameters tested per device, even a small number of eight inch wafers can produce an enormous number of data points to be studied (potentially in the millions). To aid the engineer in studying this data, two different types of analysis tools are required. Used largely during early development, a data viewer can be used on smaller subsets of data, enabling the engineer to interactively study and de-

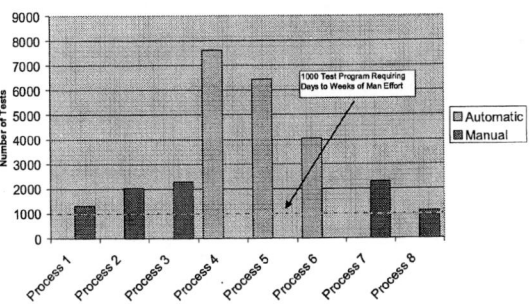

Fig. 6. Comparison of test program size between manual versus automated test program generation.

embed the effects of process changes or layout variations on the performance of a device, and to track the location of off-target die. Fig. 7 illustrates a plot produced using the data analysis tool that operates on the limited data set. In this example, process parameters coupled with the layout parameters stored in the database have been used to study the effect of different implant conditions and layout configurations on the breakdown voltage of an LDMOS transistor.

A second tool is required to work with statistically significant data sets (often involving multiple lots of

silicon) and to produce reports summarizing the result. The latter is an automated system which first filters the data to remove outliers before compiling a pre-defined set of plots and statistical results into a report. Further information on this tool is available [6].

CONCLUSION

The method outlined in this work utilizes a parametric layout solution coupled with a test program procedure that references a centralized database for information stored from layout rule entry and test chip development to automate test program generation. The automation of both test chip development and test plan generation has two primary benefits. It eliminates much of the potential for human error that comes with repetitive tasks and data re-entry, and facilitates more complete device design experimentation than was feasible in the past due to the time required for manual execution of each task. To aid in the analysis of the resulting complex experiments, two data analysis tools are available. One targets visual analysis of smaller data sets, while the other generates summary reports of statistically significant quantities of data.

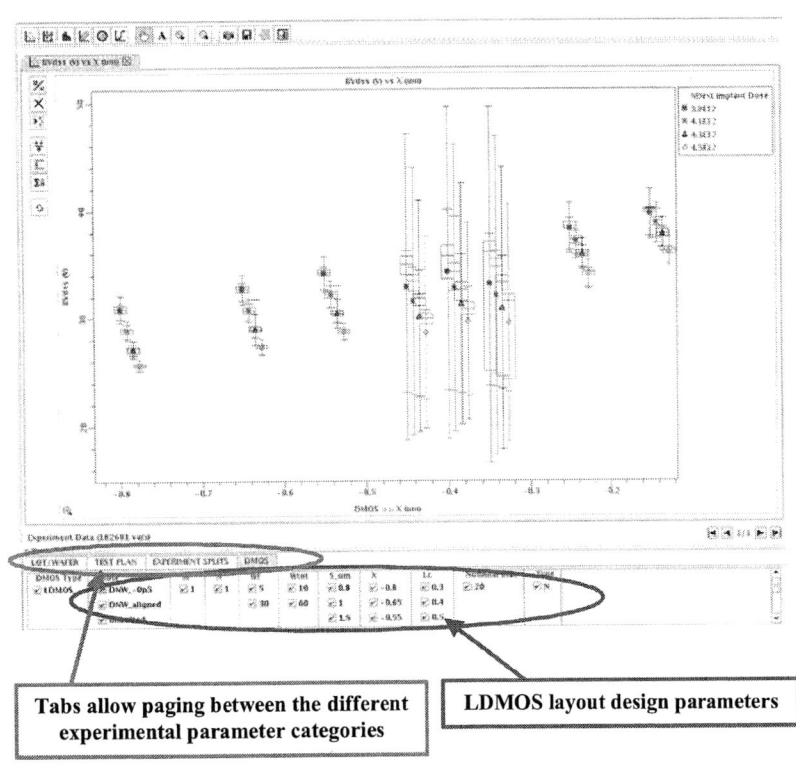

Fig. 7. An example from the data analysis utility illustrating LDMOS breakdown voltage versus the selected layout design rule; a process implant parameter is selected to color the data. As shown, process splits and device layout parameters can be considered simultaneously when studying the performance metrics of a device.

ACKNOWLEDGMENT

The authors would like to acknowledge StonePillar Technologies (www.stonepillar.com) who developed the software framework used throughout this work and to thank them for the development effort conducted during the course of this work.

REFERENCES

[1] T. Brenner et al, "A modular 0.7μm CMOS JESSI test chip for multi purpose applications," Proc. IEEE Int. Conf. on Microelectronic Test Structures, pp. 160-165, 1992.

[2] T. Ternisien d'Ouville et al, "Automatic test chip and test program generation. An approach to parametric test computer-aided design," Proc. IEEE Int. Conf. on Microelectronic Test Structures, pp. 145-149, 1992.

[3] S. Tarasewicz and R. Horner, "Development of the Test Insert Generating Expert Routine (TIGER) for BiCMOS technologies," Proc. IEEE Int. Conf. on Microelectronic Test Structures, pp. 178-183, 1994.

[4] L. Kasel, "Automated generation of SPICE characterization test masks and test databases," Proc. IEEE Int. Conf. on Microelectronic Test Structures, pp. 74-79, 1999.

[5] G. Leonardelli, "A novel system for fully automated creation of layout, documentation and test programs for electrical test structures," Proc. IEEE/SEMI Advanced Semiconductor Manufacturing Conference, pp. 205-207, 2004.

[6] Stone Pillar User Guide, v4.3.0, 2008.

[7] A. J. West, S. Mondal, D. Patra, K. Goswami and S. Sural, "A novel high speed automation layout system to place and route test structures for parametric test capability," Proc. IEEE Int. Conf. on Microelectronic Test Structures, 2008.

Circular Geometry MOS Transistor Analysis of SOI Substrates for High Energy Physics Particle Detectors

S.L. Suder,[a] F.H. Ruddell,[a] J.H. Montgomery,[a] B.M. Armstrong,[a] H.S. Gamble,[a]
G. Casse,[b] T. Bowcock,[b] P.P. Allport[b]

[a] Northern Ireland Semiconductor Research Centre, School of Electronics, Electrical Engineering & Computer Science, Queen's University Belfast, Ashby Building, Stranmillis Road, Belfast, BT9 5AH, UK.

[b] Liverpool Detector Centre, Department of Physics, University of Liverpool, UK.

ABSTRACT

SOI substrates are important for the fabrication of monolithic active pixel high energy physics particle detectors. In this work, self-aligned circular geometry MOS transistor test structures were fabricated on ion split, bonded SOI substrates to evaluate the interface between the high resistivity handle silicon and the SOI buried oxide. Pre- and post- proton irradiation transistor measurements are presented, showing an increased SOI buried oxide trapped charge of only 3.45×10^{11} cm^{-2} for a dose of 2.7 Mrad.

INTRODUCTION

Events at the Large Hadron Collider (LHC) at CERN result in huge numbers of ionising particles passing through the detectors. Interaction rates at the LHC imply unprecedented levels of radiation for both the detectors and front-end electronics, which must therefore be tolerant to such an environment. Vertex detectors use a missed distance technique to identify short-lived particles. This requires that decay product tracks are not deflected in the detector by scattering and thus it is necessary to minimise the amount of material in the tracker.

The highest dose detectors at the LHC (ATLAS, CMS Pixels and LHCb Vertex Locator) all employ hybridised pixel detectors that are bump bonded to 300 μm thick electronics readout chips. However, this technique is labour-intensive and can suffer from yield and reliability problems. Most importantly, the resulting radiation length per detector layer is usually much more than the sensing region, giving rise to undesirable multiple scattering effects. High energy particle physics therefore has a clear requirement for ultra high precision, low mass, radiation tolerant pixel detectors at low radii close to the beam interactions. Silicon On Insulator (SOI) material is very promising for such an application, as it offers an optimised

substrate without the disadvantages of flip-chip bump bonding.

MONOLITHIC ACTIVE PIXEL SOI DETECTOR

Our goal is to fabricate a monolithic active pixel detector which vertically integrates the readout electronics and the sensor device in one substrate. This solution not only eliminates the hybridisation process, but also provides thinner devices suitable for inner layers of vertex detectors. SOI technology may be used to combine reverse biased p$^+$-n$^-$ pixel diodes in a fully depleted high resistivity n-type substrate 'handle' layer with readout electronics in a low resistivity 'device' layer, both interconnected by vias through the insulating SOI buried oxide (BOX) [1]. Each ionising particle impinging on such a detector would create approximately 80 e-h pairs/μm of depleted Si. These carriers are then separated and swept to the p$^+$ pixels and n$^+$ substrate contact by the electric field. Fig. 1 shows a schematic cross-section of such a detector.

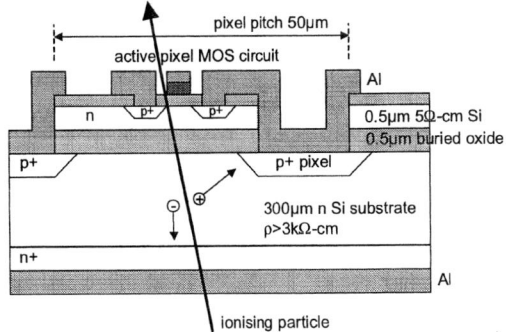

Fig. 1. Schematic cross-section of a monolithic vertically integrated active pixel ionising particle detector in a SOI substrate.

The charge sensing and amplification functions implemented in the SOI active layer need not employ fully depleted MOS technology. However, since the pixel diodes are fabricated in the SOI handle wafer, it is crucial to maintain the electrical integrity of the interface between the buried oxide (BOX) layer and the substrate in order to minimise detector dark current.

This paper describes the use of self-aligned gate circular geometry p-channel MOS transistor test structures fabricated in ion split, bonded silicon SOI substrates to examine this vitally important interface. These transistors employ the handle layer p$^+$ pixel diode diffusions as source and drain electrodes, with the BOX layer as the gate dielectric and the doped SOI active layer as the gate electrode.

SOI WAFER PREPARATION

The SOI substrates were first prepared using wafer bonding and ion split technology [2]. P-type device wafers with resistivity 10-20 Ω-cm were implanted with H$_2$ (dose: 3×10^{16} cm^{-2}) through a 100 nm thick sacrificial oxide layer. The 500 nm thick SOI BOX layer was grown on high resistivity (4 kΩ-cm) n-type FZ handle wafers at 1050°C using a hydrox process, before bonding to the device wafers. For this application it is vital that the BOX layer is grown on the handle wafers so that the bond interface will be between the BOX layer and the overlying device layer. This ensures a high quality interface between the SOI BOX layer and the substrate. The bonded wafer pairs were then ion split annealed at 500°C for 2 hrs in N$_2$ to give a 500 nm transferred SOI device layer, before a final 2 hr bond strengthening N$_2$ anneal at 1050°C.

CIRCULAR GEOMETRY TRANSISTORS

Circular geometry MOS transistor test structures with self-aligned gate lengths of 30 and 100 μm were fabricated in this SOI substrate. Fig. 2 shows the plan view layout of the transistors and Fig. 3 shows a schematic cross-section of the completed device.

First, solid source phosphorus diffusion (1000°C) was used to form a n$^+$ contact layer at the back of the n$^-$ substrate. Sacrificial LPCVD Si$_3$N$_4$ was then employed as a mask when wet etching source and drain regions in the SOI device layer and the BOX layer. Solid source boron diffusion (1000°C) was used to form the p$^+$ source and drain regions in the n$^-$ substrate. At the same time, the SOI device layer was also boron doped for use as the gate electrode. A 200 nm thick low temperature CVD SiO$_2$ layer was then deposited and patterned to form contact windows, before sputtering and patterning a 3 μm thick Al/3% Si layer as the

metallisation. A 450°C H$_2$/N$_2$ anneal completed device processing.

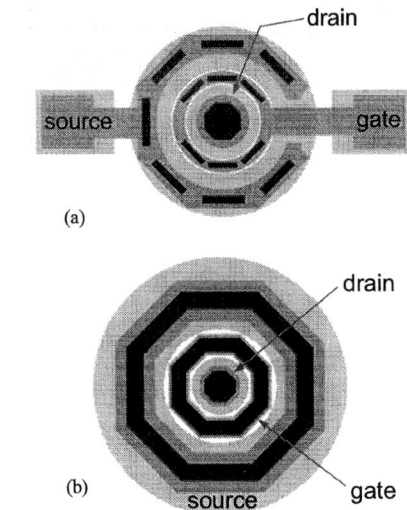

(a)

(b)

Fig. 2. Plan view layout of the circular geometry MOS transistor test structures, with gate lengths of (a) 30 μm and (b) 100 μm.

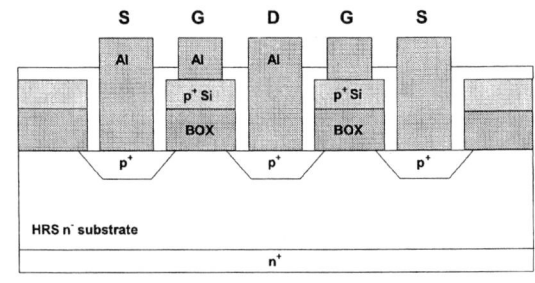

Fig. 3. Schematic cross-section of the self-aligned gate circular geometry MOS transistor.

TRANSISTOR CHARACTERISTICS

The completed MOS transistors were DC tested using a HP4155A parameter analyser in a dark, electrically screened enclosure. Output and transfer characteristics for a typical 100 μm gate length device are shown in Figs. 4 and 5. The smaller, 30 μm gate length, devices had comparable characteristics.

The threshold voltage (V$_{th}$) of all the transistors was slightly positive, as shown, at a value of +0.2 V. However, a p-channel enhancement mode MOS transistor is expected to have a negative threshold voltage, in the range -1 to -2 V, dependant on the gate oxide fixed charge density (Q_{ss}). The presence of small doses of boron in the high resistivity silicon substrate, due to trace environmental or chemical contamination, would result in the threshold voltage becoming positive. However, SIMS analysis of this substrate has been carried out and there is no evidence of such boron contamination.

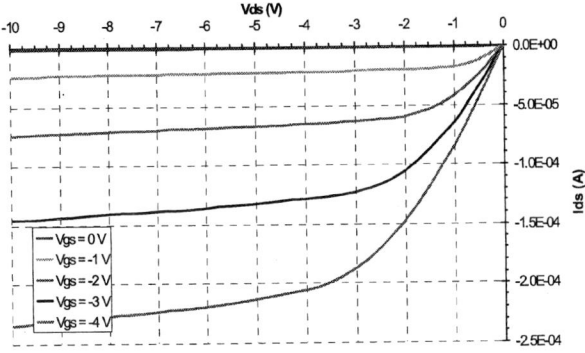

Fig. 4. Output characteristics of a typical 100 μm gate length circular geometry MOS transistor.

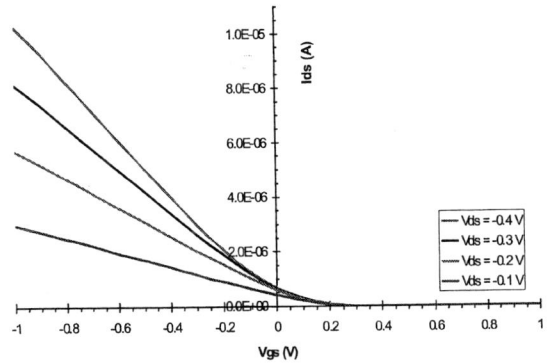

Fig. 5. Transfer characteristics of a typical 100 μm gate length circular geometry MOS transistor.

Theory suggests that for this structure a threshold voltage of +0.2 V would be achieved for an extremely low oxide fixed charge density of 1×10^{10} cm^{-2}. It has previously been observed by some of the authors, and others, that wafer bonding can lead to the creation of negative fixed oxide charge initially located at the bond interface [3]. Subsequent anneals can reduce the magnitude of this negative charge, and similar changes in charge magnitude are also observed at the handle wafer interface. This mechanism would explain the reduction in fixed charge to the ultra-low value extracted from the measurement data presented here.

TRANSISTOR IRRADIATION EXPERIMENT

The devices were then irradiated with 24 GeV protons at CERN, according to the following dose schedule:

Sample	Dose (p/cm^2)	Dose (rad)
A	7.9×10^{11}	22.57 krad
B	1.1×10^{12}	31.43 krad
C	9.5×10^{12}	271.43 krad
D	9.6×10^{13}	2.74 Mrad

After irradiation the devices were transported back to Belfast in a cool, temperature controlled container, thus ensuring a controlled annealing history. Fig. 6 shows the effect of irradiation on the transfer characteristics of a typical 100 μm gate length transistor, and Fig. 7 shows the similar results obtained on 30 μm gate length devices.

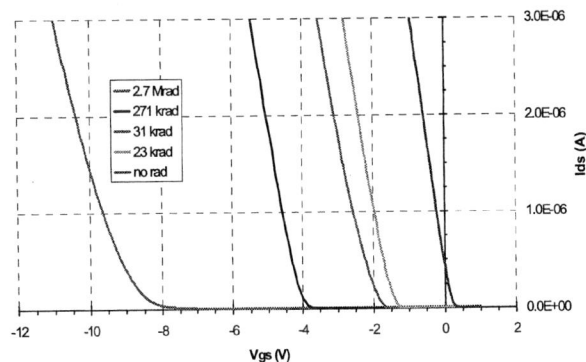

Fig. 6. Transfer characteristics for 100 μm gate length circular geometry transistors before and after proton irradiation.

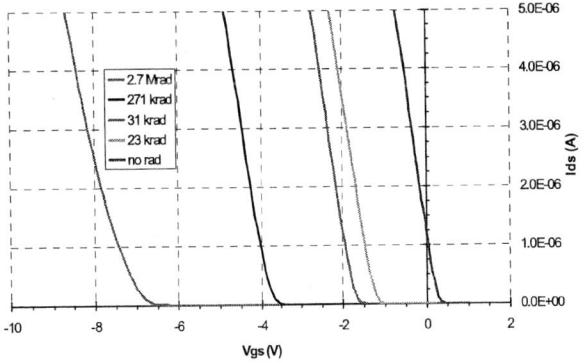

Fig. 7. Transfer characteristics for 30 μm gate length circular geometry transistors before and after proton irradiation.

It is clear from Figs. 6 and 7 that increasing radiation dose has caused an increasing negative shift in transistor threshold voltage. The approximately exponential relationship observed between radiation dose and V_{th} shift is shown in Fig. 8.

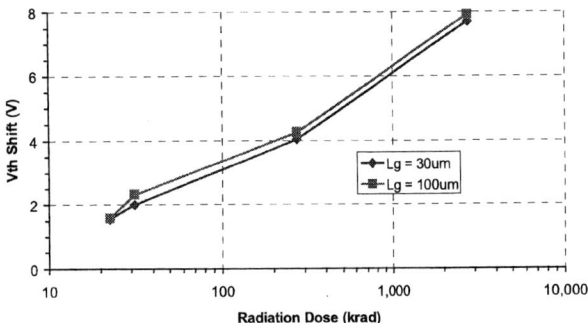

Fig. 8. MOS transistor threshold voltage shift as a function of proton radiation dose for both device geometries.

The increasing negative shift in threshold voltage with increasing dose is known to be a direct consequence of the radiation-induced buildup of ionised positive charge in the gate oxide (BOX) layer near the SiO_2/Si interface [4]. For the 100 μm gate length transistors studied in this work, the maximum proton dose of 2.7 Mrad has caused a threshold voltage shift of approximately 8 V. This indicates that the trapped charge in the 500 nm thick gate oxide has changed by only 3.45×10^{11} cm^{-2}.

Figs. 6 and 7 also show a steady decrease in the slope of the I_{ds} vs. V_{gs} transfer characteristics with increasing radiation dose. This is analogous to distortion observed in C-V characteristics [4]. In this case it may be attributed to an increase in trap density (D_{it}) at the SiO_2/Si interface between the BOX layer and the substrate. Since MOSFET transconductance (g_m) is defined by $\Delta I_{ds}/\Delta V_{gs}$, this represents a 50% reduction in 100 μm gate length transistor g_m at the highest proton radiation dose of 2.7 Mrad.

The drain-substrate p^+-n diode reverse bias leakage current was also measured, and Fig. 9 shows the characteristics obtained before and after proton radiation.

It was found for both device geometries that the reverse bias leakage current increases with increasing radiation dose. For the maximum dose of 2.7 Mrad the leakage current increased by approximately two orders of magnitude. This may also be attributed to a combination of a buildup of positive charge in the BOX layer and an increase in D_{it} at the interface between the BOX layer and the substrate. At the higher doses the radiation induced positive charge is in fact sufficient to invert the surface of the p^+ diode region [4]. Additional leakage current then results from low voltage breakdown of this field-induced n^+-p^+ junction.

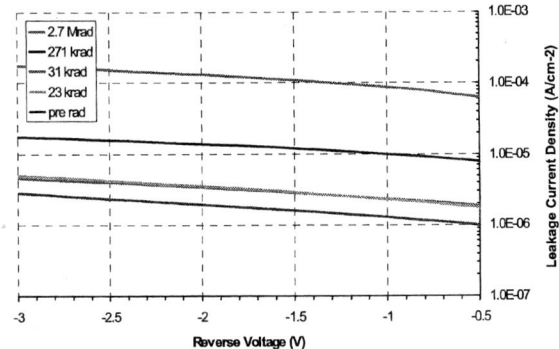

Fig. 9. Drain-substrate diode reverse bias leakage current before and after proton irradiation (30 μm gate length transistor).

CONCLUSIONS

The next generation of SOI-based high energy physics particle detectors require a high quality BOX-substrate interface. This work has shown that circular geometry transistors provide a valuable test structure to allow electrical characterisation of this crucial material interface in the high radiation environments experienced by ionising particle detectors.

ACKNOWLEDGEMENTS

The authors wish to acknowledge the financial support provided by the UK Particle Physics and Astronomy Research Council, and the technical assistance of Maurice Glaser at CERN.

REFERENCES

[1] H. Niemiec et al, *Microelectronics Reliability* 45 (2005) 1202-1207.

[2] F.H. Ruddell et al, *Electrochemical Society Proc.* 2003-05 (2003) 57-62.

[3] P.T. Bailey et al *Electrochemical Society Proc.* 95-07 (1995) 252-266.

[4] Nicollian and Brews, *MOS Physics & Technology* 1981 (repr. 2003) 549ff.

SESSION 5
3D Integration

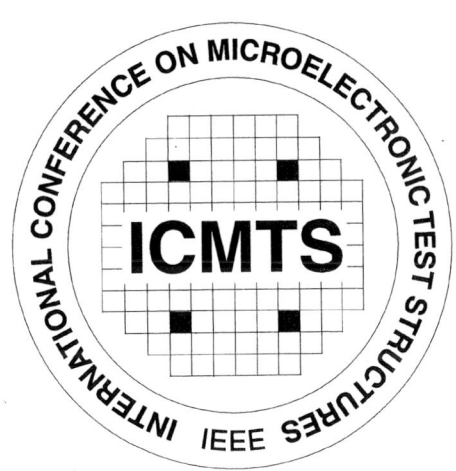

March 26, 09:00–10:20

Co-Chairs: Yoshio Mita, *University of Tokyo, Japan*

Emilio Lora-Tamayo, *Universitat Autonoma de Barcelona, Spain*

2008 IEEE Conference on Microelectronic Test Structures, March 24-27, Edinburgh, UK

5.1

Prediction of Stress-induced Characteristic Changes for Small-scale Analog IC

Naohiro Ueda, Eri Nishiyama, Hideyuki Aota, Hirobumi Watanabe
Electronic Devices Company, Ricoh Co. Ltd.
30-1 Saho, Kato-city, Hyogo, Japan
Fax: +81-795-42-6101
E-mail: naohiro.ueda@nts.ricoh.co.jp

ABSTRACT

Stress-induced parametric changes during the resin-molded packaging of a small-scale integrated circuit (IC) smaller than 1.0 mm^2 have been evaluated by a specially designed test chip. Multiple test chips with different resistor locations have been fabricated and measured by die-to-die correspondence. One contour plot was reproduced from the measurement results. The present paper shows the distribution of parametric change for the small-scale IC. In addition, a new method for evaluating the circuit performance change due to stress-induced parametric changes is presented.

INTRODUCTION

With the dramatic expansion in the use of portable electronic devices in recent years, these devices require downsizing and accurate operation, and the IC is no exception. Initial accuracy must be improved, or characteristic variation must be controlled.

One of the factors that interferes with improving the accuracy of the electrical characteristic of an IC is parametric changes due to resin-molded packaging. One of the origins of such changes is the mismatch in thermal expansion between the various assembly materials, i.e., silicon, molding compound and the copper lead frame. This introduces compressive stress on the IC surface, and the piezo effect changes the electric characteristics of individual devices [1]. Consequently, the circuit performance changes with the wafer condition [2] [3]. This parametric change is not considered in usual circuit design, because use of the SPICE model parameter is determined based on the condition of the wafer.

This report presents unique experimental results of the residual stress-induced parametric changes during resin-molded packaging of a small analog IC whose size is smaller than 1.0 mm^2.

METHODOLOGY AND MEASUREMENT

A. Test Chip

The test chip used in this evaluation is shown in Fig. 1. This test chip has the same outline and features the same

bonding pad as that of a small IC, because stress-induced parametric changes of an actual small IC have been measured accurately. The small IC has a limited number of bonding pads, which means that only a single piezoresisitive sensor can be mounted on a single test chip. The piezoresistor sensor used for this experiment is a custom p-type diffused resistor that incorporates (100) silicon.

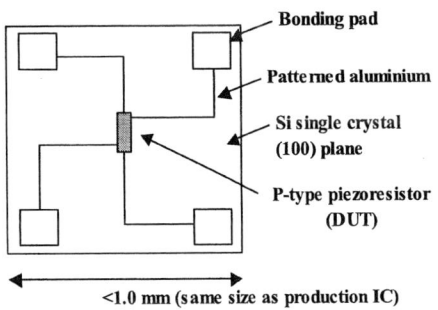

Fig. 1. Test chip.

The technique for evaluating the distribution of parametric changes on the surface of the chip is as follows.

First, multiple test chips having different piezoresistor positions were fabricated and their resistances were measured before and after packaging. The measurement data for the same test chip must be perfectly traced before and after packaging, and a single contour plot was obtained from all of the measurement results. This method enables the parametric change distribution on the surface of the small IC to be obtained with only four pins (Fig. 2).

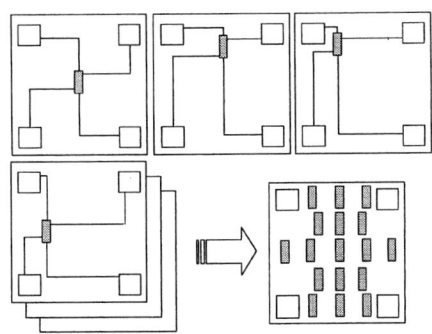

Fig. 2. Principle of multiple pointmeasuments utilizing test chips.

978-1-4244-1801-5/08/$25.00 ©2008 IEEE

Second, a high accuracy is required for each measurement in order to reproduce the dependency in the plane of a single chip from the results of different test chips. The four-point Kelvin resistance measurement was performed in order to avoid other disturbance factors.

Third, as mentioned above, since the size of the small IC is less than 1.0 mm^2, the incorporated piezoresistor should be sufficiently small compared to the entire chip. The resistor fabricated for the present study is 2.0 μm in width and 30 μm in length.

A test chip having the above-described characteristics was subjected to wafer fabrication and packaging processing under the same conditions as those for the small IC, namely, the production facility, materials, and process condition were the same.

The resistance value was measured under a 23.0±0.5ºC isothermal environment in order to reduce the temperature characteristics dependency of the resistor value and the piezoresistor coefficient.

B. Data for circuit performance estimation

In order to accurately predict changes in the electronic circuit performance, the stress dependencies of the Gm change in the N-channel MOSFET and P-channel MOSFET were measured along with the resistance value change of the N-type poly-Si resistor. Strips of silicon of 10 mm in width, 40 mm in length, and 0.4 mm thick containing an N-channel MOSFET, a P-channel MOSFET, and a Poly-Si resistor were used for this experiment.

The cantilever technique was used to calibrate the response of the device characteristic with respect to externally applied uniaxial tensile and compressive stresses [4]. The measurement was performed in such a way that the current flow in the element is parallel to the stress direction. The applied stress is in the range of 0 ~ ±150 MPa. The evaluation results are shown in Fig. 3.

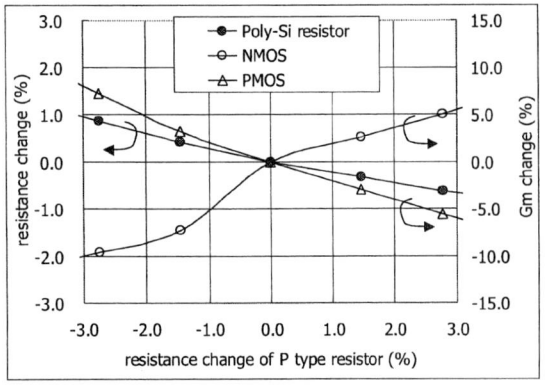

Fig. 3. Typical cantilever bending results.

RESULTS

A. Accuracy and Reproducibility of the Measurement data

High-accuracy measurement is required for each measurement result for multiple test chips in order to reproduce the dependency on the surface of a single chip.

The basic die image of 17 piezoresistor sensors (P1~17) in the 0.8×0.7 mm^2 test chip is shown in Fig. 4.

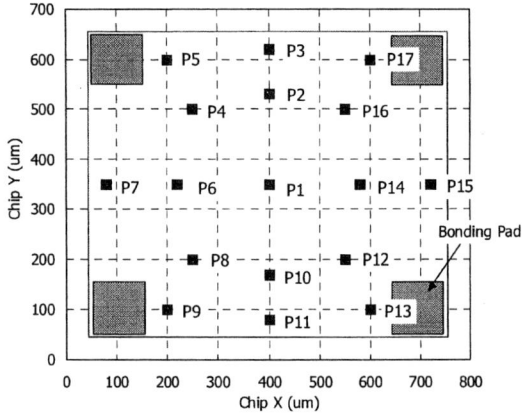

Fig. 4. Configulation of piezoresistors for a 0.8×0.7 mm^2 test chip.

Fig. 5 shows the measurement results of resistance changes for 102 piezoresistors extracted from different locations (Site 1 ~ Site 6) inside a single wafer.

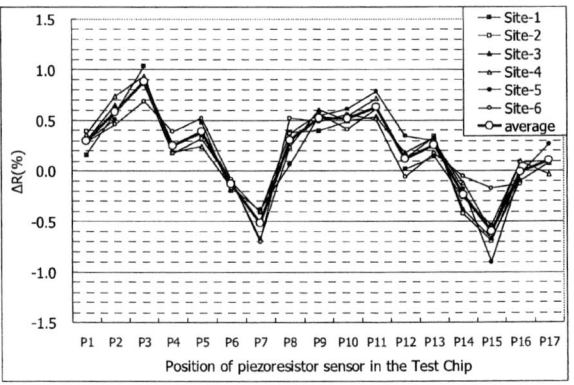

Fig. 5. Stress-induced resistance change for 102 packages of the test chip.

The measured data of Site 1 ~ Site 6 show almost the same dependency. In other words, this result shows that the method of reproducing the variation of parametric changes by using multiple test chips is effective for the small-scale IC which limits the number of bonding pads.

B. Equal stress change line (contour plot)

Table 1 shows the sample list of the present study. The measured chip sizes are 0.5×0.5 mm^2 and 0.8×0.7 mm^2 for Type A and B packages.

108

Table 1 Evaluation sample table

Chip size	0.5 x 0.5 mm^2	0.8 x 0.7 mm^2
N of sensor	20	17
Package A	Figure 6	Figure 7
Package B		Figure 8

Figs. 6 through 8 show the calculated contour plots o p-type resistance value change based on the actua measurement data. Values in the figure indicate th resistance value change rate (in arbitrary units).

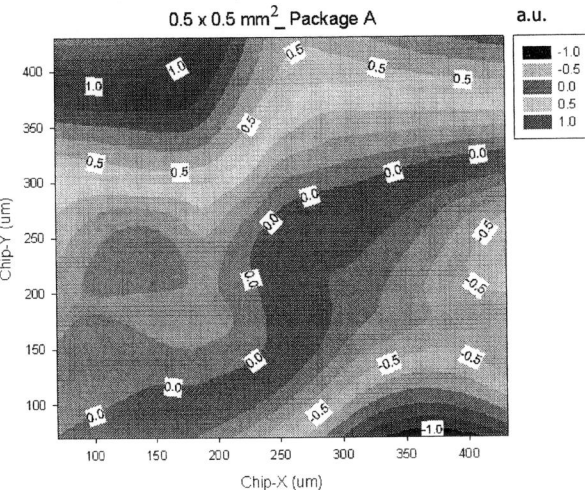

Fig. 6. Contour plot of parametric change for a 0.5×0.5 mm^2 Si chip resin-molded with package A.

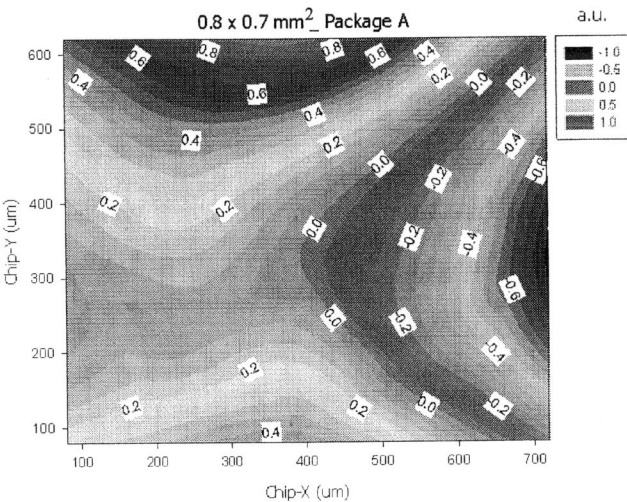

Fig. 7. Contour plot of parametric change for a 0.8×0.7 mm^2 Si chip resin-molded with package A.

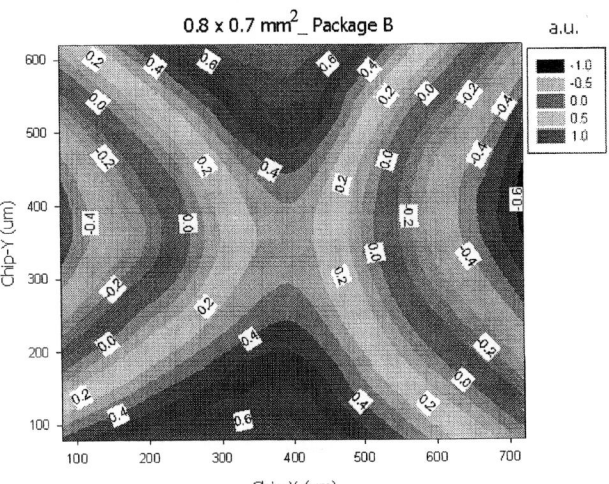

Fig. 8. Contour plot of parametric change for a0.8× 0.7 mm^2 Si chip resin-molded with package B.

As shown in Figs. 6 and 7 pertaining to package A, in both cases, the parametric change of the right region of the chip decreased, while that of the upper region increased, and a symmetrical distribution was not observed. In contrast to package A, package B shows a definite symmetrical distribution about the chip surface center.

The dependencies of the change are considered to be caused by the package structure, especially the shape of the lead frame. In addition, the asymmetrical distribution shown in package A is considered to be due to the injected resin flow to the cavity at the encapsulated step, and a more detailed analysis is expected.

C. *Estimation of performance change of the circuit*

To clarify the relationship between the parametric change distribution and the performance fluctuation of a circuit, the performance change of an Op-Amp circuit fabricated using the same process condition as the test chip was investigated. The chip size of the Op-Amp circuit was 0.5×0.5 mm^2 and the encapsulation was performed by package A. The evaluation method is as follows.

First, a square partition image was placed over the entire circuit layout in units of 0.1 mm to recognize the element included in each block (Fig. 9). Meanwhile, the same partition image was placed on the parametric change results in order to identify the parametric change value (Fig. 6). The average value inside one block was used as a representative value of the parametric change. The devices for which parametric changes were modified are N-channel MOSFET, P-channel MOSFET, and N-type Poly-Si resistor. The trans-conductance Gm for the transistor and resistance value for the resistor were respectively modified using the correlation of Fig. 3.

Second, several SPICE model parameter files were prepared. These files differ in magnitude with respect to the parametric changes in each device.

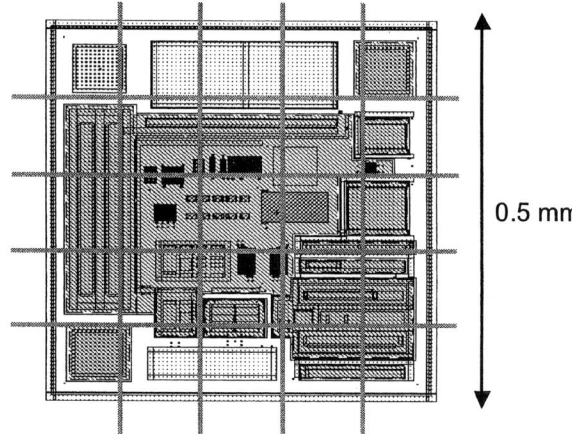

Fig. 9. Experimental circuit.

Fig. 10 shows the simulation result of the DC output fluctuation by stress-induced parametric changes. Both the best case of a layout with consideration of stress distribution and the worst case with maximum impact were summarized.

The case in which the characteristics of both the transistor and the resistor were influenced indicates the maximum shift of approximately 0.7% output fluctuation. For this circuit, the significance of the impact on the DC output due to the transistor characteristics is greater than that due to the resistor.

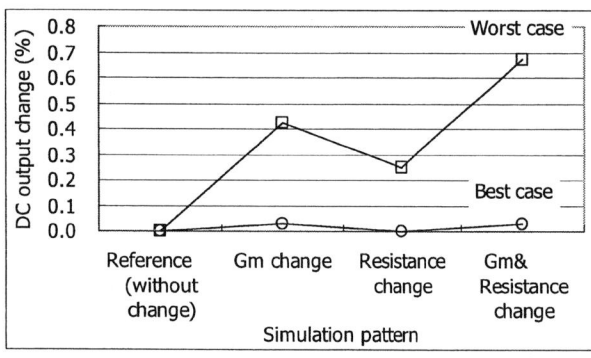

Fig. 10. Simulation results.

CONCLUSIONS

The residual stress-induced parametric changes during resin-molded packaging of a small-scale analog IC smaller than 1.0 mm^2 have been evaluated.

In the present study, we established a new method for manufacturing multiple test chips with different piezoresistor locations and reproduced the distribution on the surface of a single chip.

The present evaluation shows that the distribution of parametric change depends on the package type. In addition, these results confirmed that the performance change of an Op-Amp circuit can be predicted with higher accuracy.

The proposed method is remarkably effective for evaluating the packaging-induced parametric change for a small-scale analog IC having a limited number of bonding pads.

ACKNOWLEDGEMENTS

The authors would like to thank Mr. Kazuhiro Kubo and Mr. Hiroshi Fujiwara of Aoi Electronics, Ltd., for their helpful discussions and sample preparation.

REFERENCES

[1] H. Miura and A. Nishimura, "Device Characteristic Changes caused by Packaging Stress," ASME AMD, vol.195, pp.101-109, 1994

[2] S. Komatsu, K. Suzuki, N. Iida, T. Aoki, T. Ito, and H. Sawazaki, "Stress-insensitive diffused resistor network for a high accuracy monolithic D/A converter," in IEDM Tech. Dig., Int. Electron Devices Meet., pp.144-148, 1980.

[3] Hassan Ali, "Stress-Induces Parametric Shift in Plastic Packaged Devices" IEEE Transacitons on Conponents, Packaging, and Manufacturing Technology, Part B, vol.20, No.4, Nov. 1997

[4] Fabiano Fruett and Gerard C. M. Meijer "Low-drift Bandgap Voltage References," Proceedings of the 28th European Solid-State Circuit Conference, pp.383-386, 2002

Measurement and Optimisation of Bond Strength for Anodic Bonding of Glass to Dielectric Thin Films

G. Cummins, H. Lin and A. J. Walton
Institute of Micro and Nano Systems
Institute of Integrated Systems
University of Edinburgh
West Mains Road, EH9 3JF
Email: Gerard.Cummins@ed.ac.uk

Abstract—This paper details the optimization and characterization of an anodic bonding process for glass to dielectric thin films. Test structures suitable for non destructive, in-situ measurement of the bond strength at the interface have been used to determine the uniformity and quality of the bond.

I. INTRODUCTION

Anodic bonding is a common method used in MEMS for wafer level device packaging. This process involves an alkali glass wafer such as Pyrex 7440 being brought into contact with a wafer, usually silicon, at a high temperature while a high DC voltage is applied between the wafers. Mobile sodium ions migrate towards the electrode leaving a negatively charged region in the glass and the resulting electrostatic force pulls the glass and silicon together tightly. An electrochemical reaction takes place and bonds are formed between the glass and the silicon.

It is clearly desirable to be able to bond glass to materials other than silicon to enable the technique to be more widely applicable. However, bonding silicon nitride or oxide films to glass does not produce bond strengths comparable to those achieved by bonding silicon to glass. Previous work on anodic bonding to thin film nitride or oxide indicates that the electrostatic force decreases as the thickness of the dielectric increases causing a reduction in the bond strength [1].

Clearly a robust method of characterising the bond strength is required to enable the effect of changing the recipe parameters to be quantified for films of various thickness. One standard method for determining the bond strength is simple mechanical tensile testing, which although useful does not necessarily give an accurate measurement as the result can vary with the sample geometry and how the sample is mounted. This method also requires the usage of expensive equipment and can often result in measurement of the stress in the glass or silicon layer rather than the bond strength. Other approaches have used knife blade, chevron or blister methods [2] which are also destructive which makes them non-optimal for process control.

This paper presents a preliminary results of process optimisation of anodic bonding of glass to dielectric thin films of various thicknesses through the use of previously reported non destructive test structures.

The test structure developed by Knechtel and others [3] based on work conducted by Horning et. al [4] enable a non-destructive determination of the bond strength. These test structures consist of mesas or raised steps. How they are arranged depends on whether a multi-step or a single step structure is implemented. Multi-step structures consist of a series of steps with increasing distance between neighbouring steps. In the case of multi-step features the bond strength is determined by the minimum sized gap between adjacent structures at which bonding occurs, in single step structures it is the length of the resulting crack that allows the bond strength to be calculated. The single step structure therefore provides an analogue measurement of the bond strength whereas the multi-step structure provides a discrete measurement.

Nondestructive test structures based on cavities have also been reported, these allow the electrostatic pressure exerted during the bonding process to be characterised and not the bonding strength [5]. These test structures consist of a variety of cavities of different shapes and sizes. The bond quality is determined by whether the electrostatic pressure during the bonding process is sufficient to overcome the stiffness of the cavity and hence pull the bottom of the cavity and the surface of the glass together forming a bond. The smaller the test structure bonded the greater the bond quality as these structures would have a higher stiffness.

II. TEST STRUCTURES

The test structures reported in this paper use titanium mesas. Titanium was chosen because of its relative inelasticity which ensures that any deformation of the test structure due to the load applied during bonding is minimised. hTis differed from previously reported aluminium structures which required nitride deposition to ensure there is not excessive deformation. Another advantage of the titanium mesas is that Ti deposition is a lower temperature process than PECVD nitride. This may be advantageous for some applications where the time the wafer spends at an elevated temperature is an issue.

978-1-4244-1801-5/08/$25.00 ©2008 IEEE

Stepped Ti test structures for non-destructive in-situ evaluation of the bonding strength have been designed and fabricated on silicon wafers coated with films of oxide or nitride with different thicknesses. The thickness of the dielectric films were 0.5, 1 and 1.5 μm. Titanium with a thickness of 0.25 μm was then sputtered and patterned using a lift-off process to form the test structures. The bonding was carried out using a Suss Microtec substrate bonder.

The test structure design consists of both single and multi-step structures. These test structures are repeated across the wafer surface and the mask layout us shown in Fig. 1. This allows information regarding the bond uniformity and quality across the surface of the wafer to be determined.

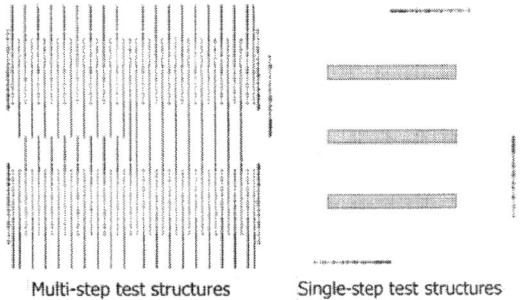

Multi-step test structures   Single-step test structures

Fig. 1. Mask design for multi-step and single step test structures

The single step test structures have a width of 500 μm. The Ti mesas that form the multi-step test structure are 10 μm in width, with an initial gap between the first Ti mesas of 10 μm, with each successive gap increasing by 2 μm up to a maximum gap of 100 μm.

III. Measurements

The test structure has been used to investigate the relationship between bond strength and the following process parameters; temperature, voltage, load pressure and film thickness. The Taguchi method is one method of reducing the number of experiments needed to determine the relationship between the bonding recipe parameters and the bond strength for films of various thicknesses. This facilitates an optimum recipe to be determined for a given film thickness. An L9 design has been used and the control factor setting are shown in Table I. The duration allowed for the bond to propagate throughout the glass/dielectric interface was set at 30 minutes for these initial experiments.

The anodic bonding was carried out between dielectric thin films on N doped (100) silicon wafers and Pyrex 7740 glass. The glass was cleaned with Piranha etch ($H_2SO_4 : H_2O_2$; 2:1) and deionised (DI)water prior to bonding, the silicon wafers were cleaned with acetone, isopropyl alcohol and DI water.

IV. Results

A. Bonding

Not all wafers were found to be fully bonded after removal from the substrate bonder. The more complete bonds were

TABLE I
A L9 TAGUCHI MATRIX WITH AND THE PARAMETER VALUES USED IN IN BOND STRENGTH OPTIMISATION

Recipe	t (μm)	T (° C)	V (V)	P (mbar)
1	0.5	350	-750	500
2	0.5	400	-1000	750
3	0.5	450	-1250	1000
4	1.0	350	-1000	1000
5	1.0	400	-1250	500
6	1.0	450	-750	750
7	1.5	350	-1250	750
8	1.5	400	-750	1000
9	1.5	450	-1000	500

found to occur when the dielectric film was thinner, air voids at the edges and weak bonds occurred in most of the thicker film wafers as shown in Fig 2. From past experience of bonding silicon to glass a wafer can be said to have been bonded fully when the resistance is such that the current applied to the wafer is 5% or less of the initial current applied.

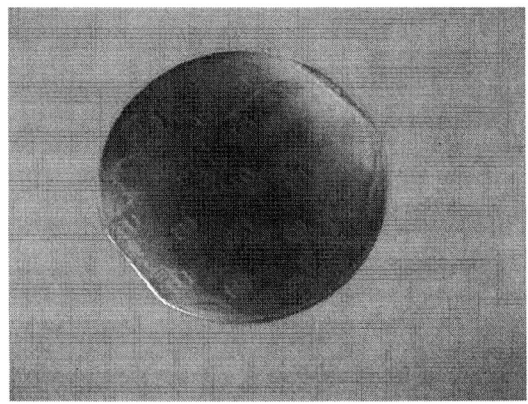

Fig. 2. Partially bonded wafer

The reason for the incomplete bonding found on several wafers is most likely due to the 30 minutes bonding time allowed not being sufficient for the depletion layer to form across the thicker dielectric layer. Future bonding of thick dielectric thin films will need to take this into account by increasing the duration for which the electric field is applied to the wafer stack or by using a higher voltage.

The bonding current as a function of time is shown in Figs 3 and 4, for most recipes it was observed that a major drop in current occurred within the first few minutes of bonding, this is consistent with what has been previously observed with the bonding of glass to silicon. It was observed that for thinner oxide films, the initial current was higher and dropped more rapidly. This can be attributed to the fact that the thicker the film on the silicon the electric field required for the ion migration necessary for bond formation is lower.

An interesting observation is that the current levels for the nitride films in Fig. 4 are significantly higher than for those observed for the oxide results in Fig. 3. Figure 4 suggest that a longer bond time would enable more ions to move with the final current level being closer to the levels obtained at the completion of the oxide bond. In these experiments time was a constant and this clearly needs to be examined.

Preliminary results suggest that the temperature plays a significant role in the rate at which current reaches steady state, this would be consistent with previously published results [6], with recipes operating at a higher temperature dropping to steady state faster than recipes operating on similarly thick films with lower temperatures. This is most likely due to the increased temperature causing increased ion mobility in the dielectric thin film.

B. Multi-step test structures

The multi-step test structures were found to only be of use when a good bond is created between the glass and film, because of this results could only be accurately obtained using the test structures found on the 0.5 μm thick dielectric films, for these the existence of the gap is easily determined by visual inspection through the presence of an obvious colour change between the bonded and unbonded area, as shown in Fig. 5. Any gaps between successive mesas could not be accurately determined by visual methods on the thicker wafers due to a diffuse colour change between areas. This made it more difficult to determine where the bonded area starts and the unbonded area begins.

Similar test structures were used to qualitatively determine the bond strength rather than quantitively. Quantitative results have been previously published using these models. However the model which was used to do this has not been [3]. The quality of the bond strength was determined by observing the smallest gap size at which bonding occurs with the smaller the gap size the higher the bond strength for mesas of similar thickness and width.

C. Single step test structures

The single step test structures were somewhat better for determining the bond strength of the wafers over the range of film thicknesses investigated. The unbonded region formed around the mesa during bonding was more discernable over a range of thicknesses and process conditions due to the larger geometries. The length of this crack (L) is a function of bond strength, the longer the crack the weaker the bond [3] [4] [7]. An example of this test structure is shown in Fig. 6

D. Recipe Optimization

The length of the crack formed around the central single step structures was determined using an optical microscope. Measurements were taken at five points across the wafer, these points were the centre, left most, right most, top and bottom squares. The mean crack length ($\bar{L}$), as well as the range (ΔL) was calculated, the results are shown for both nitride and oxide films in Tables II and III.

There was no discernable trend observed between the process parameters or film thickness and the range of measured crack lengths (ΔL).

An initial trend amongst the values for the mean crack length shown in Table II can be discerned, the mean crack length increases with the nitride film thickness. Again, this may be attributed to the nitride layer impeding the migration of ions which form the bond resulting in a surface energy. The crack length for bonding recipes 3 and 6 of the matrix for oxide to glass could not be measured due to the resulting low quality partial bond. The lack of results for these recipes may affect the validity of any Taguchi analysis performed using them. It was also noted that generally, the measured crack length for nitride films are better than that for oxide, this is unusual in that it has been previously noted that it is difficult to obtain a better bond strength for an anodically bonded nitride film than it is for oxide [1] [8] [9].

These results were used to generate main effects plots to determine any relationships between the bonding conditions and the bond strength. It can be seen from Figs 7b and 7d that the trends observed for oxide and nitride films as a function of temperature and pressure are similar. In the case of temperature, Fig 7b shows that as the temperature increases the crack length decreases. Smaller crack length result from a greater bond strength between the thin film and the glass. Increased temperatures would ensure increase ion mobility within the dielectric thin film and hence the formation of a stronger bond. Higher bond strength due to increased load pressure as shown in Fig 7d can probably be attributed to better contact between the glass and silicon wafers at higher loads.

Increasing oxide film thickness results in a lower bond strength (Fig. 7a), this is also demonstrated by the longer cracks measured at the test structures. Increasing nitride film thickness seems to show the opposite effect. The reason for this is unclear as an increase in dielectric thickness should lead to a lower bonding strength with a standard recipe due to the increased bonding time needed to ensure the migration of ions across the film. However it should be remembered that nitride films tend to exhibit good diffusion barrier properties and this may contribute for the weaker bond strengths observed for thicker films.

The relationship between voltage and the mean crack length at first seems to differ between the two dielectric materials as shown in Fig 7c. The oxide film suggests that at 1000 V a greater crack length is achieved than by either of the two experimental values used. The opposite is suggested by the nitride film. The nitride film seems to suggest that for a voltage of 1000 V there is slight decrease in the mean crack length. An increase in voltage was expected to lead to a drop in bonding time due to an increase in ion migration needed to ensure bonding. A higher voltage would be expected to lead to an increased bond strength due to a stronger electric field increasing the rate of migration of bonding ions.

V. CONCLUSION

Test structures have been used to aid in the development of a process recipe for the anodic bonding of glass to thin film dielectrics. The test structures offer advantages over previous methods used to achieve similar process optimisation, allowing

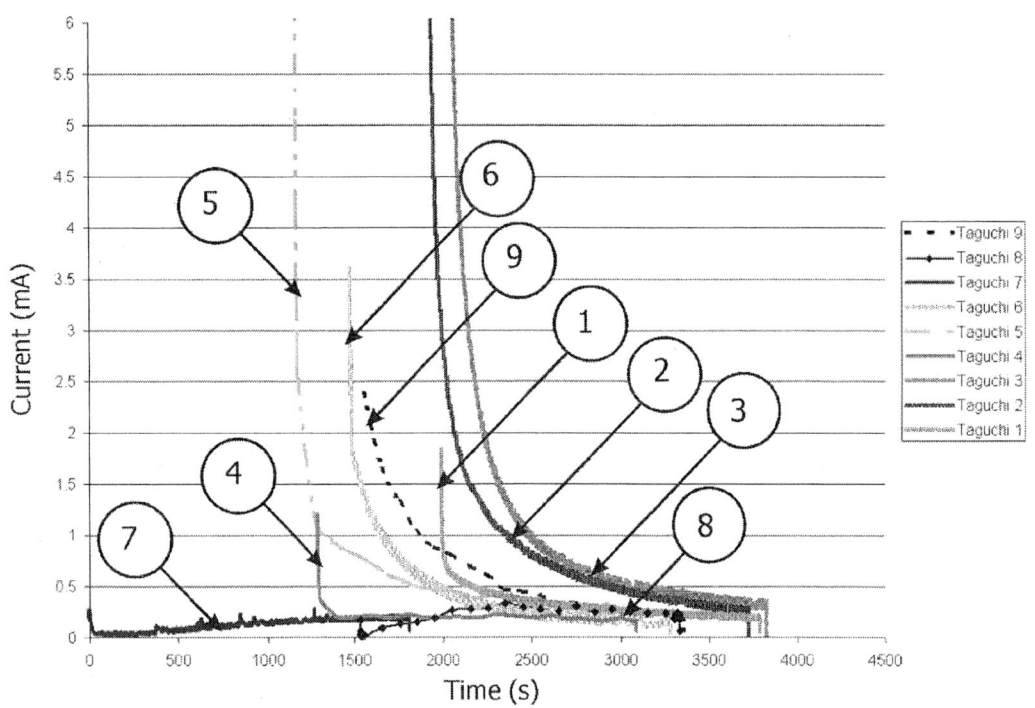

Fig. 3. Plot of current versus time during bonding of oxide thin film to Pyrex 7740 for recipe matrix

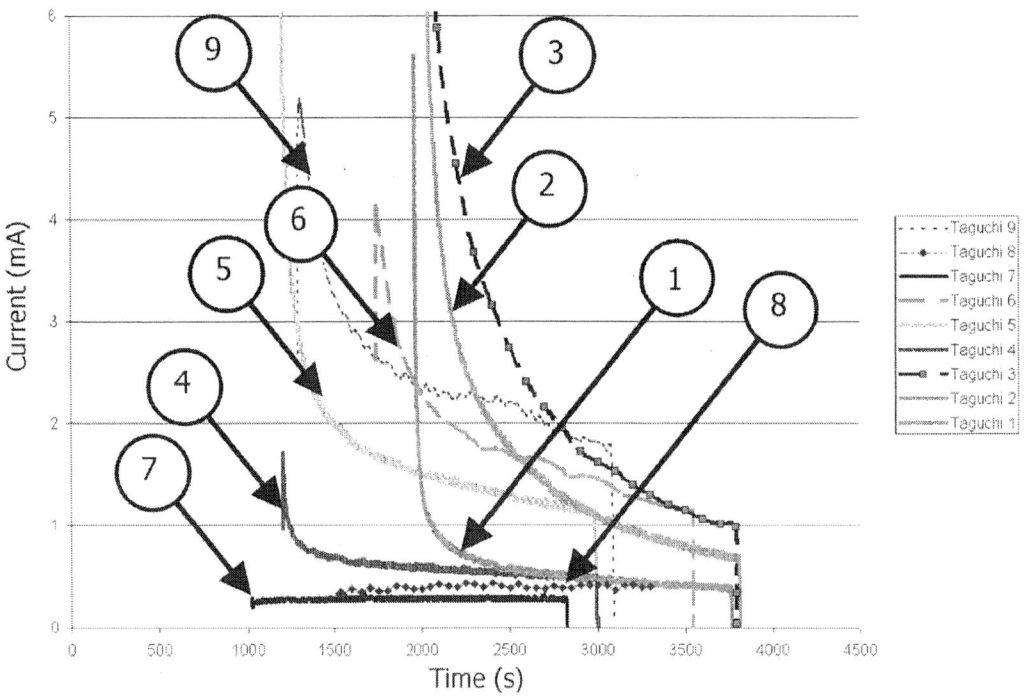

Fig. 4. Plot of current versus time during bonding of nitride thin film to Pyrex 7740 for recipe matrix

TABLE II
INITIAL RESULTS OF THE RECIPE MATRIX USED IN BOND STRENGTH OPTIMISATION EXPERIMENTS - NITRIDE FILMS

Recipe	t (μm)	T (∘ C)	V (V)	P (mbar)	$\bar{L}$ (μm)	ΔL (μm)
1	0.5	350	-750	500	5.43	2.04
2	0.5	400	-1000	750	6.05	2.28
3	0.5	450	-1250	1000	7.96	2.34
4	1.0	350	-1000	1000	11.68	11.34
5	1.0	400	-1250	500	9.06	9.94
6	1.0	450	-750	750	9.95	6.37
7	1.5	350	-1250	750	24.06	16.52
8	1.5	400	-750	1000	18.68	11.12
9	1.5	450	-1000	500	14.57	2.71

TABLE III
INITIAL RESULTS OF THE RECIPE MATRIX USED IN BOND STRENGTH OPTIMISATION EXPERIMENTS - OXIDE FILMS

Recipe	t (μm)	T (∘ C)	V (V)	P (mbar)	$\bar{L}$ (μm)	ΔL (μm)
1	0.5	350	-750	500	27.43	16.25
2	0.5	400	-1000	750	32.21	14.50
3	0.5	450	-1250	1000	*	*
4	1.0	350	-1000	1000	22.70	32.75
5	1.0	400	-1250	500	11.73	22.63
6	1.0	450	-750	750	*	*
7	1.5	350	-1250	750	6.07	0.93
8	1.5	400	-750	1000	5.54	1.06
9	1.5	450	-1000	500	14.95	19.71

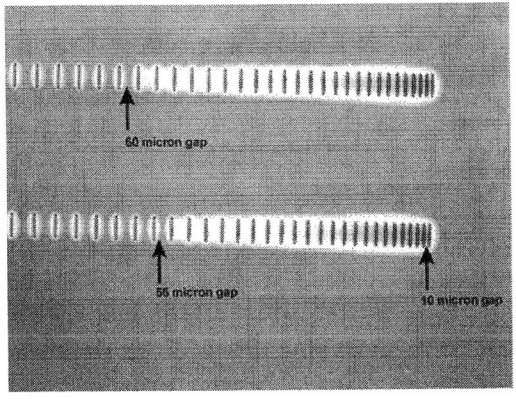

Fig. 5. Example of multi-step test structure

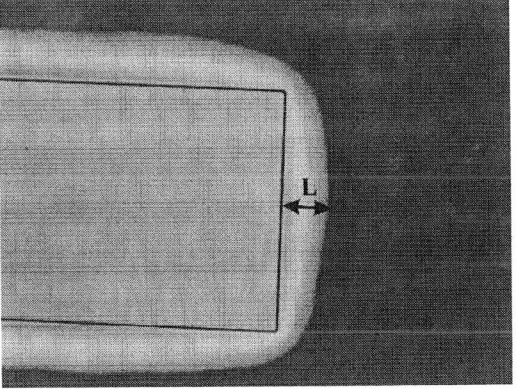

Fig. 6. Single step test structure with resulting unbonded region

non-destructive, in situ measurements without the need for external fixtures. These test structures are best used when the wafers are fully bonded.

The measurement of the crack length using the single step test structures requires the use of optical microscopy, which may lead to some error in the measurements. The multi-step test structures owing to predefined gaps between structures would have less uncertainty when measuring, the accuracy of these measurements would be limited by the photolithographic system used to define them. These multi-step test structures are discrete by their very nature which may introduce some quantization error into the measurements if suitable gap sizes are not chosen.

The effect of changing anodic bonding process parameters on bonding time and quality has been observed with the use of these test structures and it is hoped that further continuation of this work will lead to a process that can be easily altered for use with various dielectric thicknesses.

ACKNOWLEDGMENTS

The authors would like to acknowledge the support of the Edinburgh Research Partnership in Engineering and Mathematics and the Engineering and Physical Sciences Research Council.

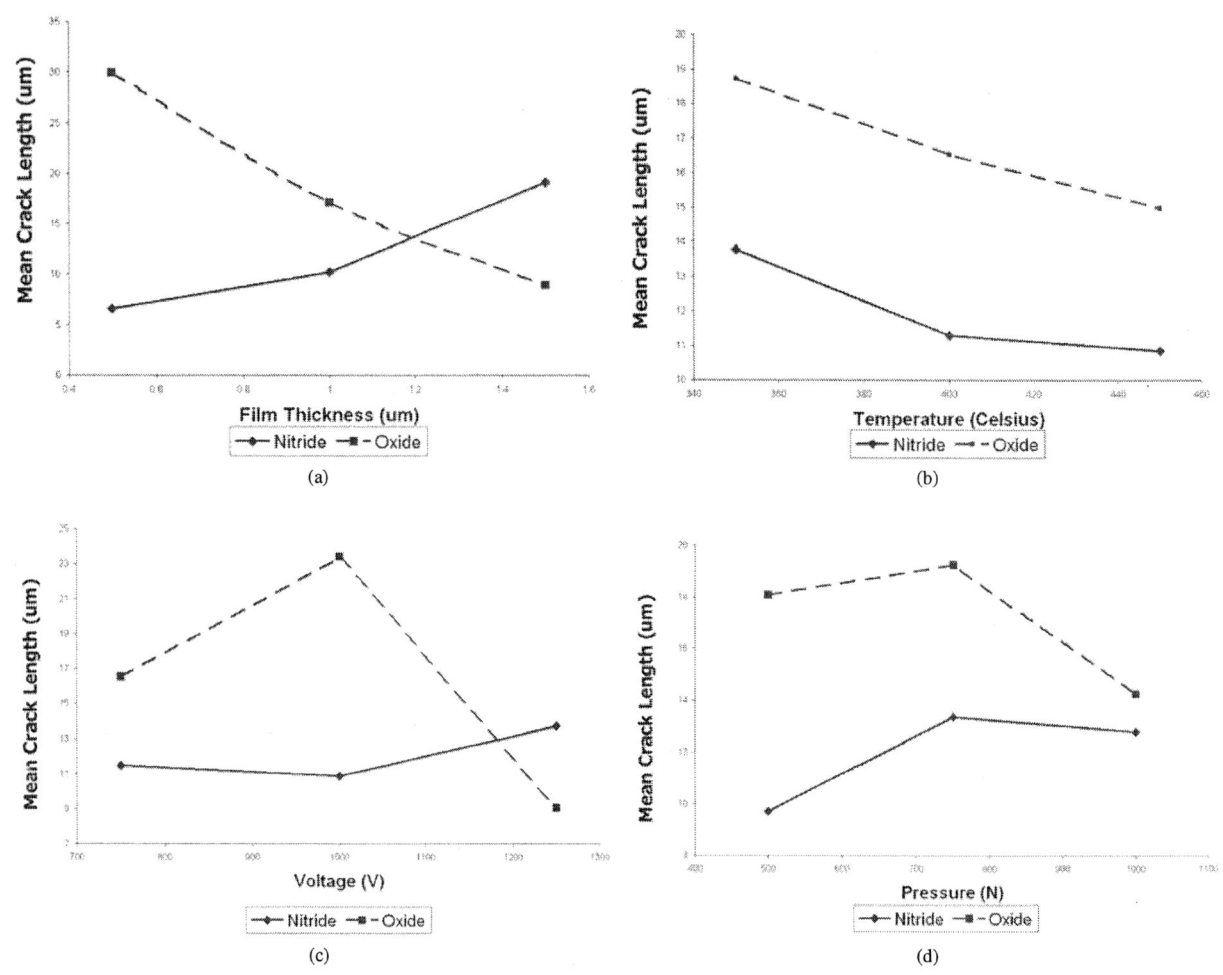

Fig. 7. Results of Taguchi Analysis

REFERENCES

[1] J. A. Plaza, J. Esteve, and E. Lora-Tamayo, "Effect of silicon oxide, silicon nitride and polysilicon layers on the electrostatic pressure during anodic bonding," *Sensors and Actuators A*, vol. 67, pp. 181–184, 1998.

[2] O. Vallin, K. Jonsson, and U. Lindberg, "Adhesion quantification methods for wafer bonding," *Materials Science and Engineering R*, vol. 50, no. 4-5, pp. 109–165, 2005.

[3] R. Knechtel, M. Knaup, and J. Bagdahn, "A test structure for characterization of the interface energy of the anodically bonded silicon-glass wafers," *Microsystems Technology*, vol. 12, pp. 462–467, 2006.

[4] R. D. Horning, D. W. Burns, and A. I. Akinwande, "A test structure for bond strength measurement and process diagnostics," in *Proceedings of the Electrochemical Society*, 1992, pp. 386–393.

[5] J. A. Plaza, J. Esteve, and E. Lora-Tamayo, "Nondestructive anodic bonding test," *Journal of the Electrochemical Society*, vol. 144, no. 5, pp. 108–110, 1997.

[6] T. M. H. Lee, D. H. Y. Leee, C. Y. N. Liaw, A. L. K. Lao, and I. Ming-Hsiang, "Detailed characterization of anodic bonding process between glass and thin-film coated silicon substrates," *Sensors and Actuators A*, vol. 86, pp. 103–107, 2000.

[7] J. S. Go and Y. Cho, "Experimental evaluation of anodic bonding process basedon the taguchi analysis of interfacial fracture toughness," *Sensors and Actuators A*, vol. 73, pp. 52–57, 1999.

[8] M. Nese and A. Hanneborg, "Anodic bonding of silicon to silicon wafers coated with aluminium, silicon oxide, polysilicon or silicon nitride," *Sensors and Actuators A*, vol. 37-38, pp. 61–67, 1993.

[9] A. Hanneborg, M. Nese, H. Jakobsen, and R. Holm, "Silicon-to-thin film anodic bonding," *Journal of Micromechanics and Microengineering*, vol. 2, pp. 117–121, 1992.

Test Structures For The Evaluation Of 3D Chip Interconnection Schemes

A.Mathewson*°, J. Brun°, R.Franiatte°, A.Nowodzinski°, R.Ancient°, N.Sillon°,
F. Depoutot[†], B.Dubois-Bonvalot[†]

* Tyndall National Institute , Prospect Row, Cork , Ireland
° CEA-Leti-Minatec 17 Avenue des Martyrs, 38050, Grenoble. France
[†]Hardware Security Research Group Gemalto, 17 Avenue des Martyrs, 38050, Grenoble. France
Corresponding Author alan.mathewson@tyndall.ie

Abstract

In this paper a test structure is described which facilitates the evaluation of interconnection schemes for chip on wafer attachment and interconnection. Microinsert technology is described and some of the characterization that the test structure permits is discussed. Thermal cycling experiments were performed on this test structure and although the resistance of the contact chain seemed not to change as a function of number of cycles, detailed investigation revealed that the metal resistance was reducing while contact resistance was increasing and the two effects were trading off against each other. Possible explanations for this behavior have been provided

Introduction

In many complex systems under development, the cost of integration and the compromises in performance that can result from the integration process can become prohibitive. Therefore, the establishment of a technology which enables the system level integration of devices from different technological families at the package level can become very attractive. In particular, since each component can be built in its optimum technology flow and interconnections within the package tend to be very short, very high performance can be achieved and many different approaches have been reported [e.g. 1,2,3,4.5,6,7,8]

The work to be presented in this paper describes results from test structures used to evaluate the electrical performance of a novel interconnection system for three-dimensional integration. There are two applications under investigation in this work which share the common theme of the inter chip connection technology.

- The creation of a crypto processor from discrete components for smart card applications (microcontroller and NVM) in which contacts between devices tend to be around the edge of the chip mounted on the wafer and
- The evaluation of the limits of the interconnection methodology for high density fine pitch applications, in which the whole surface of the stacked die can be available for interconnection.

These two application areas are constrained in different directions. In one, the number of interconnections between the two components was limited to around 40 points all of which tend to be around the periphery of the stacked die. However, this application does have a requirement for mechanically strong, high reliability, electrically conductive bonds with a small die footprint.

In the second application, very high density connections are needed in order to make a connection between complex sensor systems with regular interconnections distributed across the front surface of the device. One such application, which has a strong requirement for this type of topography involves mounting imaging devices on top of Read out Integrated Circuits (ROICs). In this application domain it is vital that all contacts are made to all relevant points in the array because any open circuits can create undesirable dead spots in the image plane. Independently of the application, the practical requirements for the interconnection system between the two die include:

- Low Resistivity Electrical Interconnection
- Mechanical Attachment
- Mechanical Stability
- Thermal Management
- Rerouting of Interconnections To Facilitate Front Surface Attachment
- CTE Compensation

In conventional flip chip approaches, where a die is attached to a printed circuit board, interposer or thin film substrate, the interconnection between the die and the board is supplemented by the use of an under fill material. This fills the void between the two surfaces and when cured matches, as closely as possible the thermal expansion coefficients of the two materials (Si and Substrate material). It also provides a degree of strength and mechanical stability to the system.

In the chip to wafer bonding approach to be described here, an appropriate glue layer is required to take the role of the under-fill. Bearing in mind the purpose of this glue layer, the choice of material for this is dependent upon several constraints including its resistance to chemical and physical attack and the temperature at which it starts to decompose or dissociate.

The test structure that permits the evaluation of this technology is described in section three, but it is important to first describe the methodology used to form the interconnections between the two die so that the test structure and what it is used to measure can be put into context.

Microinsert Technology

The approach used to form the interconnections between the two component parts of the system in this work is to adopt an approach entitled Micro-insert Technology[2,3]. In this approach, small spikes of Ni are formed at the point of interconnection between the two circuits. These could occur at either a bond pad or at the termination of a re routed interconnection. Microinserts are formed at these points by using a combination of photolithography and electrolytic plating, as shown in the sequence of diagrams in Figure 1. The approach shown here is extremely flexible because it allows individual optimization of micro-insert pitch, density and size as a function of the masks used.

Microinserts are an attractive methodology for interconnecting two die face to face for chip on wafer examples of heterogeneous integration. They permit small pads (pads as small as 20μm square have been created so far, and 2μm diameter microinsert have been achieved) and much better pitch possibilities than the more conventional flip chip or ball bumping approaches. They also form a very robust connection, which facilitates thinning of the combined system after attachment to suit the packaging scenario to be adopted. (In the case of smart card the entire system needs to be thinned to a value of less than 180μm to enable the system to fit inside the final packaged module).

Figure 1a - Deposit A Seed layer On The Wafer Surface

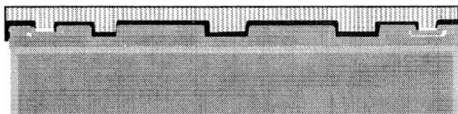

Figure 1b - Print And Expose Regions To Receive Micro-inserts Through Thick Photoresist

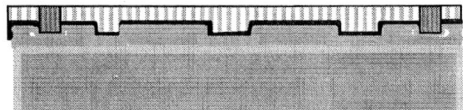

Figure 1c - Grow Nickel On Regions Where Metal Is Exposed Using Electroplating.

Figure 1d - Microinserts Located On Top Of A Pad After Thick Resist And Seed Layer Removal

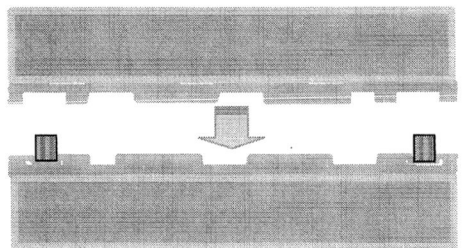

Figure 1e - The Mounting Of The Target Die To The Wafer

Figure 1f - Attachment Of Die To Wafer Showing Glue Layer

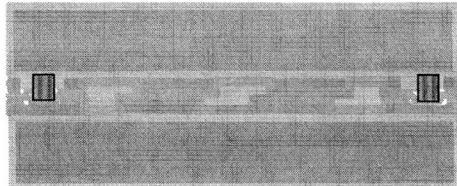

Figure 1g System Assembly After Thermo-Compression Bonding.

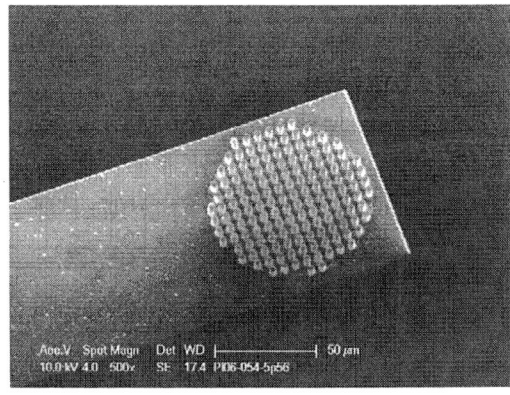

Figure 2 - Microinserts Formed On A Metal Pad On Wafer

In addition there is no additional processing required on the top die and rerouting that interconnects the system at appropriate points can be incorporated into the process. High aspect ratio bumps which can alleviate CTE issues can simply be created as a function of the time required for the plating technology and resist thickness. However, in all instances the choice of polymer/glue layer is crucial. This is particularly true in the case of smart cards, which have the additional constraint (above those for more conventional face to face bonding approaches) of requiring a high destruction temperature and very good mechanical behaviour. This feature is not particularly significant for the assembly of imaging systems but the choice of glue layer remains one of the most significant technological parameters, which requires adjustment for optimum performance and reliability. In this work, non photo imageable, Cyclotene BCB[9],has been used as the glue layer because of its useful viscosity properties as a function of temperature in its pre polymerisation state.

Microinsert Test and Ealuation Structures
The test structures used to evaluate the micro-insert technology in this work are created by joining two components together. For the sake of clarity, the two parts are called the wafer cell and the die cell. The wafer cell is shown in Figure 3 a. This circuit represents the base of the test structure which is defined at wafer level onto which the

second part, the die cell, is aligned and attached using a DATACON 220APM+ 'die on wafer' flip-chip tool and then bonded using a Süss Micro-Tec SB8 Thermo-compression Bonder.

The Datacon system enables the dispense of dots of polymer glue and then aligns and mounts the components under a combination of heat and pressure. This process can be performed at a rate of approximately 2000 units per hour with a placement accuracy of better than five microns in 3 Sigma. Figure 4 shows a typical wafer with approximately 500 die chips mounted on it. This wafer is an experimental structure with large spacing between wafer cells to facilitate the use of different sized bonding heads and to minimise die utilisation during the evaluation experiments. Figure 5 shows a close up an assembled system.

Figure 4 Assembled Wafer and Die Components Connected On A 200mm Wafer

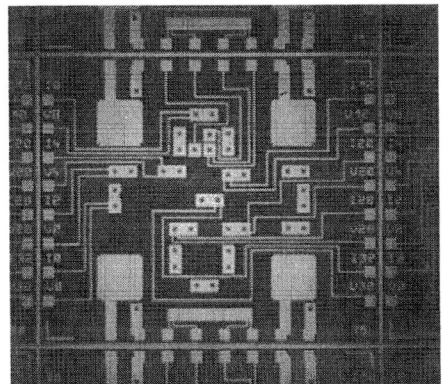

Figure 3a - The Wafer Cell

Figure 5 Close Up Of An Assembled System

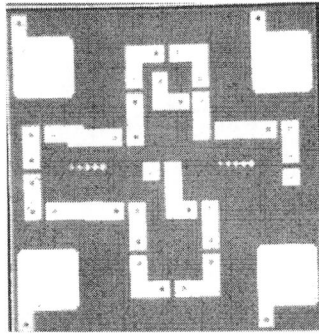

Figure 3b -The Die Cell

When these two components are joined together the assembled test structure enables the measurement of contact resistance, local dielectric thickness, die planarity and reliability. The ability to measure these parameters enables considerable process optimisation to be performed.

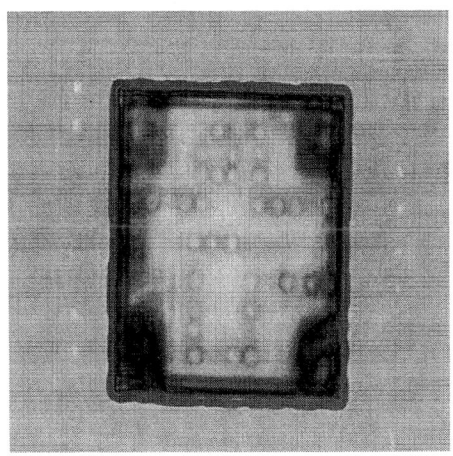

Figure 6 An Acoustic Micrograph Of An Assembled System Showing Contacts To All Connections Between Chip And Wafer

System Interconnect Characterisation

Figure 6 shows an acoustic micrograph of an assembled system. It shows the contacts between the two die and the metal interconnect that permits measurement of the resistance of the daisy chain. In the acoustic micrograph the visible circular regions represent regions where physical and electrical contact is made. Figure 7 shows the acoustic micrograph of a poorly connected device. In this micrograph the dark region in the top right hand corner of the system has been found to exhibit higher resistance than the well connected system shown in Figure 6. This test structure permits the electrical detection of this type of anomaly and can aid in process optimisation.

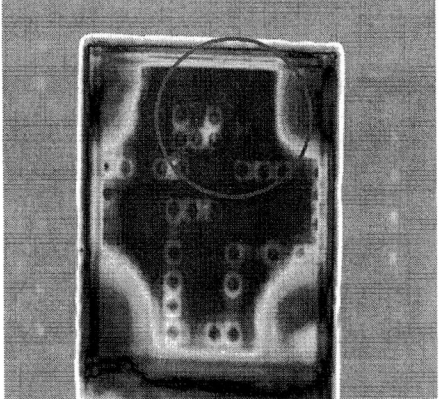

Figure 7 An Acoustic Micrograph Of A Poorly Connected System

Both the wafer cell and die cell contain metal structures which, when assembled, constitute a 32 contact daisy chain, four capacitors, a metal reference resistor, and a Kelvin completed contact resistance measurement structure. The potential applications of this technology add a further constraint to the system in that the impact of different sized contact regions need to be evaluated and optimised for each application domain and current carrying regime. Therefore, several different sized contact regions have been defined and have been evaluated in the course of the work. All of the contact points lie are on the same centres and the metal portions of the contact chain were defined to be 220 μm. This is larger than the largest contact region used in the study (200um square) and its boundaries lie well within the placement capabilities of the Datacon system.

By evaluating the resistance of the 1300 μm long metal resistor R1 on each chip, the local series resistance of the interconnection regions can be extracted from the structure using simple Kelvin completed IV measurements. All resistance measurements are made using this type of four terminal measurement because they remove the series resistance of the wires to the test structure. However, since the contacts and their associated interconnect are all lumped together as one effective resistance, they do not remove the resistance of the metal in the test structure.

A thirty two element daisy chain was chose to represent the largest number of contacts required for smart card applications whilst also addressing the requirements of larger, more surface intensive application domains, by evaluating the performance of the interconnection mechanism at various locations on the interfacial plane between the wafer and die surfaces. It can be seen from figure 3a and b that the contact chain facilitates the evaluation of peripheral and internally located connections between chips

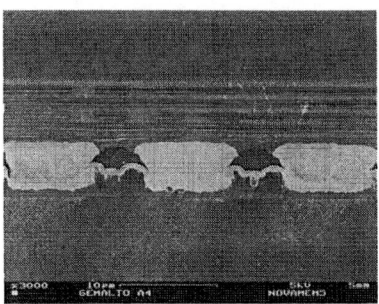

Figure 8 A Cross Section Through A Single 50 μm Contact Point.

. This contact chain is also divided into segments that permit the localisation of any failure point in the daisy chain. The test chip is configured in such a way that all of the interconnections between die and wafer cells are at the same locations on both circuits and the area of interconnection region can be varied, on different die on the system wafer. Figure 6 shows the different tapping points on the daisy chain, the location, and designation of the capacitors in each corner of the die and the 1300mm long resistor (R1). In this experimental structure tap points, metal lengths and typical values of resistance measured between connections for a 50 μm square contact region are shown in Table One.

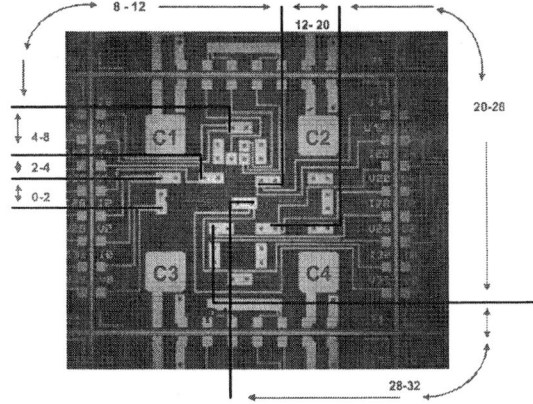

Figure 9 – The Locations Of The Tapping Points On The Contact Chain, The Capacitor Structures And The 1300 mm Resistor (R1)

Once the metal interconnection resistance has been extracted from the measurements an estimation of the resistance introduced by each contact can be made. The mean value of a single contact extracted from the 32 contact daisy chain was found to be 8mΩ and the resistance of the contact extracted from measurement of individual contributions was found to be 7mΩ. These values are consistent with each other. Since there is only one 32 contact daisy chain per chip and seven measurements per chip of the individual contributions, it is possible that the availability of more data for the individual contributions has reduced the mean value slightly.

120

Contact Point	Length (mm)	No Contacts	Mean Measured Resistance (Ω)
0-32	15069	32	1.33
0-2	677	2	0.06
2-4	1097	2	0.07
4-8	1759	4	0.15
8-12	2153	4	0.21
12-20	3310	8	0.29
20-28	4058	8	0.38
28-32	2015	4	0.16

Table 1 Number of Contacts Measured In Daisy Chain

Thermal Cycling Experiments

Figure 10a and 10b show a wafer map as its associated value histogram for the resistances measured of die mounted in a 10 x 10 matrix on a 200 mm silicon wafer. In the case of interconnections between chips mounted face to face an acceptable definition of parametric yield which is one which provides an acceptably low resistance for the contact. Not shown in Figure 10a there are several structures which exhibit resistances in excess of 10 ohms and these need to be filtered out to provide a good representation of the results and enable interpretation of the observed data. Figure 10a Shows the histogram of the data using a 10Ω filter and Figure 11a represents it using a 3Ω filter. Table Two shows the yield, mean and standard deviation of the data presented in figure 10a with different values of filters which permit the definition of data taken into account. Figure 11a and 11b show the same data as that presented in Figure 10 with the 3Ω filter in place. These figures indicate that the resistance of the 32 contact daisy chain inside the 3 ohm window is quite well represented using a Gaussian distribution.

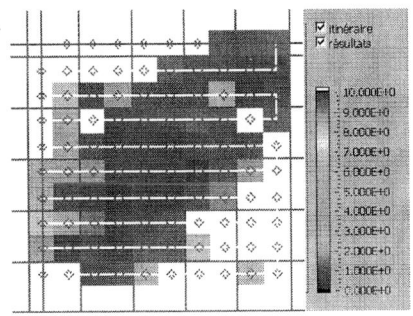

Figure 10a A Wafer Map With 10Ω Filter In Place

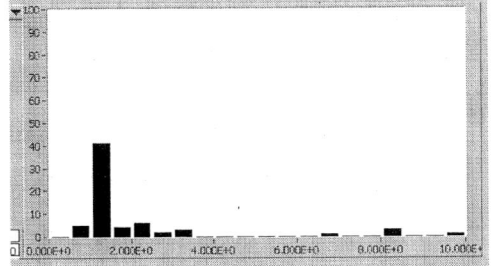

Figure 10b The 10Ω Histogram Showing a Skewed Distribution Of Resistances Towards ~2Ω

Max Resisitance (Ω)	Yield (%)	Mean (Ω)	Std. Deviation (Ω)
<10	66	1.95	1.92
<9	62	1.83	1.67
<8	62	1.52	0.92
<7	61	1.44	0.63
<6	61	1.43	0.63
<5	61	1.43	0.63
<4	61	1.43	0.63
<3	58	1.33	0.47
< 2	50	1.17	0.21

Table Two Mean Measured Values Of Contact Chain Resistance With Different Value Filters

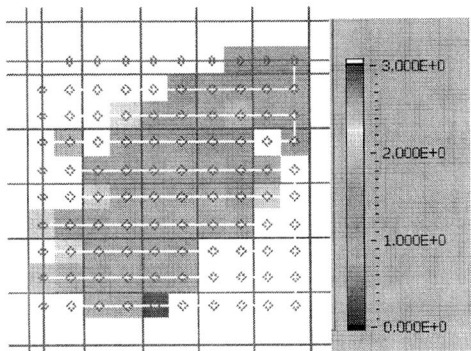

Figure 11a The Wafer Map With 3Ω Filter In Place

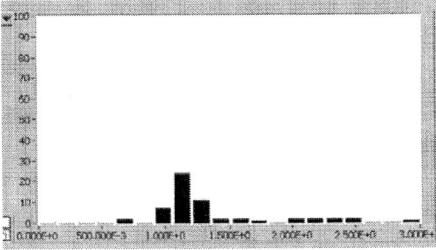

Figure 11b The 3 Ω Filtered Data Histogram Indicating A Gaussian Distribution of Data

This wafer was then cycled between -40°C and + 85 °C (which is the standard required for smart card applications [10]) with the 3Ω data map being measured at intervals during the cycling period. Table 3 presents the resistance of the 32 contact daisy chain (Res 32), the observed standard deviation of structures measured (SD32) as well as the extracted resistance of the metal interconnections between all of the contacts (RM) and the average contact resistance of a single contact in the daisy chain (CR1) as a function of number of thermal cycles that the wafer was exposed to

From Table Three it is evident that the mean value of resistance of the 32 element contact chain is not changing significantly as a function of thermal cycling. However, the standard deviation of these resistance measurements is increasing. This indicates that, while, on a macroscopic level, the contact chain is not particularly sensitive to thermal cycling, there is an effect in place which seems to create some dispersion in the results as a function of temperature. Inspection of the extracted metal interconnect resistance shows that the resistance of the tracks are, in the main, decreasing.

No cycles	Res 32 Ω	SD32 Ω	RM Ω	CR mΩ
0	1.33	0.43	0.98	8
5	1.34	0.45	0.98	11
13	1.29	0.45	0.90	9
33	1.35	0.51	0.87	13
486	1.35	0.60	0.89	14

Table Three Measurements Of Average Resistances From Thermal Cycling Measurements

However, the contact resistance at each contact is increasing. The decrease in resistance of the metal tracks is possibly a function of metal annealing and changes of state (possibly grain size growth) in the Al interconnect[11] while the increase in contact resistance can be attributed to changes in state of the plated Ni inserts or Al/Ni inter diffusion or the formation of higher resistance $NiAl_3$ intermetallics at the interface[12]. It is interesting that there is a reduction in all of the resistances after 13 cycles , this could be an artifact of the measurement, a statistical effect or a real physical phenomenon. At this stage, it is not clear which is the case. However, additional wafers are currently being tested to reveal whether this is really a physical phenomenon or an artifact of the measurement or analysis.

Conclusions

Microinsert technology is a very interesting approach to use as the interconnection medium between two separate systems connected together face to face. It is reliable and has few of the constraints of conventional flip chip mounting including the fact that no under bump metallization (UBM) is required and only one die needs to be treated at wafer level using a commonly accepted electrolytic plating process to provide the bumps and the re routing on the chip surface.

Limited thermal cycling of interconnected structures does not seem to have a significant impact on the 'ensemble' resistance behavior of the interconnected devices. However, there are changes observed in both interconnect resistance and contact resistance which seem to trade off against each other. This increased value of contact resistance is considered to be acceptable within the scope of chip to chip interconnection when coupled with the reduction in interconnection metal resistance although the reason for this is unclear at the moment. It seems to be a genuine effect and more thermal cycling experiments are underway to elucidate the reasons for this behavior. A broader range of temperatures (-40 to 150°C) is also under evaluation to determine whether these observed phenomena are a function of thermal exposure or if they are specific to the test regime used in these experiments.

Acknowledgments

The Pidea+ project 04.160 SM@RTPACK is gratefully acknowledged by the authors for their support of this work.

References

1] Eric Laine, Klaus Ruhmer, Eric Perfecto, Hai Longworth, David Hawken C4NP as a High-Volume Manufacturing Method for Fine-Pitch and Lead-Free FlipChip Solder Bumping Proc. Electronics Systemintegration Technology Conference Dresden 2006 pp 518-524]

2] A.Mathewson, et al Microstructured Interconnections For High Security Systems. Proc IEEE Electronic Systems Integration Conference (ESTC) Dresden 2006.

3] A.Mathewson°, J. Brun°, G. Ponthenier°, R.Franiatte°, A.Nowodzinski°, N.Sillon° G;Poupon°, F. Depoutot†, B.Dubois-Bonvalot† Detailed Characterisation of Ni Microinsert Technology For Flip Chip Die on Wafer Attachment Proc. 57 th IEEE Electronic Components and Technology Conference (ECTC) Reno NV. 2007

4] Low-Cost Wafer Bumping. P. A. Gruber, L. Be´ langer,G. P. Brouillette,D. H. Danovitch,J.-L. Landreville,D. T. Naugle,V. A. Oberson,D.-Y. Shih,C. L. Tessler, M. R. Turgeon IBM J. RES. & DEV. VOL. 49 NO. 4/5 JULY/SEPTEMBER 2005 pp622-639

5] Lou Nicholls, Robert Darveaux, Bernt Hansen, Chuck Carey, Toyohiro Aoki, Toshiharu Akimoto, & Jason Chang Fine Pitch (150µm) Pb-free Flip Chip Bumping & Packaging. Electronic Components and Technology Conference 2006 ECTC pp 131-138

6] Rao R. Tummala, P. Markondeya Raj, Ankur Aggarwal, Gaurav Mehrotra, Sau Wee Koh, Shubhra Bansal, Tan Teck Tiong , CK Ong, Jimmy Chew, Kripesh Vaidyanathan, Vempati Srinivasa Rao. Copper Interconnections for High Performance and Fine Pitch Flip-Chip Digital Applications and Ultra-miniaturized RF Module Applications IEEE ECTC 2006 pp 102-111

7] Naoya Watanabe and Tanemasa Asano A Large Number of I/O Connections Using Compliant Bump. Electronic Components and Technology Conference ECTC 2006 pp 125-130

8] P.Triboulet Vision IR: du composant à l'image - HgCdTe Technology In France. CR Physique 4 2003 pp 112-1131

9] DOW Cyclotene (BCB) http://www.dow.com/cyclotene/

10] MIL – STD -883D – 1011.6

11] K.Kawakita Annealing Effects In Aluminum Films Vacuum-Deposited On Room-Temperature Substrate Japanese Journal of Applied Physics, Volume 25, Issue 3, pp. 366 (1986)

12] E.Ma and M-A Nicolet andM.Nathan $Nial_3$ Formation In Al/Ni Thin Film Bilayers With And Without Contamination J.Appl Phys 65 (7) 1April 1989 pp 2703-2709

2008 IEEE Conference on Microelectronic Test Structures, March 24-27, Edinburgh, UK

An Evaluation of Test Structures for Measuring the Contact Resistance of 3-D Bonded Interconnects

H. Lin, S. Smith, J.T.M. Stevenson, A.M. Gundlach, C.C. Dunare, and A.J. Walton

Institute for Integrated Micro and Nano Systems, Part of the Institute for Integrated Systems
School of Engineering and Electronics, The University of Edinburgh
Kings Buildings, West Mains Road, Edinburgh, EH9 3JF, U.K. Email: h.lin@ed.ac.uk

Abstract—This paper evaluates test structures designed to characterise electrical contacts between interconnect on bonded wafers. Both simulation and experimental measurements are used to explore the capability of a stacked Greek cross type test structure to extract the contact resistivity (ρ_c) between two bonded conductive layers. It is concluded from the simulations and actual electrical measurements of the benchmark Kelvin structures that the stacked Greek cross can only be used where there is a relatively high specific contact resistivity. For the structures evaluated in this study, this was found to be greater than $\rho_c \geq 9.0 \times 10^{-7} \Omega \cdot \text{cm}^2$.

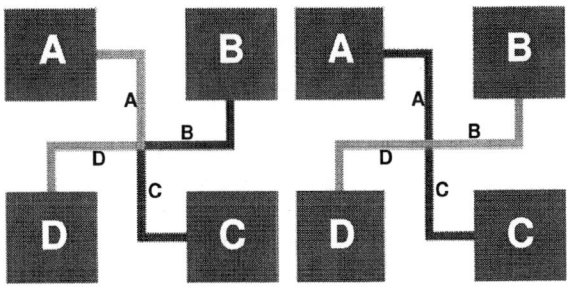

(a) Kelvin test structure   (b) Stacked Greek cross

Fig. 1.   Layouts of test structures for the electrical characterisation of the wafer bonding process

I. Introduction

Electrical interconnect between wafers for the integration of MEMS and CMOS devices or the construction of three-dimensional ICs [2] can be achieved by using a low temperature wafer bonding process [1]. Low contact resistivity (ρ_c) is clearly desirable for all applications, therefore, appropriate test structures for ρ_c measurement are essential for characterising and optimising different fabrication technologies.

This paper presents test structures designed to characterise the electrical interconnects between bonded wafers. Two types of test structures: Kelvin structures and stacked Greek cross test structures have been fabricated using a two-level aluminium metallisation process. These two architectures have also been evaluated using simulation software, and the results compared with electrical measurements.

II. Test structures

Kelvin test structures [3], [4] (Fig. 1(a)) and stacked Greek cross structures [5] (Fig. 1(b)) have been considered as candidates for measuring the contact resistivity between two metal layers connected by the wafer bonding process.

A. Kelvin test structure

An ideal Kelvin structure consists of two L-shaped lines (with line-width W) in separate conducting layers. When these layers are perfectly aligned, they form a contact area of A_c ($A_c = W^2$), as illustrated in Fig. 1(a). The contact resistance is determined by forcing a current from arm C to arm A (I_{AC})

and measuring the potential difference between B and D. The specific contact resistivity is then calculated using

$$R_c = \frac{(V_B - V_D)}{I_{CA}}, \qquad \rho_c = A_c R_c \qquad (1)$$

The accurate extraction of ρ_c depends upon the contact area A_c being known precisely. However, this is not the case in practice because undesirable misalignment will always be present during the lithography step.

In order to avoid the contact area being a function of misalignment, the basic Kelvin test structure has been modified to minimise the inevitable misalignment problem from the lithography process [3], [4]. Depending on the arm width and the via size of the test structure, modified Kelvin structures can be categorised into L-shaped and D-shaped, as shown in Fig. 2. The addition of collars solves the misalignment issue during the fabrication, but they cause lateral current spreading in the structure during the electrical measurement [6]. Fortunately, these errors can be calculated through simulation and the universal error correction curves are available in the literature [7], [8]. However, the misalignment issues associated with wafer bonding are significantly greater than with conventional IC technology and this makes the Kelvin structure less attractive. Another issue of the L-shaped and D-shaped Kelvin structures is that the contact area is defined by the via size, which is less straightforward to define in a wafer bonding process.

978-1-4244-1801-5/08/$25.00 ©2008 IEEE   123

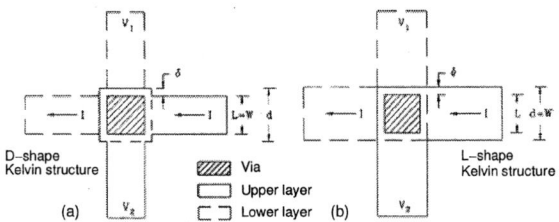

Fig. 2. Kelvin structures for the measurement of specific contact resistance between two metal layers [7], [8]: (a) D-shaped Kelvin structure, and (b) L-shaped Kelvin structure.

B. Stacked Greek cross structure

In order to overcome the issues associated with the use of Kelvin structures to extract contact resistance between interconnect in a wafer bonding process, a test structure, which will be referred to as the stacked Greek cross, has been proposed [5] and is illustrated in Fig. 1(b). This paper does not justify the measurement and associated extraction of contact resistance, which is identical to that used for the Kelvin structure. However, the fabrication sequence of the stacked Greek cross structure is suitable for the wafer bonding process [1].

In this structure, two metal lines (with line-width W) are orthogonally bonded to create a contact area $A_c = W^2$. As the two tracks are orthogonal, which can normally be satisfied in fabrication, the resulting contact area A_c is constant regardless of the lateral and vertical misalignments. When measuring this test structure, the "contact resistance" is determined by forcing a current from pad C to pad D (I_{CD}) and measuring the potential difference between B and A. The specific contact resistivity can be calculated using

$$ R_c = \frac{(V_B - V_A)}{I_{CD}}, \qquad \rho_c = A_c R_c \qquad (2) $$

However, unlike the Kelvin structure which essentially has a 1-D current flow, in this structure the forced current turns 90° when it passes through the contact interface from one metal layer to another as shown in Fig. 1(b). Hence there is current crowding at the corner of the contact region, which results in a non-uniform current distribution at the contact interface. Consequently it can be expected that the resistance derived from this test structure will not extract the same contact resistance value as the Kelvin structure.

C. Test chip

Structures with contact line widths of 3, 5, 7, 10 and $14\mu m$ have been implemented on the test chip, which results in contact areas of 9, 25, 49, 100 and $196\mu m^2$, respectively. For the L-shaped and D-shaped Kelvin structures, a fixed collar of $2\mu m$ has been used.

III. FABRICATION OF THE TEST STRUCTURES

A two-level Al metallisation process illustrated in Fig. 3 which included CMP (chemical mechanical planarisation) and a thick Al lift-off technique was used to evaluate the above test structures for contact resistivity measurements.

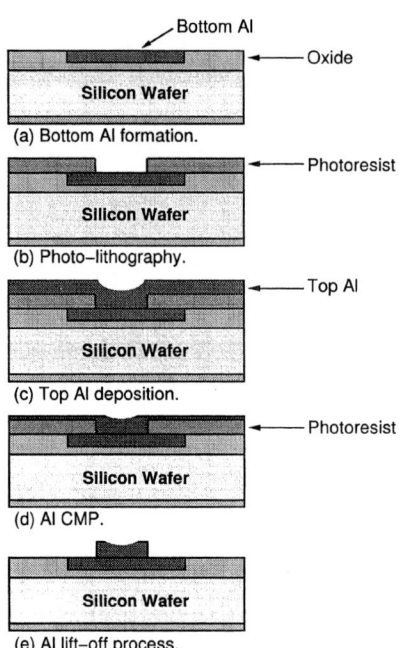

Fig. 3. Two-level Al metallisation process for the fabrication of Kelvin structures and stacked Greek cross structures.

A standard Al damascene process has been used to create the bottom metal tracks and a planar wafer surface, Fig 3(a). The following lithography step defined contact vias and the patterned photoresist is used as a shadow mask for the subsequent Al deposition, as shown in Fig. 3(b). The top Al layer ($\sim 1.0\mu m$) is then deposited (Fig. 3(c)) and followed by Al CMP (Fig. 3(d)). The Al CMP is completed when the Al thickness on top of photoresist becomes thin enough ($\sim 0.5\mu m$) for the lift-off process to work. Assisted by megasonic agitation, the photoresist dissolves in the acetone solvent, and the thin Al film on top of the photoresist is removed, as shown in Fig. 3(e).

N-type (100) silicon wafers with a resistivity of 1-10Ωcm were used in this study. In the Al damascene process, a 2.2μm thermal oxide layer was grown at 1100°C, followed by a 1.1μm deep trench etching using PlasmaTherm PK 2440 RIE (reactive ion etching) etcher. An Al layer with a thickness of 1.2μm was deposited on the patterned wafers using Balzers BAS450M sputterer. The Al damascene process was then performed using a Presi E460 wafer polisher which generated a surface with Al dishing of 20-40nm and bottom Al thickness of 0.9-1.0μm after the polishing step. Negative photoresist AZ 5214E was spun onto the polished wafers for the subsequent lift-off process. The top Al layer (1.0μm) is then deposited and followed by Al CMP. The challenge was to reduce the thickness of the Al layer without degrading the soft photoresist underneath. A delicate, low-down-force and high-rotation-speed Al CMP process has been developed for reducing the thickness of the Al layer at a reasonable polishing rate, while keeping the underlying photoresist intact. Once the thickness of the top Al is reduced to $\sim 0.5\mu m$, lift-off becomes relatively straightforward using acetone and megasonic agitation. The fabricated D-shape Kelvin structure and a stacked Greek cross structure after the lift-off step are shown in Fig. 4.

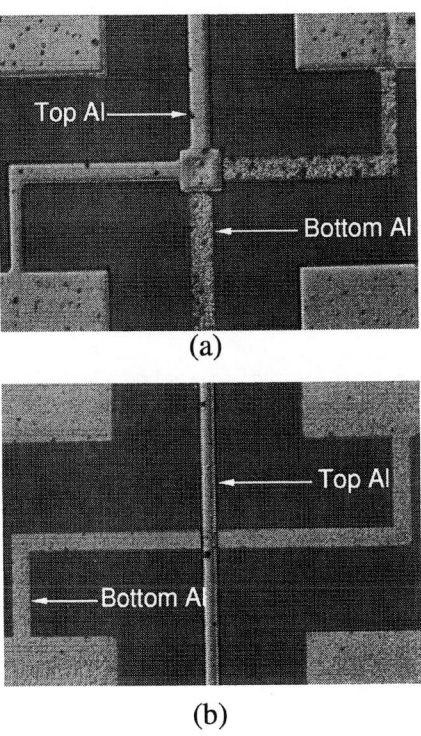

(a)

(b)

Fig. 4. Optical microscope images of (a) D-shape Kelvin structure and (b) stacked Greek cross structure after the Al lift-off process.

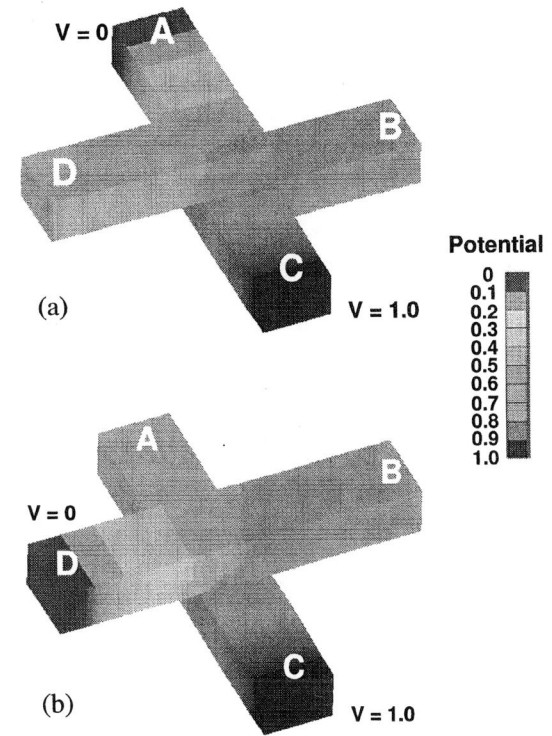

(a)

(b)

Fig. 5. Potential distribution in a stacked Greek cross structure during the simulation of contact resistance extraction. (a) ideal Kelvin test structure, and (b) stacked Greek cross structure.

IV. SIMULATION OF THE TEST STRUCTURES

The contact resistance between two conductive layers has been modelled using the numerical analysis software, Raphael [9], from Synopsys. Raphael is a three-dimensional simulation program for solving Poisson's equation, and is based on the finite-difference method with an automatically adjustable rectangular mesh. It has been used to model both the ideal Kelvin structure and stacked Greek cross structure, as shown in Fig. 5. The attraction of performing these simulations is that it enables the input of a known contact resistance and, if this value can be extracted from the simulated voltages and currents, this confirms the measurement approach and the associated extraction strategy.

For the Kelvin structure (Fig. 5(a)), a potential of 1V was applied to the lower arm C and 0V set on the upper arm A, which results in a current flowing from C to A through the contact area. The potential difference between arms B and D can be extracted from the simulation and this enables the contact resistance (R_k) to be calculated using

$$R_k = \frac{(V_B - V_D)}{I_{CA}} \quad (3)$$

A similar procedure was applied to the stacked Greek cross structure shown in Fig. 5(b) in order to extract the contact resistance R_g using

$$R_g = \frac{(V_B - V_A)}{I_{CD}} \quad (4)$$

It should be pointed out that there is an interface with *infinitesimal* thickness and variable contact resistivity when the conductive layers contact to each other. However, this resistive interface has not been included in the simulation presented above, and also, it can not be directly set within Raphael. Hence, a resistive layer with a *finite* thickness has to be defined as the contact interface, which, inevitably introduces a potential error to the result. To minimise this error, the thickness of the resistive layer was set to be as thin as possible within the simulation program's limitation.

In Raphael, the resulting contact resistivity between two conductive layers is determined by the thickness and bulk resistivity (ρ) of the contact resistive layer. Since the thickness of this layer was set to be the thinnest within the program's limitation (0.05μm), the ρ value became the sole parameter to determine the contact resistivity in the Raphael simulation. The ability to control the contact resistivities between two conductive layers in the simulation enables many useful studies.

V. ELECTRICAL MEASUREMENTS AND SIMULATION RESULTS

The fabricated Kelvin structures and stacked Greek cross structures were both measured using a HP4062B and a Solartron 7065 voltmeter by forcing a current of 100mA from the top probe pad to bottom pad and measuring the voltage drop across the contact area. The contact resistances as a function of contact area for each structure before the Al sintering process are plotted in Fig. 6. It should be noted that the contact resistances of Kelvin structures were the average values from both L-shape and D-shape structures and revised using error correction curves. The average specific contact resistivity of $2.008 \times 10^{-8}\Omega\cdot\text{cm}^2$ was extracted from the Kelvin structures and $3.556 \times 10^{-8}\Omega\cdot\text{cm}^2$ from the stacked Greek cross structure. Hence, before the Al sintering process, the contact resistances extracted from the stacked Greek cross were 70% higher than the results from the Kelvin structure.

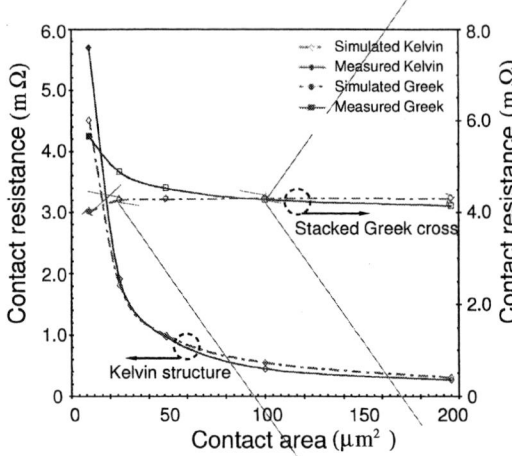

Fig. 7. Contact resistance measurements and Raphael simulation results for the Kelvin structures and stacked Greek cross structures after the Al sintering process.

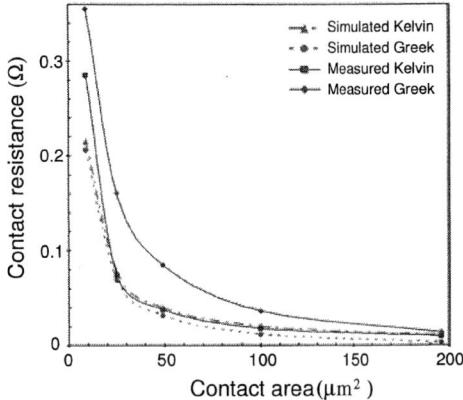

Fig. 6. Contact resistance measurements and Raphael simulation results for the Kelvin structures and stacked Greek cross structures before the Al sintering process.

The specific contact resistivity (ρ_c) between two Al layers can be reduced by employing a standard Al sintering process at 435 °C with H_2 (5%) and N_2 (95%) for 15 minutes. The measured contact resistances after this process are plotted in Fig. 7. The average ρ_c of $4.850 \times 10^{-10}\Omega\cdot\text{cm}^2$ was extracted from the Kelvin structures, which is almost two orders of magnitude smaller than that before the Al sintering process. However, it can be observed that for the stacked Greek cross structure the extracted contact resistance became independent of the contact areas. This indicates that this structure is not providing information about the contact resistance, but is measuring the sheet resistance of the interconnect at the contact region. Hence, in this paper we have identified this structure as a "stacked Greek cross" as in this regime it measures the sheet resistance of the stack of the two layers of metal in the contact region.

It is worth noting that the increase of the contact resistance value for the small geometries for the stacked Greek cross test structure could be caused by self-heating during the measurement or Al dishing during the processing of the bottom metal. Further work is required to confirm if either of these factors have any effect on the extracted value.

The electrical measurement data from Kelvin structures was used as a reference to determine input parameters of the resistive contact layer in the simulation. The same parameters were used for the stacked Greek cross structure. Following the simulations, these results and the actual electrical measurements can be compared, and this data is presented in Fig. 6

and Fig. 7.

It is well-known [3], [4], and has been once again confirmed by this work, that a Kelvin structure can be used for measuring ρ_c in all ranges when error correction curves are used. However, the electrical measurements and simulation results have suggested that the stacked Greek cross structure provides some information about the contact interface when the contact resistivity is high (e.g. $\rho_c \approx 2.0 \times 10^{-8}\Omega\cdot\text{cm}^2$), but is unable to do so at low contact resistive conditions (e.g. $\rho_c \approx 5.0 \times 10^{-10}\Omega\cdot\text{cm}^2$).

VI. DISCUSSION

A. Validation of the stacked Greek cross structure

The comparison between an ideal Kelvin structure and stacked Greek cross structure at various contact conditions is also useful for providing information about the usefulness of the stacked Greek cross test structure for different interfacial specific contact resistivity. In the Raphael program, resistive interfaces are created to mimic various contact conditions by varying the resistivity of the insertion layer. The respective contact resistivity ρ_k and ρ_g can then be extracted by using an identical contact resistance for both structures. The comparison is presented in Fig. 8 using a normalised contact resistivity (ρ_g/ρ_k).

It can be seen from Fig. 8 that ρ_g/ρ_k is close to 1 when ρ_k is in the range of 9.6×10^{-6} to $9.6\times10^{-7}\Omega\cdot\text{cm}^2$. This indicates that the stacked Greek cross structure measurement extracts the same R_c as the Kelvin structure when it is used for measuring a high resistive interface. However, as ρ_k decreases, there is a significant difference between the R_c values extracted from the two structures. Since the ideal Kelvin test structure is capable of extracting contact resistivity for all ρ_c, it is clear that the stacked Greek cross structure is not suitable for measuring interfaces with low contact resistivity. For the stacked Greek cross structure, the simulated R_g eventually becomes independent of the contact area when the interface contact resistivity is in the range of $5.0\times10^{-10}\Omega\cdot\text{cm}^2$. From these Raphael simulation results, it can be concluded that the stacked

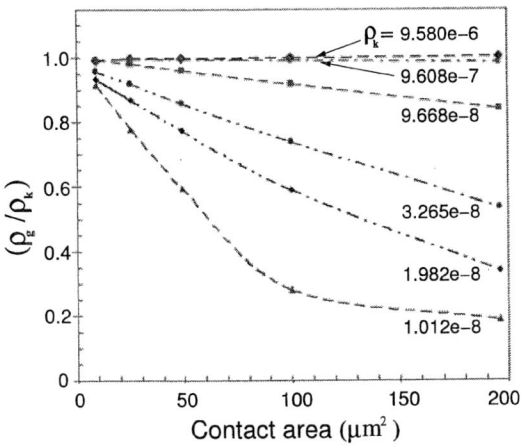

Fig. 8. Comparison of extracted contact resistance of the ideal Kelvin and a stacked Greek cross structures using Raphael.

Greek cross can be used for contact resistivity measurement when ρ_c is greater than $9.0 \times 10^{-7} \Omega \cdot cm^2$. However, it can only be used as an indicator of the interconnect quality when the ρ_c is less than that critical value.

B. Stacked Greek cross vs planar Greek cross structure

For the stacked Greek cross structure with a low resistive interface (e.g. $\rho_c < 5.0 \times 10^{-10} \Omega \cdot cm^2$ in Fig. 7), the extracted contact resistances (R_g) are independent of the contact area, but there is a relationship with the film thickness. More simulations were conducted for stacked Greek cross structures with $R_c = 0$, and the results compared with those from the planar Greek cross structure, in Fig. 9. The inverse proportional of the resistance and film thickness indicates that the stacked Greek cross shows the same characteristics to that of a planar Greek cross structure which is widely used for the sheet resistance measurement of metal films [10]. This is not surprising as the stacked Greek cross approximates to a planar Greek cross with double the thickness in the centre of the cross.

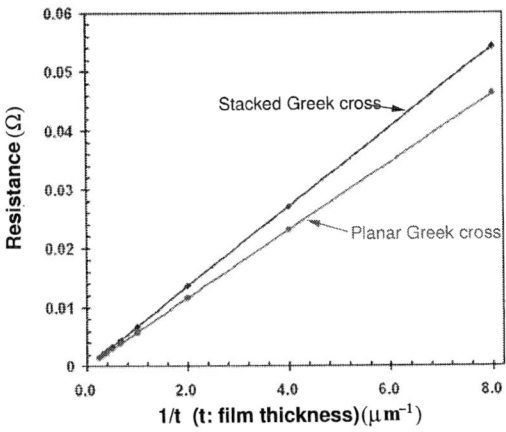

Fig. 9. Comparison of extracted sheet resistance for the stacked Greek cross ($\rho_c = 0$) and planar Greek cross structures. These results were simulated with Raphael.

VII. CONCLUSIONS

Kelvin and stacked Greek cross test structures have been compared using both simulation and experimental measurements. In particular their abilities to extract contact resistance between two conductive layers have been examined. It can be concluded that the stacked Greek cross can only be used to identify the existence of a highly resistive contact interface, such as $\rho_c \geq 9.0 \times 10^{-7} \Omega \cdot cm^2$, when Al is used as a conductive material.

VIII. ACKNOWLEDGMENTS

The authors would like to acknowledge Applied Materials, EPSRC, and the Edinburgh Research Partnership for financial support, and the staff in Scottish Microelectronics Centre for their technical assistance.

REFERENCES

[1] H. Lin, A.J. Walton, C.C. Dunare, J.T.M. Stevenson, A.M. Gundlach, S. Smith and A.S. Bunting, "Test Structures for the Characterisation of MEMS and CMOS Integration Technology", in *Proc. IEEE International Conference on Microelectronic Test Structures*, USA, 2006, pp. 143-148.

[2] R. S. Patti, "Three-dimensional integrated circuits and the future of system-on-chip designs", *Proceeding of the IEEE*, Vol. 94, pp. 1214-1224, 2006.

[3] S. J. Proctor, L. W. Linholm, J. A. Mazer, "Direct measurements of interfacial contact resistance, end contact resistance, and interfacial contact layer uniformity", *IEEE Transactions on Electron Devices*, Vol. 30, pp. 1535-1542, 1983.

[4] W. M. Loh, S. E. Swirhun, T. A. Schreyer, R. M. Swanson, K. C. Saraswat, "Modelling and measurement of contact resistances", *IEEE Transactions on Electron Devices*, Vol. 34, pp. 512-524, 1987.

[5] K.N. Chen, A. Fan, C.S. Tan, R. Reif, "Contact Resistance Measurement of Bonded Copper Interconnects for Three-Dimensional Integration Technology", *IEEE Electron Device Letters*, Vol. 25, no. 1, pp. 10-12, 2004.

[6] R. L. Gillenwater, M. J. Hafich, G. Y. Robinson, "The effect of lateral current spreading on the specific contact resistivity in D-resistor Kelvin devices", *IEEE Transactions on Electron Devices*, Vol. 34, pp. 537-543, 1987.

[7] J. Santander, M. Lozano, C. Cané, "Accurate extraction of contact resistivity on Kelvin D-resistor structures using universal curves from simulation", *IEEE Transactions on Electron Devices*, Vol. 40, pp. 944-950, 1993.

[8] J. Santander, M. Lozano, A. Collado, M. Ullán, E. Cabruja, "Accurate Contact Resistivity Extraction on Kelvin Structures with Upper and Lower Resistive Layers", *IEEE Transactions on Electron Devices*, Vol. 47, pp. 1431-1439, 2000.

[9] Raphael - Interconnect Analysis Program - Reference Manual, Synopsys.

[10] A.J. Walton, "Microelectronic Test Structures", *Semicon '97*, 1997.

SESSION 6
Yield and Reliability II

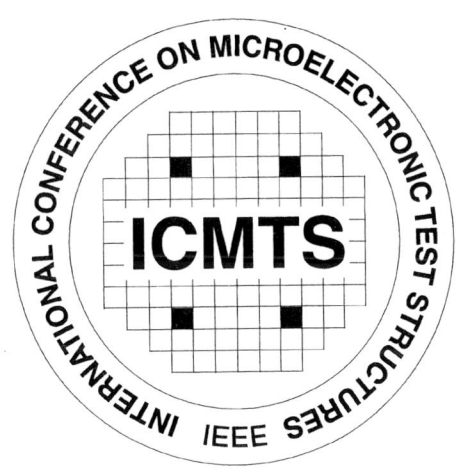

March 26, 10:50–12:10

Co-Chairs: Kiyoshi Takeuchi, *NEC Corp., Japan*

Alan Mathewson, *Tyndall National Institute, Ireland*

High Density Test Structure Array
for Accurate Detection and Localization of Soft Fails

Christopher Hess, Michele Squcciarini, Shia Yu, Jonathan Burrows
Jianjun Cheng[1], Ron Lindley[1], Andrew Swimmer[1], Steven Winters[1]

PDF Solutions Inc., San Jose, CA 95110, USA
Phone: +1-408-938-6436, FAX: +1-408-280-7915, Email: hess@pdf.com
[1]PDF Solutions Inc., San Diego, CA 92128, USA

Abstract – **To resolve performance yield issues it is required to detect and localize soft fails such as a contact having 500ohms instead of its nominal 50ohms. Soft fails can only be detected within very small test structures, which requires an array design to efficiently use the area of test chips. Here we present a novel High Density Test Structure Array, which will enable accurate 4 terminal measurements of 1000 or more very small devices under test (DUT) within each array. On average, only 2 selection devices are required per DUT, which will provide outstanding utilization of the test chip area. Experimental results reveal that within this array traditional test structures can be used beyond their intended purpose to detect additional defect types, which opens the door to significant reduction of overall mask and wafer consumption.**

1 Introduction

Test structures such as via or contact chains, snakes and combs are commonly used to detect defects [IpSa77], [Bueh83]. Four Terminal Resistance Measurements are used to measure the resistance or leakage most accurately [Keit04]. To do so, these test structures are usually connected to individual terminals or pads for testing. However, more advanced design rules require far more than 1000 differently designed test structures per layer to achieve yield and performance characterization, which would require thousands of pads making this impractical.

To keep the pad area overhead low, a typical method is to use large area test structures to detect completely failing test structures since its measured resistance is many orders of magnitude different from its expected value. However, at advanced technology nodes so called "soft failing" events have become more severe in semiconductor manufacturing. A soft fail is present, if only a small fraction of a test structure changes its local resistance by as little as one order of magnitude compared to its expected value. Detecting such soft fails requires access to smaller test structure segments, since a soft fail will not be detectable within a single large area test structure.

It is difficult if not impossible to place all those thousands of small test structures on a single test chip if they are each connected to individual pads for testing, since there is not enough area to place all those pads. Several methods of pad sharing are used to address and access individual test structures or devices under test (DUT) by significantly reducing the number of pads required for testing.

Purely passive arrays have been described, which do not need devices to address different DUTs. To do so, single vias have been placed in a passive array where each via is connected to a unique set of two lines as described in detail at [WWRH89]. A similar array is described by [HSWM02], which places long via chains in an array to break down the huge chain resistance for testing. However, all test structures within those arrays have to be identical and the array size is relatively small (typically around 8 rows by 8 columns) to maintain a reasonable signal to noise ratio during testing. Also this method is restricted to detect open fails. A method to permute pairs of lines has been described at [HeWB97] making it possible to place many test structures within a reduced number of pads to investigate short circuits. Unfortunately, it is limited to short circuits only and there are problems in disentangling multiple fails within the same array as reported at [HeWe98].

The passive arrays mentioned above can be expanded by adding transistors, which are placed in serial connection to each test structure as described at [WGMS90]. Still, only open fails can be detected. The method of permuting lines has been combined with a diode array [HeWe94] or a simple transistor array [HWLS96] to enable the detection open and short circuits in test structures. Nevertheless, disentangling multiple short circuit fails is still a problem, which limits the number of DOE levels within an array.

To overcome such uncertainties, more complex multiplexer circuitry is required to address passive test structures as described at [WWGH92]. However, this array does not allow one to apply any analog resistance and leakage measurements through the multiplexing circuitry, which is required to characterize so-called soft fails. A more sophisticated multiplexer array is introduced by [DBCH06], using addressable transmission gates. Transistor devices as well as complex test structures with up to 6 terminals can be used as DUTs. Analog measurements can be executed per DUT. However, the large number of addressable transmission gates as well as the complex logic to address them requires a relatively large area compared to the DUT. Typically, the area per addressable DUT is in the order of 100um by 100um. Thus, neither localization nor analysis of soft fails is supported.

Finally, SRAM, ROM or other memory based arrays are used to detect defects [FNMO07], [DGGY05], [KMGS94]. Such arrays are a derivative of existing memory based layouts – in other words they typically start from a well understood circuit design and are then modified for the usage as defect monitor. The periphery circuits like decoders and sense amps are used more or less without changes, while the bit cells are modified in a way that they can be used to detect defects and failures. However, the modification of the bit cell has its limits. For instance, DUTs barely include FEOL layers, since they are used for the bit cell itself. ROM based arrays provide more area for DUTs by only using one selection device per DUT [KFCY06] [KGCY07]. However, even they cannot overcome the ambiguity whether the selection device or the DUT cause a single bit to fail. The ability to assign a measured fail to a specific layer is essential for process characterization and successful yield and performance improvement. Finally, memory based arrays do not provide accurate analog resistance and leakage measurements per DUT.

In summary, existing addressable arrays fail to provide accurate analog measurements of very small DUTs within large arrays. Thus, this paper will introduce a novel High Density Test Structure Array, which will enable accurate 4 terminal measurements of 1000 or more very small devices under test (DUT) within each array.

2 Array Design

The goal is to place many small DUTs in an array as illustrated in Figure 1. In principle there are two approaches to implement such an array. The top down approach starts from a known circuit design modifying it to insert DUTs. Memory based arrays are an example for such top down designs. The bottom up approach starts from best known test structures to detect hard and soft fails (e.g. snakes, chains, combs) and the best known measurement method (Four Terminal Resistance Measurement). Then, circuitry is build around it to reduce pad overhead. In this paper, the bottom up approach was selected to ensure high accuracy measurement results. A total of three basic design steps are implemented to design the High Density Test Structure Array for accurate detection and localization of soft fails:

STEP 1: Starting point is a typical DUT connected to 4 pads to enable a Four Terminal Resistance Measurement as shown on the left of Figure 2. Transistor based transmission gates will be added to the routing path between the DUT and the PADs. This step is a known starting point of existing DUT arrays e.g. at [DBCH06].

STEP 2: Existing arrays typically use large transmission gates and large DUTs. Here we aim to fit much smaller DUTs down to 5um^2, which require less and much smaller selection devices to minimize the area overhead of the selection logic. To be area efficient, the area required for the selection logic should be smaller than the area of the DUT. Thus, design step two will cut the number of selection devices in half by replacing the dual gate transmission gates with just a single transistor as illustrated in Figure 3. This step limits the direction of the measurement current, which is an acceptable trade off for detecting defect.

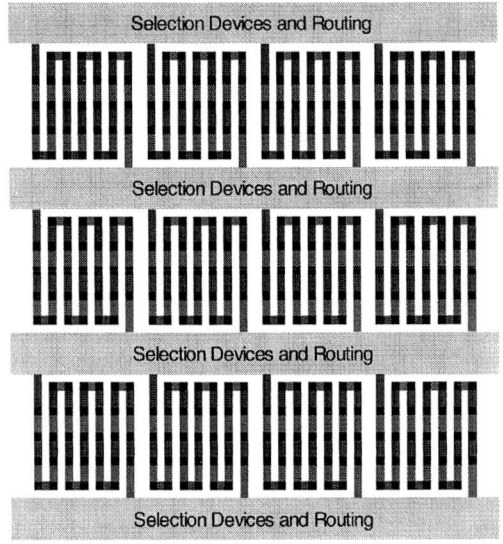

Figure 1: Concept to place small DUTs in an array.

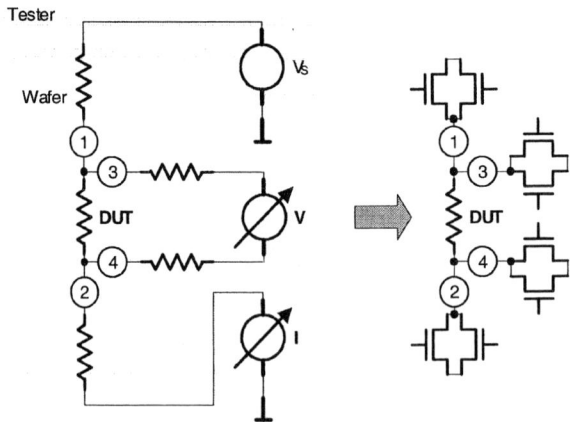

Figure 2: Step 1: Adding transmission gates in the routing path of a 4 terminal resistance measurements.

132

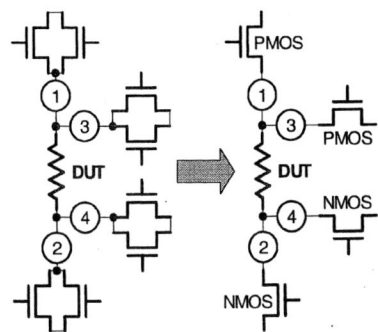

Figure 3: Step 2: Replacement of transmission gates with just ONE transistor.

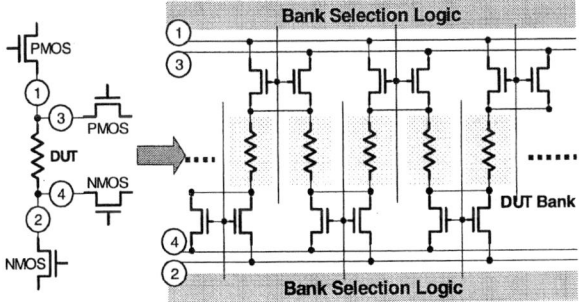

Figure 4: Step 3: Sharing of transistors between neighboring DUTs.

STEP 3: To further reduce the number of selection transistors, always two neighboring DUTs will share two selection devices among them as illustrated in Figure 4. It will connect all DUTs in a large chain, where the individual pairs of selection transistors are taps along the chain of DUTs. Thus, on average, only 2 selection transistors are used per DUT. As comparison, a transmission gate based array requires 8 transistors per DUT and an SRAM based array requires 6 transistors per DUT.

Finally, multiple banks will be arranged within an array in a way that each bank will become one row of the High Density Test Structure Array. Figure 5 shows one DUT within the array. Here a contact chain has been implemented as DUT. In general, any 2 terminal test structures such as snakes, chains or combs can be implemented as DUT.

Combinatorial decoder circuits are implemented to address a particular DUT within the rows and columns of the array. Only simple gates like NAND, NOR and inverters are being used to be robust against any process variability that may cause performance or timing issues. Dependent on the dimensions of the standard cells that are used within the decoder circuitry, each DUT can be as small as "standard cell height" by "standard cell height".

3 Defect Detection Method

In general, one resistance measurement value will be recorded for each DUT within the array. The following steps will interpret the measurement results to detect and classify defects.

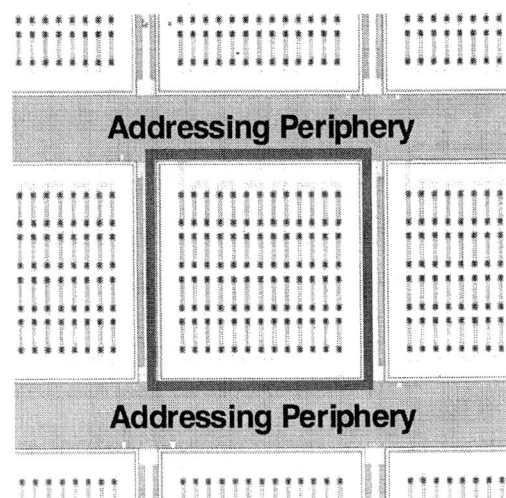

Figure 5: Zoom into the core of the High Density Test Structure Array.

1. Detect all hard fails: Initially, all DUTs with hard fails are detected. A DUT will be flagged as a hard fail, if its measurement value is at least one order of magnitude away from the target measurement value of the DUT. Dependent on the process type, even larger gaps of two or more orders of magnitude can be used.

2. Detect all soft fails: Soft fails are detected among those measurements values that have not previously been flagged as hard fails. Within each array we will group all measurement values coming from identical DUT layout cells. It is expected that the measured resistance follows a normal distribution. A soft fail is defined as a measurement that deviates substantially from the main population. Typically, a gap of 10% is a good starting point to extract random soft fails. Physical failure analysis will help to further optimize the gap value, which depends on factors like DUT type (layout), experiment type, and the overall maturity of the process technology.

3. Separate out circuitry fails: Since the overall real estate is not exclusively dedicated to DUTs only, finite probability of fails among the supporting circuitry exists. They need to be separated from DUT fails to avoid ambiguity to localize them. Usually, circuitry fails have distinct spatial pattern involving multiple DUTs. Thus, non single bit fails of neighboring DUTs are a strong indicator for circuitry related fails. To flag circuitry fails, a bitmapping algorithm is executed over all detected hard and soft fails to identify any non single DUT fails such as row & column fails, partial row & column fails, alternating row & column fails, address block fails etc. In addition, for single bit fails occurring within the array proposed here, it is possible to clearly separate circuitry fails from DUT fails without ambiguity. Since selection devices of one specific DUT are shared with its neighbors, their functionality can always be confirmed by simply checking the resistive path along the force/sense pin pairs 1&3 as well as 2&4 (ref. Figure 4). Accordingly, a local circuitry fail usually causes a dual bit fail, which can be easily flagged by a bitmapping algorithm.

133

Figure 6: Pad group with 4 Arrays (left) and zoom in to one array (right).

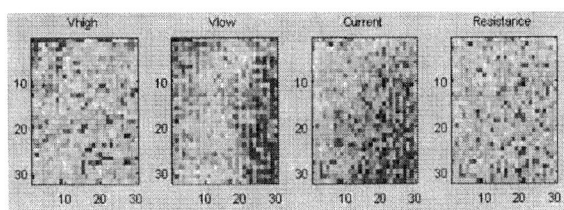

Figure 7: Distribution of measurement values within one 32 by 30 array. Each pixel represents the terminal characteristics of a cell

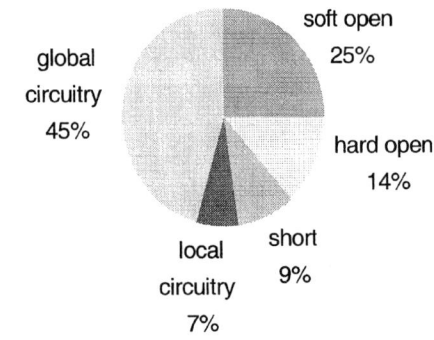

Figure 8: Distribution of the different types of electrically detected fails.

4 Experimental Results

A High Density Test Structure Array has been implemented in a 0.13um technology to detect failing contacts. For that, 9um by 9um DUT cells have been designed to best accommodate given design rules. 960 DUTs have then been arranged within an array of 32 rows and 30 columns. Finally, 4 arrays sharing the same addressing pads are placed within one pad group as shown on the left side of Figure 6. The right side zooms into the lower left corner of a single array. The box marks the DUT, which serves as reference for physical failure analysis (PFA) activities.

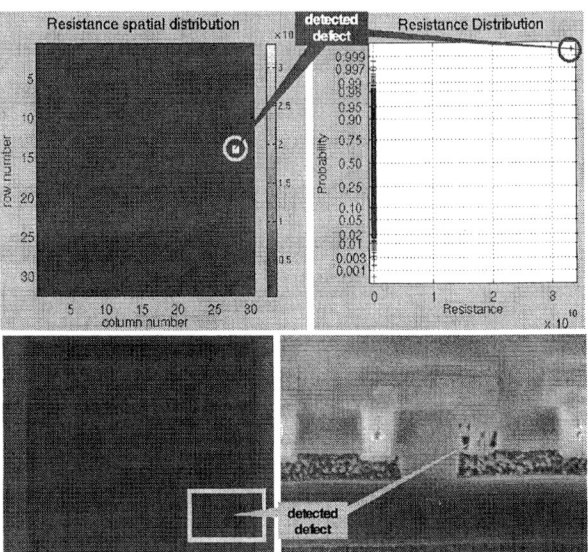

Figure 9: Detected defect causing a hard open fail within one array.

In general, testing can be executed as early as after the first 2 metal layers have been manufactured. Figure 7 shows left to right the spatial distribution of the analog measurement values for the Vhigh voltage (Pin 3 of Figure 4), the Vlow voltage (Pin 4 of Figure 4), the current (Pin 2 of Figure 4) as well as the resulting resistance of the DUT within an array. While a trend can be seen in the distribution of the measurement values of Vhigh, Vlow and the current, the distribution of the DUT resistance is random, which indicates the expected gain in accuracy of 4 terminal measurements over simple 2 terminal measurements.

Figure 8 shows the distribution of electrically detected fails. The percentage of circuitry fails is as expected, taking into account the presence of selection logic and pass gates. 45% of all fails are global circuitry fails, which impact rows or columns of the arrays. However, 7% of all detected fails are pure dual bit fails. Here, at least one of the two local selection devices failed, which are shared among two neighboring DUTs. Those findings confirm that a solid method to separate circuitry fails from DUT fails is essential to report reliable data about defect densities for yield and performance predictions. If such distinction cannot be made the 7% device fails would erroneously be accounted as hard open fails.

48% of all detected fails are DUT fails. 14% of all fails are classified as hard open fails within a DUT. Figure 9 shows an example. The upper left picture shows the resistance distribution within the array. The upper right graph shows the cumulative distribution function (CDF) plot of the measured resistance values. One hard fail can be clearly seen and located. PFA can be executed knowing the exact location of the failing DUT within the array. The top view of the DUT is presented in the lower left picture, while the lower right picture shows the cross section of the DUT at the failing contact. Both the contact plug and the liner are missing, which causes the hard open fail.

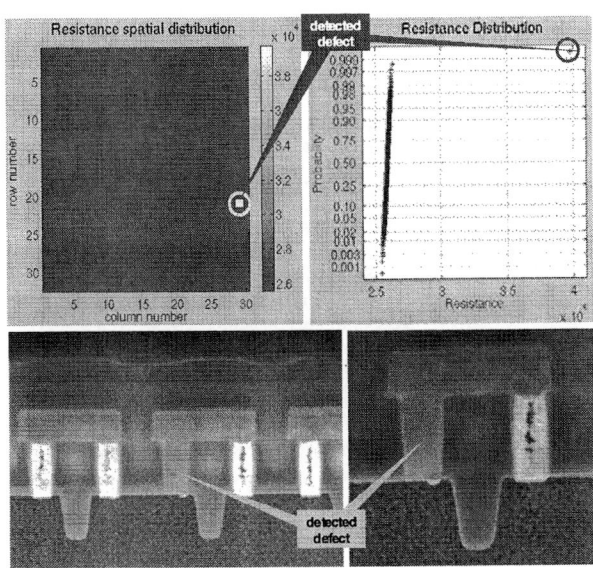

Figure 10: Detected defect causing a soft fail within one array.

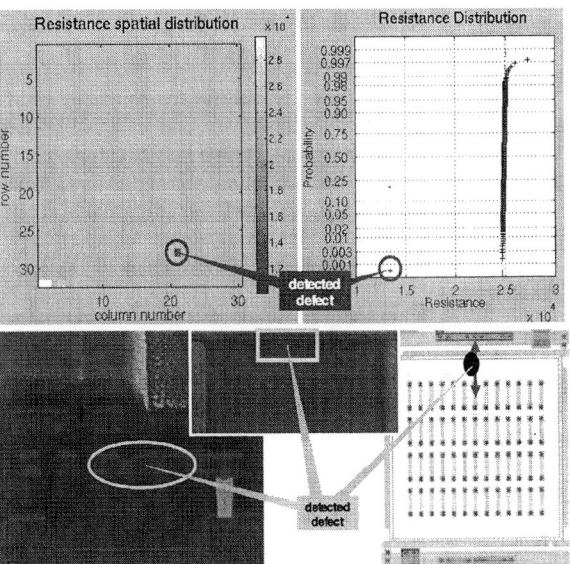

Figure 11: Detected defect causing a short fail within a contact chain.

25% of all fails are classified as soft open fails within a DUT, which is almost twice the amount of hard open fails. Figure 10 shows one of those soft opens. The upper left picture shows the resistance distribution within the array. The upper right CDF plot indicates one measurement value being about 50% above the median. PFA reveals that the Tungsten plug is again missing. However, the cross section indicates that the hole is completely filled with Aluminum and there is a small portion of liner at the bottom, which still provides "some" resistive connection.

It was expected to see hard and soft open fails within contact chains. However, we also found a group of fails with measurement values below the median as indicated in the CDF plot on the upper right of Figure 11. The cross section in the lower left picture shows a short circuit between one link of the chain and the routing line of the 2 selection pass gates. The lower right shows the layout view as well as the defect location and the direction of the cross section. Since half of the chain is cut off, the measured value of around 50% below the median is accurately reflecting the scenario.

Figure 12 shows another example of a detected short circuit. The measured resistance value is about 15% below the median. The cross section reveals a metal defect between two adjacent chains. Given the location of the defect as shown in the layout of the lower right, only 11 of the 13 links are tested, which explains the measured drop of the resistance value.

Overall, 9% of all fails are actually classified as short circuits, which is more than a quarter of all detected DUT fails and about 40% of all detected hard fails. It may initially come as a surprise to actually find short circuit fails within contact chains, simply because it is practically impossible to find those in large area test structures. However, the proposed smaller footprint provides enough resolution to detect such fails.

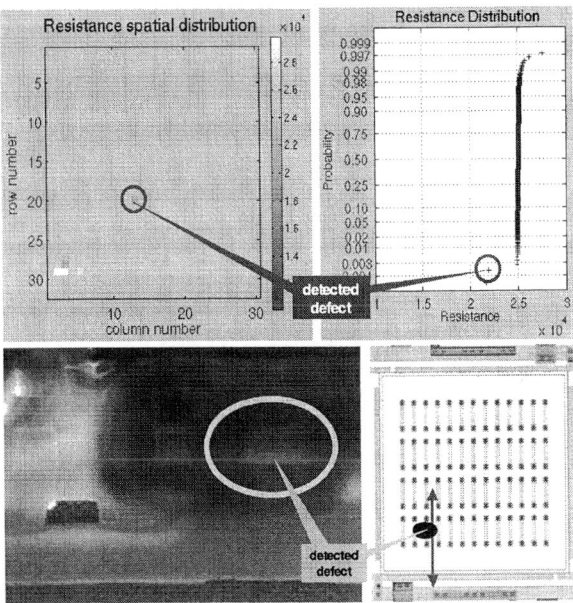

Figure 12: Detected defect causing a short fail within a contact chain.

While some memory based test structures have reported localization of soft opens in the past, even they missed those short circuits, because they cannot provide the accuracy of analog resistance measurements required to identify them [KFCY06] [KGCY07]. The capability to use a test structure beyond its intended purpose opens the door to significantly reduce mask and wafer consumption. For instance, it is possible to reduce the test chip area covered by comb test structures targeting short circuits if the proposed test structure array is implemented for contact and via chains.

5 Conclusion

A High Density Test Structure Array has been introduced to detect and localize hard and soft fails. The starting point is using known 2 terminal test structures such as chains, snakes and combs, which are arranged in an array, connected through newly developed circuitry. Each array can have 1000 or more DUTs as small as "standard cell height" by "standard cell height", which translates to $5um^2$ and less for 65nm technologies and below. Accurate 4 terminal resistance measurements are available for each DUT, which enables easy distinction of hard and soft fails as well as circuitry related fails. Even short circuits within chains of contacts can be detected. Each failing DUT can be assigned to a specific cell within the array, which greatly supports physical failure analysis. Any FEOL and BEOL layers can be used as DUTs. On average just 2 selection devices are required to address each DUT, which provides outstanding utilization of the test chip area. Addressing of DUTs is done through simple combinatorial logic to ensure robustness to process variability as well as easy separation of DUT and circuit fails.

Acknowledgement

The authors would like to thank Michael B. Schmidt and Larry A. Dworkin for providing FIB based physical failure analysis of the electrically detected defects at FEI Company, Hillsboro, OR 97124, USA.

References

[Bueh83] Buehler, M. G., "Microelectronic Test Chips for VLSI Electronics", VLSI Electronics Microstructure Science, pp. 529-576, Vol 9, Chapter 9, Academic Press, 1983

[DBCH06] Doong, K., Y., Y., Bordelon, J., Chang, K.-J., Hung, L., J., Liao, C., C., Lin, S., C., Ho, P., S., Hsieh, S., Young, K. L., "Field-Configurable Test Structure Array (FC-TSA): Enabling design for monitor, model and manufacturability", International Conference on Microelectronic Test Structures, pp. 98-103, Austin (USA), 2006

[DGGY05] DeBord, J. R. D., Grice, T., Garcia, R., Yeric, G., Cohen, E., Sutandi, A., Garcia, J., Green, G., "Infrastructure for Successful BEOL Characterization and Yield Ramp at the 65 nm Node and Below", Proc. IITC 2005

[FNMO07] Fuji, M., Nii, K., Makino, H. Ohbayashi, S., Igarashi, M., Kawamura, T., Yokota, M., Tsuda, N., Yoshizawa, T., Tsusui, T., Takeshita, N., Murata, N., Tanaka, T., Fujiwara, T., Asahina, K., Okada, M., Tomita, K., Takeuchi, M., Shinohara, H., "A Large Scale, Flip-Flop RAM imitating a logic LSI for fast development of process technology", International Conference on Microelectronic Test Structures, pp. 131-134, Tokyo (Japan), 2007

[HeWB97] Hess, C., Weiland, L. H., Bornefeld, R., "Customized Checkerboard Test Structures to Localize Interconnection Point Defects", Proc. VLSI Multilevel Interconnection Conference (VMIC), pp. 163-168, Santa Clara (USA), 1997

[HeWe94] Hess, C., Weiland, L. H., "Drop in Process Control Checkerboard Test Structure for Efficient Online Process Characterization and Defect Problem Debugging", Proc. International Conference on Microelectronic Test Structures (ICMTS), pp. 152-159, San Diego (USA), 1994

[HeWe98] Hess, C., Weiland, L. H., "Strategy to Disentangle Multiple Faults to Identify Random Defects within Test Structures", Proc. International Conference on Microelectronic Test Structures (ICMTS), pp. 141-146, Kanazawa (Japan), 1998

[HSWM02] Hess, C., Stine, B. E., Weiland, L. H., Mitchell, T., Karnett, M., Gardner, K., "Passive Multiplexer Test Structure For Fast and Accurate Contact and Via Fail Rate Evaluation", Proc. International Conference on Microelectronic Test Structures (ICMTS), pp. 163-167, Cork (Ireland), 2002,

[HWLS96] Hess, C., Weiland, L. H., Lau, G., Simoneit, P., "Control of Application Specific Interconnection on Gate Arrays Using an Active Checkerboard Test Structure", Proc. International Conference on Microelectronic Test Structures (ICMTS), pp. 55-60, Trento (Italy), 1996

[IpSa77] Ipri, A. C., Sarace, J. C., "Integrated Circuit Process and Design Rule Evaluation Techniques," RCA Review, pp. 323-350, Volume 38, Number 3, September 1977

[Keit04] "6th Edition: Low Level Measurements Handbook", Keithley Instruments Inc., pp. 2-39 & 3-18, 2004

[KFCY06] Karthikeyan, M, Fox, S., Cote, W, Yeric, G, Hall, M, Garcia, J., Mitchell, B., Wolf, E., Agarwal, S., "A 65nm Random and Systematic Yield Ramp Infrastructure Utilizing a Specialized Addressable Array with Integrated Analysis Software", International Conference on Microelectronic Test Structures, pp. 104-109, Austin (USA), 2006

[KGFY07] Karthikeyan, M., Gasasira, M., Fox, S., Yeric, G, Hall, M, Garcia, J., Mitchell, B., Wolf, E., "Development and Use of Small Addressable Arrays for Process Window Monitoring in 65nm Manufacturing", International Conference on Microelectronic Test Structures, pp. 135-139, Tokyo (Japan), 2007

[KMGS94] Khare, J., Maly, W., Griep, S., Schmitt-Landsiedel, D., "SRAM-based Extraction of Defect Characteristics", International Conference on Microelectronic Test Structures, pp. 98-107, San Diego, USA, 1994

[WGMS90] Walton, A. J., Gammie, W., Marrow, D., Stevenson, J. T. M., Holwill, R. J., "A novel Approach for Reducing the Area Occupied by Contact Pads on Process Control Chips", International Conference on Microelectronic Test Structures, pp. 75-80, San Diego, (USA), 1990

[WWGH92] Ward, D., Walton, A. J., Gammie, W. G., Holwill, R. J., "The Use of a Digital Multiplexer to Reduce Process Control Chip Pad Count", International Conference on Microelectronic Test Structures 1992, pp. 129-133, San Diego (USA), 1992

[WWRH89] Walton, A. J., Ward, D., Robertson, J. M., Holwill R. J., "A Novel Approach for an Electrical Vernier to Measure Mask Misalignment", 19th European Solid State Device Research Conference ESSDERC '89, pp. 950-953, Springer Verlag, 1989

2008 IEEE Conference on Microelectronic Test Structures, March 24-27, Edinburgh, UK

6.2

New method for non destructive snap-back characterization in multi-finger power MOSFETs

François Dieudonné, Aurore Constant [(*)], Julien Rosa, Benoit Gautheron, Jean-François Revel

STMicroelectronics, Crolles Site
850 rue Jean Monnet, 38926 Crolles, France
[(*)]: now with SID/CNM, Bellaterra, Spain

ABSTRACT

A non destructive conductance-based electrical characterization method of the snap-back phenomenon has been implemented to investigate multi-finger power MOSFETs. The context of our study is presented, then the specific test structures and the measurement methodology are shown. The robustness and repeatability of our approach is demonstrated on a variety of power MOSFETs regarding to some technological parameters. Comparisons between our results and the ones issued from a destructive characterization are also drawn. The temperature's influence on snap-back is evidenced as well as the measurement's repeatability.

INTRODUCTION

Multi-finger power MOSFETs are of a huge and growing interest for D-Class Amplifiers & Switch Mode Power Supplies (SMPS) used for dedicated Analogue Applications [1,2]. In these circuits, large currents, associated to elevated switching frequencies and parasitic inductances due to interconnections & bonding can induce some problematic drain bias overshoots.

These circuit overshoots are often critical at t=0 as regards device (or circuit, or product) operating lifetime. Indeed, those overshoots can be obtained in some cases for bias values below the nominal operating voltage of a given technology node. It results in the onset of the snap-back phenomenon [3] at device level that leads to structure's breakdown. Such an initial failure must be avoided, this is why providing some design guidelines (particularly regarding to device's immunity to snap-back) is a crucial goal to achieve. Moreover, implementing a fast non destructive characterization method at parametric test level in worst-case conditions is very topical in order to ensure a good monitoring during production ramp-up.

This dual need led us to develop a new simple approach to characterize the snap-back phenomenon based on a detection of sharp conductance variations as close as possible to the destructive snap-back onset. Some authors previously used the normalized I_{well}/I_d ratio in worst-case conditions to determine a "bipolar induced breakdown" voltage [4] in conventional MOS

structures. We also found a reference dealing with a comparison of base current reversal and bipolar snap-back in advanced NPN bipolar transistors [5]. Our method only relies on the drain current measurement and its calculated derivative (the conductance), without measuring any other transistor's current. Indeed, in our case, the drain current's knowledge is necessary and sufficient to calculate the conductance.

SNAP-BACK DEDICATED TEST STRUCTURES

Our study is performed on both N-& P-MOSFETs from a 0.13µm CMOS technology platform developed for analogue applications. The dedicated test structures are double gate oxide (85Å) large area multi-finger power MOSFETs in which the snap-back can be easily triggered due to the significant involved currents and the large well resistances induced by the multi-finger geometry [3]. A cross-section and a layout top view are shown in Fig. 1.

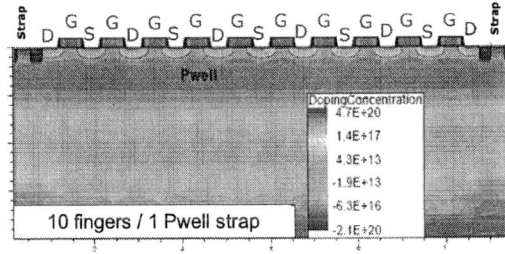

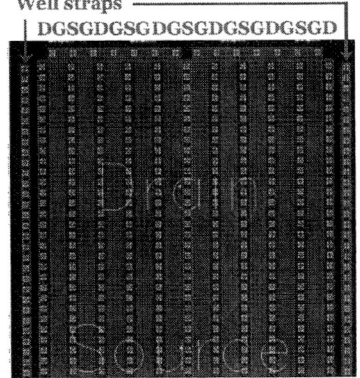

Fig. 1. Cross-section & layout top view of a 10 fingers N-MOSFET (L=0.5µm W_{tot}=1mm).

The following layout variations are considered in this paper: poly gate length, L_g: 0.5 to 1µm; finger numbers between each well access strap, N_f: 1 to 10.

978-1-4244-1801-5/08/$25.00 ©2008 IEEE

Other structures with variations of the well access contact rows' number have also been studied. Electrical characterization results are presented in this paper only for N-MOSFETs biased at Vgs=1.5V (worst-case gate bias provided by designers). A similar approach has been performed in P-channel MOS devices, and we found no interest of "duplicating" test results in this paper. Moreover, snap-back is a less critical issue in such devices at identical transistor's dimensions.

TEST IMPLEMENTATION & RESULTS

Involved physics and snap-back illustration

The snap-back phenomenon appears for elevated drain biases that result in large lateral electric fields between source and drain. Channel electrons (in the case of N-MOSFETs) are thus accelerated, and impact ionization occurs at the drain edge. Holes from electron/holes pairs are pushed away into the transistor's well [3] and electrons are collected by the drain. Due to the large hole current, the well potential rises up, and for a certain level of impact ionization, reaches the forward-biasing condition of the well-to-source diode, activating the source (emitter) – well (base) – drain (collector) lateral parasitic NPN bipolar transistor. This bipolar transistor being switched on, an additional electron current is injected into the drain leading to avalanche and then snap-back (see Fig. 2) occurrence.

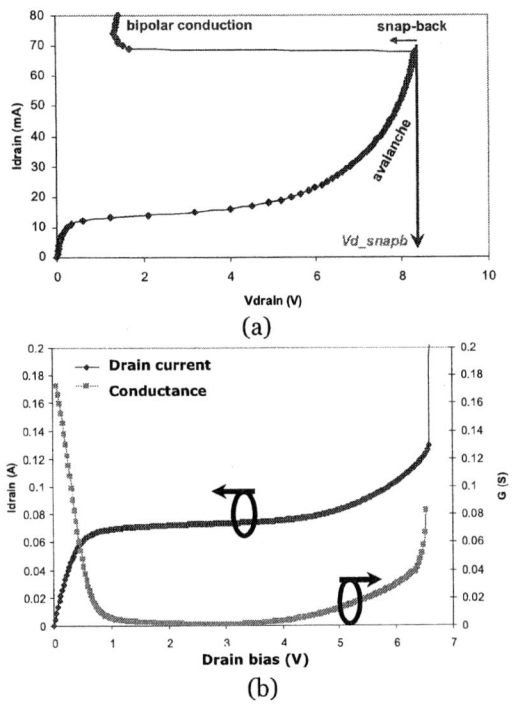

Fig. 2. Snap-back triggering on a W_{tot}/L=1000/0.5µm N-MOSFET with V_{gs}=1.5V: (a) V-I measurement (Nf=1) (b) I(V) measurement (N_f=10) along with the conductance G.

As illustrated in Fig. 2 (a) on a "Force I_d, Measure V_d" measurement, snap-back is characterized by the onset

of a dynamic negative resistance on the transistor drain pad. Indeed, for the "*Vd_snapb*" drain bias, the drain current goes straight to a lower drain bias while it keeps increasing. Once the snap-back point is reached, devices got broken and measurements are not repeatable. This is a critical issue and justified the need of introducing a non destructive method to detect snap-back triggering.

Introducing a new detection approach

Fig. 2 (b) also shows on the right-hand Y-axis the conductance G defined by $\dfrac{\partial I_d}{\partial V_d}$. G is more sensitive to drain current variations before snap-back as illustrated in this "Force V_d, Measure I_d" approach. Contrary to the "Force I_d, Measure V_d" approach, the reversal of the characteristics is not measurable due to the measurement system configuration. Since we found no quick and efficient mean to stop the measurement before device's destruction using the $V_d(I_d)$ approach, our work relies on $I_d(V_d)$ characterizations. Using the conductance to detect snap-back in this configuration is reliable enough, and very quick to implement. Moreover, we do only acquire some data points for drain current, freeing ourselves from all other transistor currents (source, gate, well) which are not measured. Source & well device pads are grounded. Some experiments also show measuring well current induced perturbations on the snap-back onset by introducing a parasitic resistance on the well pad.

A barycentric smoothing technique is used to get rid (and strongly reduce!) the noise amplification effects observed when « poor quality » experimental data points are obtained [6]. Measurements are done regularly each ΔV_d step, and a pondered calculation is used with 5 points averaging the different derivatives.

The limiting factor of this method is the resolution given by the drain bias step. After a few trials (not shown here), the optimal step (in terms of measurement time & results precision) we considered was ΔV_d=25mV. Other aspects such as the system resolution or initial drain bias *Vd_start* used for starting the detection (usually above the nominal drain voltage of the technology) are among the studied aspects. Two items are also to be considered to obtain a good detection. Firstly, to get rid of the pronounced increase of G in the linear regime, a *Vd_start* value is fixed to only consider G variations in the saturation regime of the MOSFET. Moreover, a *G_threshold* is introduced to avoid any premature stop of the measurement caused by some instability. Once the detection starts i.e. when both the *G-threshold* and the *Vd_start* are reached, a *G-variation criterion* between two consecutive measurements is then used to obtain the drain snap-back voltage value (*Vd_snapb*).

A synoptic diagram summarizing our implemented algorithm in HP Basic is shown in Fig. 3. The conductance variation criterion was fixed over a wide

138

range of tested devices (fingers' number per active from 1 to 20, channel length from 0.5 to 1µm), depending on the device family (N- or P-MOSFET) or the gate bias value.

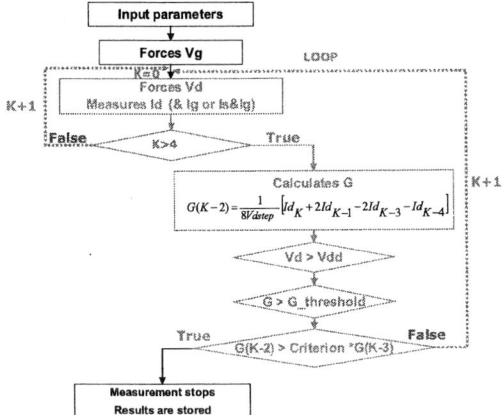

Fig. 3. Synoptic diagram of the snap-back detection protocol implemented in HP Basic.

Validation of the test methodology

The snap-back non destructive detection is illustrated in Fig. 4 for a N-MOS device, along with a comparison with destructive $I_d(V_d)$ measurements. Comparable snap-back voltages are obtained with our non destructive approach that can be therefore validated.

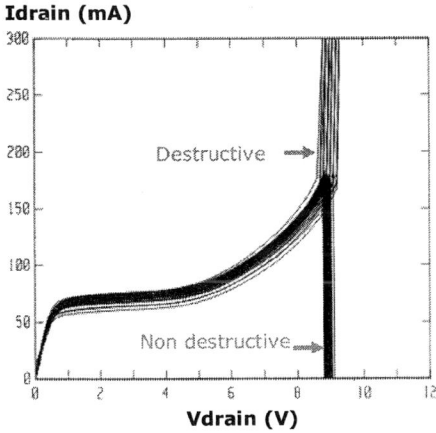

Fig. 4. Illustration of the snap-back voltage detection curve (both NON DESTRUCTIVE & DESTRUCTIVE ways) for a N-MOSFET (L=0.5µm, W_{tot}=10*W_{finger}=1mm) @ Vgs=1.5V.

To check for the non destructive side of the method, the linear drain current is measured before and after snap-back characterization. Drain current distributions are shown in Fig. 5 for 30 dice. No drastic evolution of the current is observed, giving assessment to the non destructive aspect of the test method. The tiny variations come from measurement instability, more precisely from contact resistance fluctuations during measurements. In Fig. 5, the right-hand top corner inlet explains the as plotted data (see also Fig. 7 to 10): 50% of the population is located within the box, and 80% within segments. The median value is in full

line, and the average one in dotted line in this boxplot representation.

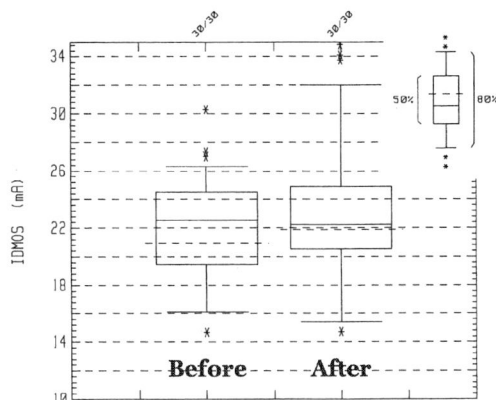

Fig. 5. Linear drain current for Vgs=4.8V and Vds=0.1V for a N-MOSFET (L=0.5µm, W_{tot}=10*W_{finger}=1mm). 30 measured dice.

Repeatability in terms of contact sensitivity during measurements is now discussed.

Snap-back characterization has been repeated five times on a N-MOS transistor. The distribution of the obtained snap-back voltages is plotted in Fig. 6. We can clearly notice that the measurements are reproducible. Taking into account the chosen ΔV=25mV measurement step, it is not possible to get below. Contact resistances are also measured on dedicated test scribes, and very slight variations are observed. We assume these variations are responsible of the ones obtained on snap-back voltage values because of the elevated drain currents measured just before snap-back. They do not find their origin in any snap-back measurement induced electrical stress. Similar results are also shown on P-MOSFETs.

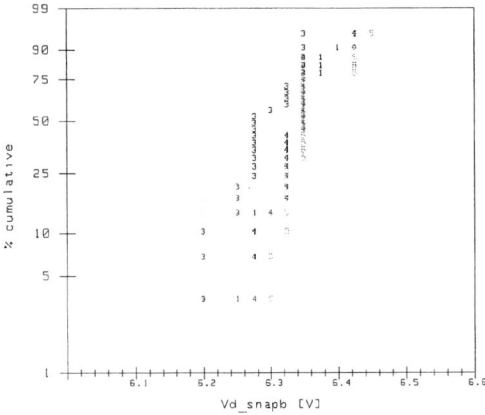

Fig. 6. 5-times repetition of snap-back detection for a Nf=10 - L=0.5µm N-MOSFET with Vgs=1.5V. ΔV=25mV.

Snap-back electrical characterization

Prior to some electrical characterization results linked with technological parameters (see *Snap-back dedicated test structures*), Vgs' impact on snap-back voltage is quickly shown as illustrated by Fig. 7. *Vd_snapb* strongly decreases then saturates while increasing gate bias. Increasing Vgs enhances impact ionization, and thus induces an easier turn-on of the well-to-source junction. Then, once the MOSFET is in strong inversion regime, no more influence appears and snap-back voltage remains almost constant. In our case, worst-case gate bias for circuit operation is Vgs=1.5V which corresponds to *Vd_snapb* starting to saturate.

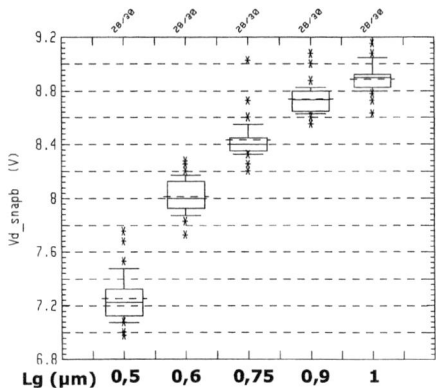

Fig. 9. Snap-back voltage (V) versus channel length. Nf=5 Vgs=1.5V.

Another technological parameter is the width of the Well strap in such transistors. 1, 2 and 3 contact rows per strap have been studied for Nf=10 / L=0.5µm N-MOSFET. The strap widening modifies snap-back conditions by making easier the holes evacuation through the strap. An illustration of Well strap design by increasing the contact rows number is shown in Fig. 10 (a). Associated measured snap-back voltages are plotted in Fig. 10 (b). A larger value is obtained with 2 contact rows compared to one row, and then saturation is observed for 3 contact rows. The holes evacuation being enhanced, a larger drain bias is needed to forward bias the well-to-source junction. Above 2 contact rows, no more improvement is obtained in our case.

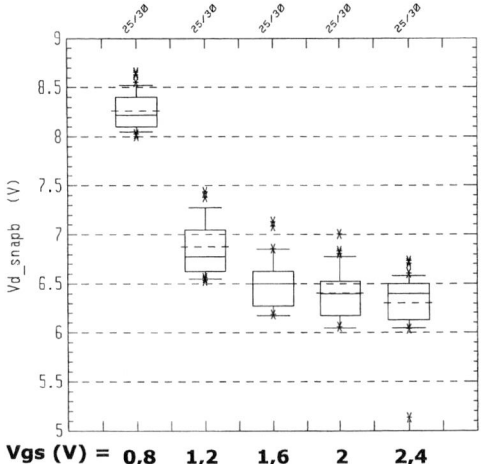

Fig. 7. Snap-back voltage (V) versus gate bias. Nf=10 - L =0.5µm.

Two examples are then respectively presented in Fig. 8 and 9 in which the influence of the fingers' number per active region and channel length is presented. Increasing the finger's number per active region induces an increase of the well resistance. This gives rise to an easier switch-on of the well-to-source junction, and consequently to an earlier snap-back onset (Fig.8).

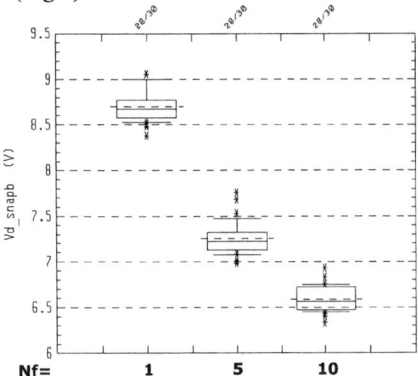

Fig. 8. Snap-back voltage (V) versus fingers' number / active. L=0.5µm.

Regarding Fig. 9 and the channel length variation, a decrease of the snap-back voltage is shown for the shortest lengths. Reducing the channel length enhances impact ionization at the drain edge, resulting also in an earlier onset of the parasitic bipolar transistor in the MOS structure. In other words, a lower equivalent drain voltage is needed to result in the snap-back triggering.

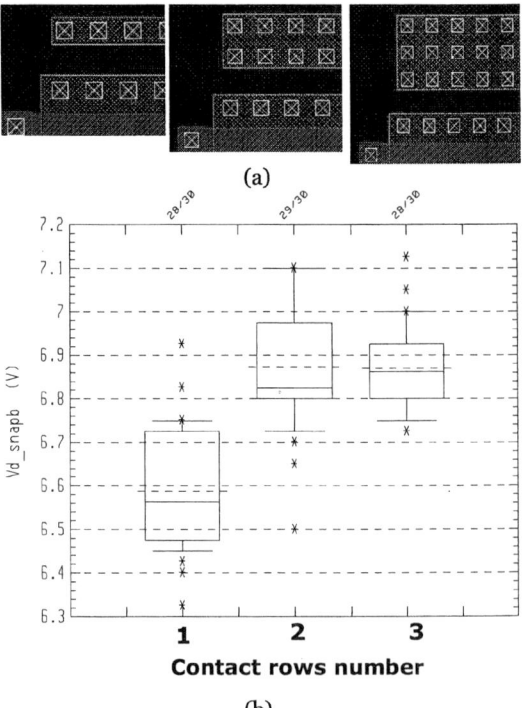

Fig. 10. (a) Illustration of contact rows variations (b) Snap-back voltage (V) versus contact rows number. Nf=10 - L=0.5µm.

The temperature dependence of the effect [7] can also be presented. Since circuits operate at elevated temperatures, it is important to evaluate such a behaviour. Fig. 11 reports the snap-back voltage evolution versus temperature for a Nf=10 - L=0.5μm N-MOSFET. A reduction of the snap-back value is noticeable in this temperature range. This is the result of the competition between the impact ionization and the parasitic bipolar transistor action. Indeed, on the one hand, impact ionization is reduced, and, on the other hand, the conduction threshold of the source-well junction is also decreased. Finally, a lower snap-back voltage value is obtained.

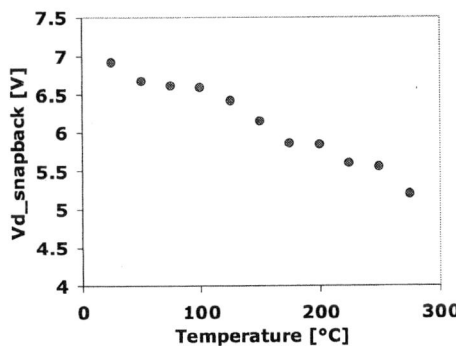

Fig. 11. Snapback voltage (V) versus temperature. Nf=10 L=0.5μm.

CONCLUSION

We have implemented a non destructive conductance-based electrical characterization method of the snap-back phenomenon on dedicated multi-finger power MOSFETs. The context of the study is firstly presented, then specific test structures along with the measurement methodology are shown. The robustness and repeatability of the methodology is demonstrated on a variety of power N- & P-MOSFETs regarding to some technological parameters. Comparisons between our results and the ones provided by a destructive characterization are also drawn. The temperature's influence on snap-back is also demonstrated.

REFERENCES

[1] J. Cerezo, Application Note AN-1070, International Rectifier, www.irf.com.

[2] J. B. V. Reddy, Proc. INDICON'05, December 11-13, 2005, pp. 585-589.

[3] P. Rossel, H. Tranduc, D. Montcoqut, G. Charitat, I. Pagès, Proc. 21st ICM, Vol. 1, September 14-17, 1997, pp. 371-381.

[4] G. J. Gaston, B.S. Bold, J.B. Mason, Proc. ICMTS'94, Vol. 7, March 1994, pp. 28-32.

[5] J.D. Hayden, IEEE EDL, Vol. 12, N°8, pp. 407-409, 1991.

[6] INRP-TECHNE, « Some basic numerical tools », 2000.

[7] D. Uffman, Microelectronics Reliability, Vol. 38, pp. 1133-1138, 1998.

Test Circuit for Measuring Pulse Widths of Single-Event Transients Causing Soft Errors

Balaji Narasimham, Matthew J. Gadlage, Bharat L. Bhuva, Ronald D. Schrimpf,
Lloyd W. Massengill, W. Timothy Holman, Arthur F. Witulski and Kenneth F. Galloway

Dept. of Electrical Engineering and Computer Science, Vanderbilt University, Nashville TN 37235 USA
balaji.narasimham@vanderbilt.edu

ABSTRACT

A novel on-chip test circuit to measure single event transient (SET) pulse widths has been developed and implemented in IBM 130-nm and 90-nm processes for characterizing logic soft errors. Test measurements with energetic ions show transient widths ranging from 100 ps to over 1 ns, comparable to legitimate logic signals in such technologies. Design options to limit the SET pulse width and hence to mitigate soft errors were evaluated with the test circuit to demonstrate the effectiveness of such design techniques. [*Keywords:* soft error, single event, single event transient, SET, SER, pulse width, combinational logic]

INTRODUCTION

Soft errors caused by the interaction of radiation with integrated circuits (IC) have been predicted by the SIA roadmap to be the most dominant failure mechanism for advanced technologies [1]. In fact, failures rates for single-events in sub-micron technologies are expected to be greater than for all other failure mechanisms combined. With each new sub-micron technology node, the reductions in nodal capacitances and drive currents result in increased vulnerability to charge deposited directly by alpha particles or indirectly by neutrons [2]. When these particles are incident on a memory cell (such as a latch or an SRAM cell) and deposit sufficient charge on a node, a soft error or single event upset (SEU) occurs, flipping the state of the memory element [2, 3]. A single event transient (SET), on the other hand, is a voltage perturbation that is caused by the collection of radiation-induced charge at a combinational logic node. Such transients may propagate through the circuit and become latched, resulting in data corruption or system failure [3]. Previous work has shown that errors due to SETs increase linearly with clock frequency [4]. At higher frequencies, or for advanced technologies, the error rates for combinational logic begin to dominate [4]. This phenomenon can be explained by the increased number of clock edges for latching SETs and the lower charge needed to represent a logic HIGH state in smaller transistors (resulting in more SETs).

For advanced technologies, a large fraction of observed soft failures are estimated to be latched SET events. The width of the SET is one of the main factors that determine whether an SET will result in an upset (fault) or not. The width of a transient voltage pulse is determined by many factors, including the nature of the ionizing particle, the linear energy transfer (LET) of the particle, the circuit characteristics, the technology used, the location of the strike, and the incident angle of the strike [3 – 5]. As the circuit vulnerability to alpha particles and neutrons is directly related to SET pulse widths, precise measurements of SET pulse widths are required for failure predictions. Additionally, soft error mitigation techniques eliminate SET pulse widths smaller than a predetermined threshold value. SET pulse width measurements are necessary for determination of such threshold pulse width for optimized circuit performance. Thus measuring the widths of transient voltage pulses is critical for prediction and mitigation of logic soft errors [5].

The characterization of single event transient pulses has been accomplished with a variety of techniques in prior research. In some cases the transient current pulses that are responsible for initiating SETs have been measured directly using oscilloscopes [6]. However, such direct off-chip measurements are difficult to perform because of pulse distortion due to the stray capacitances of the measurement system (i.e., loading and line capacitance effects). Other researchers have characterized voltage pulse widths as transients propagated through logic gates using multiple latches with delayed signal paths and/or delayed clock signals [7, 8]. Such techniques characterize the number of SET pulses greater than a specific delay, but do not measure the width of individual SET pulses.

An autonomous or self-triggered on-chip SET pulse measurement circuit that addresses the above concerns for measuring the widths of SET pulses has been developed [9] and implemented in IBM 130-nm and 90-nm technologies. This test circuit can characterize the width of each SET voltage pulse as it propagates through logic gates without the need for an external trigger. The circuit technique can also be used for characterizing other spurious signals such as noise or cross-talk pulses. Measurements with energetic ions indicate that SET pulse widths range from 100 ps to over 1 ns and are comparable to legitimate logic signals in such technologies, posing serious reliability concerns. Results from the test circuit also show that certain layout techniques such as the use of multiple well contacts and guard rings

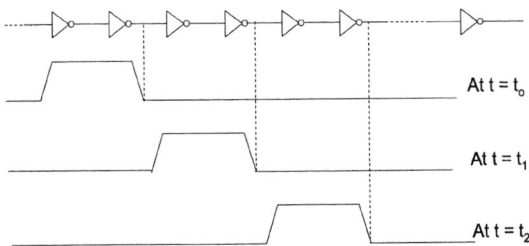

Fig. 1. Pulse propagation through a series of inverters. Time instances t_0, t_1, t_2 are 2 inverter-delays apart.

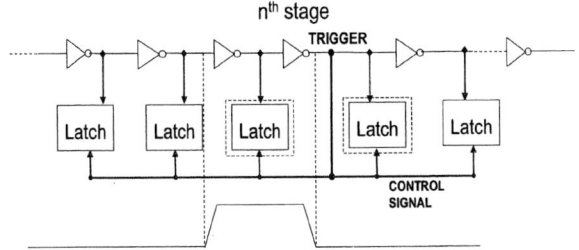

Fig. 2. The output of the nth stage can be used to provide hold signal for latches to freeze the data and the SET pulse.

mitigate the SET pulse width, and hence can be used effectively to reduce the soft error rates.

AUTONOMOUS PULSE CAPTURE CIRCUIT

A common parameter for specifying the performance of a digital IC is the propagation delay associated with an inverter, designated as one inverter-delay. The test circuit demonstrated here characterizes SET pulse widths in such units of inverter-delays. The circuit is based on the principle that if a pulse of sufficient duration is input to an inverter chain, it will affect a certain number of inverter outputs at any instant. As the pulse propagates through the inverter chain, it will affect different inverter outputs; however, the number of affected inverter outputs will remain the same at any given instant of time. This number of affected outputs is proportional to the input pulse width. Fig. 1 illustrates an example of pulse propagation through a series of inverters when the SET pulse is two inverter-delays long. If the number of inverters whose outputs are affected by the SET pulse is determined at any instant, the pulse width can be estimated as an integral number of inverter delays. Simulations showed that for all pulse widths between $[(n-0.5) \times$ stage delay$]$ to $[(n+0.5) \times$ stage delay$]$, the number of affected stages are n. Thus the pulse width determined will be accurate to within $\pm$ one half of the individual inverter stage delay.

To measure the SET pulse originating from (for example) a target combinational logic circuit, it should first be fed to the measurement circuit composed of a chain of inverters. Next, the number of inverter stages that are affected by this SET pulse at any given instant of time must be measured. This can be accomplished if the SET pulse is frozen when it is within the measurement chain of inverters. Latches can be used to freeze the state of the inverter outputs at any given instant. Thus, to capture the affected outputs from a chain of inverters, the output of every inverter is connected to an asynchronous latch as shown in Fig. 2. As the SET pulse propagates through an inverter, the data stored in its respective latch will change. However, once the SET pulse passes, the inverter output and latch data will revert to their original states. (Note that the additional loading due to the latch at the inverter output will alter the pulse characteristics. Hence, capacitance at the latch input must be minimized and accounted for when determining the inverter delay for accurate measurement of pulse width.) If the latches are placed in a *hold* mode while the SET pulse is within the inverter chain, each latch will retain the logic state of its respective inverter.

For ion testing, information regarding the hit time and hit node is usually not available. To address autonomous operation in such cases, the output of one of the inverter stages can be used as a trigger signal. To make this circuit *self-triggering*, a transition at the output of the nth stage (or trigger stage) can be used to trigger the latches to hold the states of the inverters as shown in Fig. 2. As the output of the nth stage triggers the *hold* signal internally, precise information regarding the hit time (or location) is unnecessary. Note that a hit on any stage prior to the trigger stage will result in an SET that gets captured.

The instant when the SET pulse is latched, the initial hit stage may or may not have recovered to its original state. The initial stage will not recover to its original state if the charge collection is not completed when the SET pulse is latched. If the initial stage has recovered fully when the pulse is latched, the pulse width measured is the actual pulse width (to within the accuracy of the one inverter delay). However, if the initial stage has not recovered, it is possible that the charge collection is still continuing and the actual pulse width could be longer than the measured width. For ion tests, the hit stage is not identifiable, and hence it cannot be ascertained whether the hit stage has fully recovered or not. To address this uncertainty, a delay is introduced in the trigger signal, and extra inverter stages are added beyond the trigger stage to allow the complete SET pulse to propagate well past the trigger stage before latching occurs. The distance the SET pulse travels along the inverter chain is proportional to the trigger signal delay, which is equal to the maximum SET pulse width expected during measurement. If the complete SET pulse has moved beyond the trigger stage, one can safely assume that the estimated pulse width is the actual pulse width (to within the resolution of one inverter delay), irrespective of the hit node.

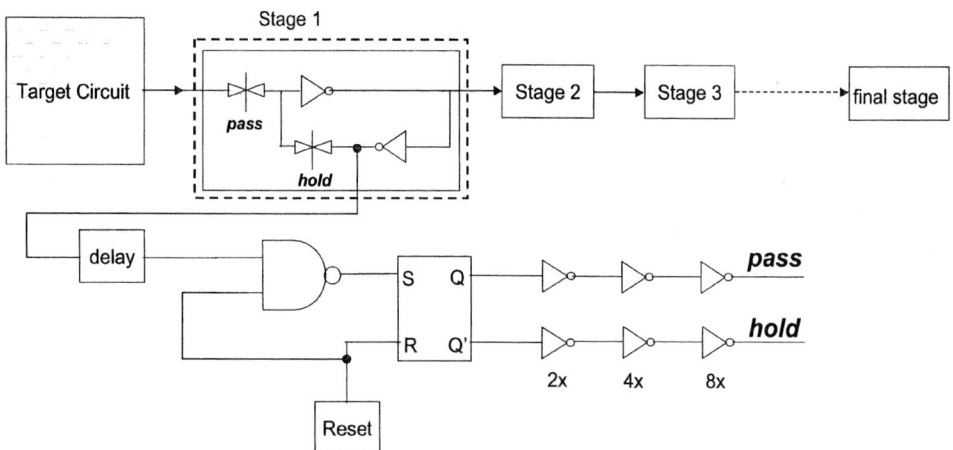

Fig. 3. SET pulse measurement test circuit showing pulse capture stages along with the trigger/reset circuit. Outlined region shows the internal circuitry of individual stages in the measurement circuit. The pulse capture stages are preceded by a sufficiently large array of target inverter circuits.

To increase the probability that an SET will be created in a given test environment, an array of target circuits, which functions as the source of SETs, precedes the measurement circuit as shown in Fig. 3. The target circuit also allows the trigger signal to be taken from the 1st stage of the measurement circuit and delayed in time to allow the SET to propagate completely into the measurement chain of inverters. Depending upon the designer's requirement, the target circuit can be composed of any combinational logic network. In this work, a minimum drive-strength inverter chain was used for the target since it yields SETs similar to those in standard ICs and such a chain will propagate SETs with little attenuation.

The design of the complete test circuit (composed of the target circuit and the pulse measurement circuit) is shown in Fig. 3. To simplify the circuit and reduce loading effects, the individual inverter stages in the measurement circuit were implemented using the inverting outputs of standard CMOS pass-gate latches. The operation of the test circuit is straightforward. An ion hit in the target circuit creates an SET pulse that propagates to the measurement circuit. The measurement circuit essentially forms a series of latches that freeze the SET pulse for measurement (Fig. 3). The latches in the measurement circuit are initially in the SET-propagate phase. During the SET-propagate phase, the pass signal is ON and the hold signal is OFF which allows a pulse to propagate through the measurement chain of inverters and pass-gates. When the leading edge of the SET pulse reaches the 1st stage of the measurement circuit, it creates a trigger signal which is delayed in time and hence the SET pulse continues to propagate through the inverters and pass-gates. When the trigger signal reaches the SR flip-flop, it turns off all pass-gates by inverting the pass signal and freezing the data in the latches by turning on the hold signal. The SET pulse width is directly proportional to the number of latches whose output is affected. Once the latch outputs have been read out, a reset signal is used to initialize the pass and hold signals and make the circuit ready for measuring the next pulse.

Latch upsets due to direct ion hits on the latches can corrupt the measurement. However, only latch upsets that occur after an SET is latched, but before the data are read out, are a concern. Since the frequency of operation of this circuit is much higher than the rate at which events are created, the probability of a latch upset during the time interval when an SET is captured and before the circuit is reset is very low. Also, most latch upsets can be identified by analyzing the data pattern and hence can be discarded.

Custom integrated circuits with the pulse capture circuit and a target circuit composed of an array of inverters were designed and fabricated in 90-nm and 130-nm CMOS processes. Thirty-two latch stages were used in the pulse measurement circuit. A ring oscillator consisting of pulse measurement circuit latch stages was fabricated to obtain the precise delay of an individual latch stage, as shown in Fig. 3. This delay was measured to be about 120 ps for the 130-nm process when operating at the nominal supply of 1.2 V. When the trigger signal was enabled in the measurement circuit, the leading edge of the pulse was latched at the 22nd stage. This enabled pulse width measurements from 120 ps (1 stage) to about 2520 ps (21 stages, excluding the first stage), with an accuracy of ±60 ps for the 130-nm process. For the 90-nm process, the individual stage delay was found to be about 100 ps, with a measurement range of 100 ps to 2 ns.

EXPERIMENTAL RESULTS AND DISCUSSION

Tests were performed using the cyclotron facilities at Texas A&M University and at Lawrence Berkeley National Laboratory with energetic ions incident at different angles

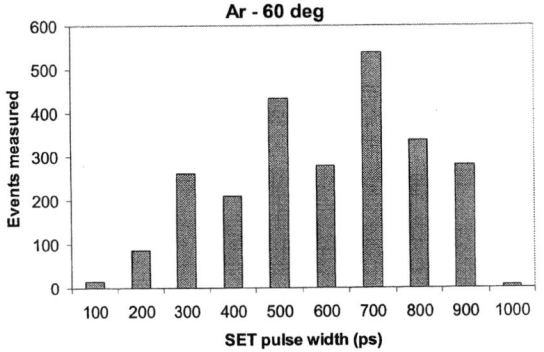

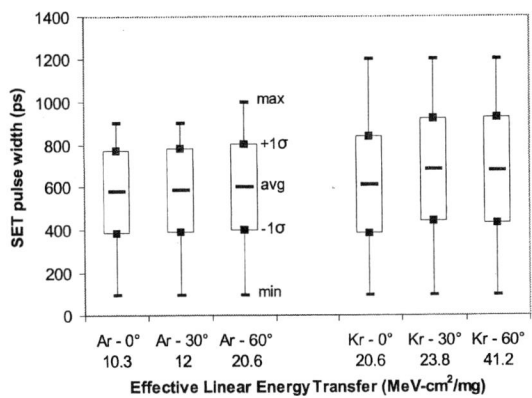

Fig. 4. Histogram of distribution of SET pulse widths in 90-nm process with Ar ions incident at 60 degrees from the normal.

Fig. 5. Heavy-ion induced SET pulse widths comparing data for normal, 30°, and 60° incidence angles for two ions in 90 nm. This is a box plot indicating the average, min, max, and ±1σ in SET pulse widths and is based on about 12,000 measured SET events.

on the test IC. Ions with linear energy transfer (LET), a measure of the charge deposited in silicon, comparable to LETs of neutron reaction products in Si were chosen for the experiments. Similar pulse widths can be expected in ICs exposed to terrestrial neutrons. Fig. 4 shows a histogram of the distribution of SET pulse widths in the 90-nm process with Ar ions incident at 60°. The Ar ion used has effective LETs of 10.3 MeV-cm²/mg when incident normal to the IC and 20.6 MeV-cm²/mg when incident at 60° from the normal. The SET pulse width for a given LET is not a constant and varies over a wide range. This is because the pulse width varies with the collected charge, which depends on the location of the strike with respect to the sensitive

drain. It is possible to measure such precise distributions, as the autonomous pulse characterization technique measures each individual pulse for most of the SETs that are created. Fig. 5 is a box plot comparing the SET pulse widths for two ions at different angles of incidence for 90-nm technology. This data is based on measurements of about 12,000 SET events. The SET pulse widths were found to range from about 100 ps to over 1 ns.

Another set of pulse measurement test circuits was designed in the 130-nm process. Each test circuit consisted of the pulse capture circuit, along with the target circuit composed of either inverters with a single well contact or inverters with multiple well contacts and guard rings. A guard ring is an n+ (or p+) diffusion region surrounding a device in the n-well (or p-substrate) (see Fig. 6). 3-D TCAD simulations have shown that the charge collection (and hence the SET width) in advanced technologies is enhanced due to a parasitic bipolar effect [10]. This parasitic transistor turns on due to collapse of the well potential following a radiation event. Thus, the presence of additional well contacts and guard rings should mitigate the bipolar effects and reduce the transient pulse width. Fig. 7 compares the SET pulse widths for a conventional circuit with only one well contact per transistor to those of a circuit with multiple well contacts per transistor. As expected, the charge collection decreases for the heavily contacted circuit, resulting in a reduction of the maximum transient pulse width. Guard rings and

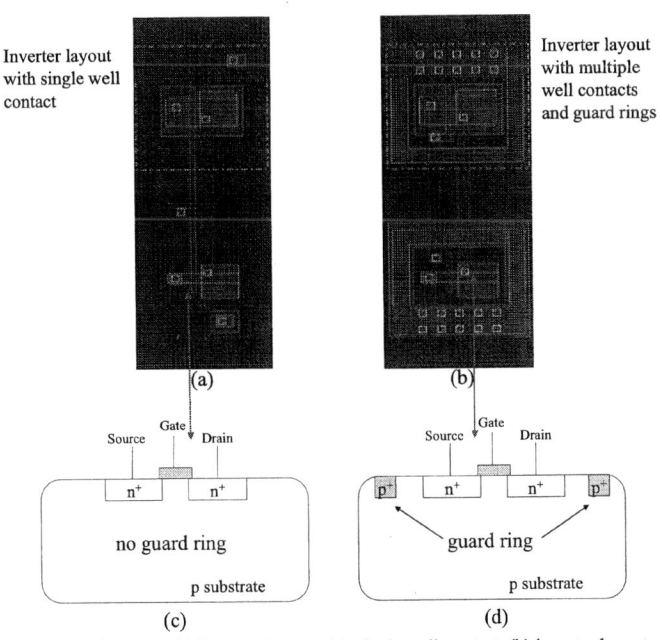

Fig. 6. Test circuit layouts (a) inverter layout with single well contact, (b) inverter layout with multiple contacts and guard rings. Cross section of layout (c) without guard rings and (d) with guard rings.

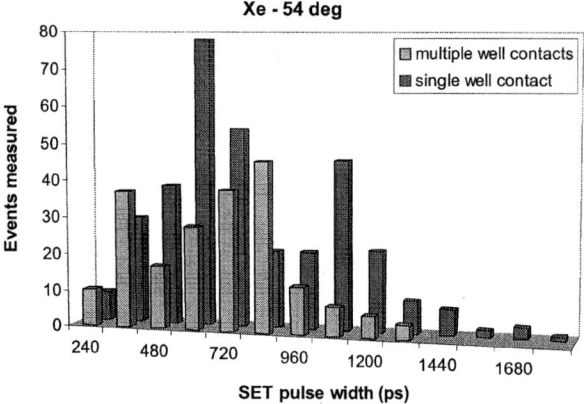

Fig. 7. Histogram of distribution of SET pulse widths for structures with single and with multiple well contacts in 130 nm. The maximum SET width reduces for the structure with more well contacts.

multiple well contacts also help reduce the sensitive area over which an SET is created. This is evident as the total number of events is considerably lower for the circuit with guard rings compared with the circuit with single well contacts.

CONCLUSION

Combinational logic soft errors are becoming one of the major radiation based reliability issues. One of the main factors that determine logic error rates is the width of the SET pulse. A novel self-triggered test structure for measuring SET pulse width in increments of an inverter delay has been developed and used to measure the SET pulse widths in 130-nm and 90-nm CMOS processes. Results with energetic ions incident indicate that SET pulse widths can range from about 100 ps to over 1 ns for the 90-nm process. Such pulse widths are comparable to legitimate logic signals in such a process and indicate that as technology is scaled to lower operating voltages and higher operating frequencies, SETs may become a serious reliability problem.

Transient widths for different layouts in the 130-nm process were measured. Heavily contacted well areas were found to decrease the maximum SET pulse widths as they are able to mitigate the parasitic bipolar effect more effectively as compared to the SETs in a conventional circuit with only one well contact per transistor. The results indicate the need for developing mitigation techniques to limit logic soft errors and suggest that certain layout approaches for logic circuits can be used to limit error rates.

REFERENCES

[1] Semiconductors Industry Association (SIA), International Technology Roadmap for Semiconductors, 2003.

[2] R. C. Baumann, "The impact of technology scaling on soft error rate performance and limits to the efficacy of error correction," *IEEE IEDM Tech. Dig.,* pp. 329-332, 2002.

[3] P. E. Dodd and L. W. Massengill, "Basic Mechanisms and Modeling of Single-Event Upset in Digital Microelectronics," *IEEE Trans. Nucl. Sci.,* vol. 50, pp. 583-602, 2003.

[4] S. Buchner, M. Baze, D. Brown, D. McMorrow and J. Melinger, "Comparison of error rates in combinatorial and sequential logic," *IEEE Trans. Nucl. Sci.,* vol. 44, pp. 2209-2216, 1997.

[5] S. Buchner and M. Baze, "Single-event transients in fast electronic circuits," in *Proc. IEEE Nuclear and Space Radiation Effects Conf. Short Course Text,* 2001.

[6] V. Ferlet-Cavrois, P. Paillet, A. Torres, M. Gaillardin, D. McMorrow, J. S. Melinger, A. R. Knudson, A. B. Campbell, J. R. Schwank, G. Vizkelethy, M. R. Shaneyfelt, K. Hirose, O. Faynot, G. Barna, C. Jahan, and L. Tosti, "Direct Measurement of Transient Pulses Induced by Laser and Heavy Ion Irradiation in Deca-Nanometer SOI Devices," *IEEE Trans. Nucl. Sci.* vol. 52, pp. 2104-2113, 2005.

[7] P. Eaton, J. Benedetto, D. Mavis, K. Avery, M. Sibley, M. Gadlage, and T. Turflinger, "Single event transient pulsewidth measurements using a variable temporal latch technique," *IEEE Trans. Nucl. Sci.,* vol. 51, pp. 3365-3368, 2004.

[8] M. P. Baze, J. Wert, J. W. Clement, M. G. Hubert, A. Witulski, O. A. Amusan, L. Massengill, D. McMorrow, "Propagating SET Characterization Technique for Digital CMOS Libraries," *IEEE Trans. Nucl. Sci,* vol. 53, pp. 3472-3478, 2006.

[9] B. Narasimham, V. Ramachandran, B. L. Bhuva, R. D. Schrimpf, A. F. Witulski, W. T. Holman, L. W. Massengill, J. D. Black, W. H. Robinson, D. McMorrow, "On-chip characterization of single event transient pulse widths", *IEEE Trans. Dev. and Mat. Rel.,* vol. 6, pp. 542-549, 2006.

[10] B. D. Olson, O. A. Amusan, S. Dasgupta, L. W. Massengill, A. F. Witulski, B. L. Bhuva, M. L. Alles, K. M. Warren, and D. R. Ball, "Analysis of parasitic PNP bipolar transistor mitigation using well contacts in 130 nm and 90 nm CMOS technology," *IEEE Trans. Nucl. Sci.,* pp. 894-897, 2007.

2008 IEEE Conference on Microelectronic Test Structures, March 24-27, Edinburgh, UK

6.4

CMOS Latch Metastability Characterization at the 65-nm-Technology Node

Manjul Bhushan, Mark B. Ketchen* and Koushik K. Das*

IBM Systems and Technology Group, 2070 Route 52, Hopewell Junction, NY 12533, USA
*IBM Research, T. J. Watson Research Center, Route 134, Yorktown Heights, NY 10598, USA

ABSTRACT

A new test structure utilizing a differential technique for measuring CMOS latch delay with sub-ps time resolution is described. The latch delay and error count in the metastability region are measured as a function of clock-data delay which can be incremented in 0.1 ps steps. This compact test structure is configured to be placed in the scribe line for characterizing different latch designs and correlating their behavior with model predictions.

INTRODUCTION

A fairly extensive high frequency characterization of CMOS circuits can be carried out in the manufacturing line by using test structures that employ only DC I/O's [1]. However, there are some applications, for example characterization of individual circuits with multiple high speed inputs, where sub-ps time resolution is required, and such tests are best conducted off-line using special high frequency equipment. Measurement of CMOS latch metastability is one such application. Metastability occurs in latches when the data arrival time with respect to the clock trailing edge is sufficiently delayed that the time to resolve the data input increases resulting in erroneous output [2]. In asynchronous clocking schemes, metatstability introduces random errors in the data output. It is therefore important that the latch behavior be fully characterized and correlated with model predictions. Earlier work has experimentally demonstrated such metastability effects with time resolution in the ns range [3, 4, 5]. Recently this effect has been utilized in random number generation applications [6].

In this work we describe a differential technique and test structure for measuring latch delay with resolution in the sub-ps regime, suitable for present day technologies. The test structure has been implemented for characterizing level-sensitive latches at the 65-nm-technology node. The clock-data delay can be incremented in steps as small as 0.1 ps using a mechanically adjustable external delay line. The latch delay is measured as a function of clock-data delay for the falling edge of the data signal and compared to model predictions. The output error count is recorded as the latch is tuned through the metastability region. With straightforward modifications in the present test circuit design, the full range of data and clock input waveforms for both level-sensitive and edge-triggered latches and flip-flops can be investigated.

DESCRIPTION OF TEST STRUCTURE AND METHODOLOGY

The output of a latch circuit depends on the sequence of data (D) and clock (C) signals that it has recently received. As an example, consider the case of the level sensitive latch shown in Fig. 1. Initially the clock input is "1", the data input is "1" and the latch output is "0". If a data input goes to "0" while the clock is at "1", followed by the clock going to "0" at a substantially later time, the latch output will switch to "1" after some characteristic latch delay time τ_L. It will remain at "1", independent of what the data subsequently does, provided the clock remains at "0". In this scenario, as the magnitude of the clock-data time delay, $\Delta t = t(D) - t(C)$, between the falling clock signal and the falling data signal is decreased, the latch delay first remains at τ_L, but eventually begins to increase when $|\Delta t|$ becomes sufficiently small, as indicated in Fig. 2.

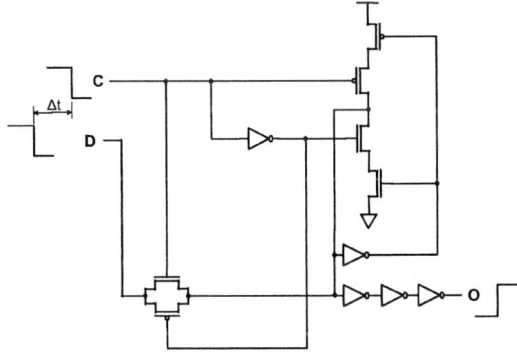

Fig. 1. Circuit schematic of the level sensitive latch.

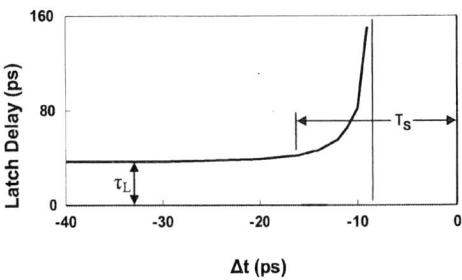

Fig. 2. Simulated latch delay vs. clock-data timing, Δt, at a power supply voltage VDD of 1.0V. The rightmost end of the curve is within 1 ps of the metastability point (vertical line).

978-1-4244-1801-5/08/$25.00 ©2008 IEEE

147

As $|\Delta t|$ is further decreased the latch delay increases exponentially, theoretically reaching infinity at a singular metastability point [2]. To the right of this point the latch does not change state as the data falls to "0". In 65nm technology, at a power supply voltage VDD of 1.0 V, the anomalous latch delay may occur over a range of 10 ps in clock-data delay, with the metastability point being of order 10 ps from the falling clock edge. The concept of set-up time is also frequently introduced in the discussion of latch operation. This set-up time, T_S, is defined as the minimum delay between the arrival of data before the fall of the clock to ensure both that the data is captured by the latch and that the latch delay has not increased significantly beyond τ_L. The value of T_S for the case shown in Fig. 2 is about 15 ps, although its exact value depends upon the specification of a "significant increase in latch delay".

To characterize the latch in this critical region of operation we employ the circuit shown in Fig. 3. This circuit uses two high speed inputs, C and D, and two low speed inputs, a1 and a2, which select the output of the circuit as specified in Table 1. The input and output waveforms are as shown schematically in Fig. 4(a) and 4(b) respectively. Holding a2 at "0" and toggling a1 between "0" and "1", the output of the circuit stimulated by

TABLE 1
Settings a1 and a2 for selecting the output signal

a1	a2	Path
0	0	latch
1	0	data
1	1	clock

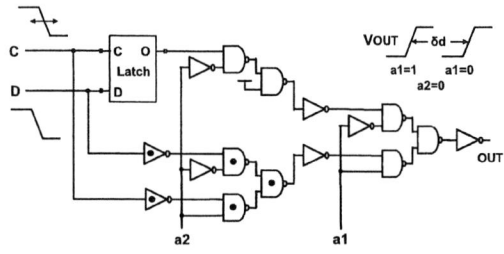

Fig. 3. Differential circuit for measuring various time delays. The waveforms in the upper right correspond to measuring the latch delay.

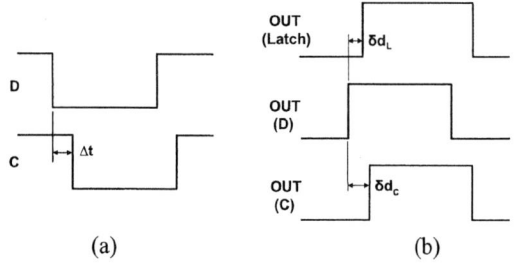

Fig. 4. Schematic (a) input and (b) output waveforms associated with the circuit in Fig. 3.

properly timed falling C and D edges will be rising edges at two locations, the time displacement, δd_L, being the difference in delay between the latch path and the data path, which by design is one inverter delay less than the delay through the latch itself. Holding a1 at "1" and toggling a2 between "0" and "1" the output of the circuit stimulated by falling C and D edges will be rising edges at two locations. The time displacement, δd_c, is the difference in delay between C and D at the input of the latch, assuming that the delays through the physically separate but nominally identical sections of the C and D paths from the latch input through their first common NAND gate are identical. Typically the input waveforms as shown in Fig. 4(a) take the form of nominally identical complementary pulses 2 ns in width, with a variable time delay between them. The output waveforms shown in Fig. 4(b) are all true pulses approximately 2 ns in width, and the pulse repetition rate is typically 1 μs. Note that at the end of the active 2 ns interval, the latch resets to its initial value and remains there until the next pulses arrives.

Fig. 5 shows the configuration of the measurement set up. The 2 ns wide C and D pulses are derived from a single pulse stream from an HP 8131A pulse generator at a frequency of, for example, 1 MHz. A second pulse from the complementary output of the same channel of the pulse generator serves as the trigger for an HP 54120A sampling scope that is used to measure the waveform at the output of the experiment. The relative timing between the C and D pulses is set with a mechanically adjustable delay line DLC with a resolution of 0.1 ps. We use a manually operated phase adjuster, Part# P1607DRE, from Advanced Technical Materials, with an LED readout directly calibrated in ps over a total range of 250 ps. For a given set of data unidirectional travel is used to avoid backlash error in the delay

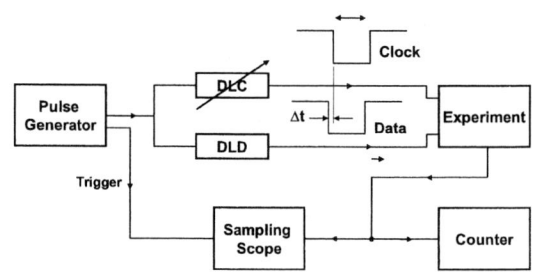

Fig. 5. Configuration of apparatus for measuring latch characteristics. The clock and data pulses have a common source with relative timing set by mechanically adjustable delay line DLC with respect to fixed delay line DLD.

line. The experiment to measure δd_L then consists of measuring δd as a function of the setting of DLC. The resolution of δd_L is ultimately limited by the jitter between the leading edges of the C and D pulses, the jitter between the true and complement outputs of the pulse generator, and the intrinsic jitter of the sampling scope itself, all of which have standard deviation values ≤ 1 ps, as will be subsequently discussed. The output from the experiment is simultaneously routed to an Agilent 53131A counter and the count rate measured as a function of the DLC setting. The count rate transitions from the drive frequency to zero as the metastability point is passed. Alternatively, by setting a1 =1 and toggling a2 at a fixed DLC setting, DLC_{ref}, the δd_c value can be used to directly measure Δt. The Δt value at any DLC setting, DLC_L, is given by

$$\Delta t = DLC_{ref} - DLC_L - \delta d_c \quad (1)$$

Fig. 6 depicts the test structure in which the latch metastability experiment is placed. This structure is 200 μm x 2500 μm in size and incorporates up to eight experiments, each using up to five high speed inputs and multiplexed to a common high speed output. There is a 1 x 25 padset with 10 ground pads. The six high speed I/O's are all wide bandwidth ground-signal-ground probes that share adjacent pads. The set of experiments and the I/O circuitry are powered by different power supplies but share a common ground. Both power supply sectors are provided with substantial decoupling capacitance within the test structure itself and also on the probe card (from GGB Industries) in close proximity to the silicon surface.

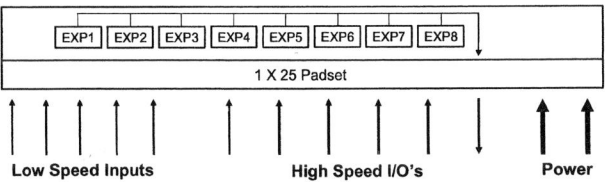

Fig. 6. Configuration of test structure. Physical dimensions are 200 μm x 2500 μm.

The differential techniques featured in the latch metastability circuit and experimental configuration provide a very high degree of common mode rejection with many potential sources of inaccuracy being eliminated by subtraction. The output waveform shapes for clock, data and latch are nearly the same with small differences in timing of the leading edges of these isolated pulses. The I/O driver which switches 300 mV into the 50 Ohm output line is on only 0.2% of the time, and it fires after the critical timing path segments are completed. Any I/O power bus disturb that may carry over from one pulse to the next has the same effect on C, D, and latch output and is subtracted in the determination of δd_L and δd_c.

EXPERIMENTAL RESULTS

The measured results shown here are from experimental 65 nm CMOS PDSOI (Partially Depleted Silicon on Insulator) hardware. The same test structure design can be used for bulk Si CMOS technology also. All measurements shown are made on a level-sensitive latch design. The latch delay, δd_L, is measured as a function of DLC setting. Calibration for clock-data delay is obtained by setting a1 = "1", switching a2 from "1" to "0" and measuring δd_c at a setting DLC_{ref}. This calibration is then applied to get the clock-data delay, Δt, for all values of DLC_L as described in Equation (1).

Fig. 7 shows the measured δd_L vs. Δt for different values of the power supply voltage, VDD. All waveform positions are measured with the sampling scope set with a voltage widow of 5mV centered on the rising transient. In all cases 100 samples are acquired and averaged to obtain the mean value. For the clock and data paths and also for the latch path well away from the metastability region, the histogram of these points as presented by the sampling scope is approximately Gaussian with a standard deviation, σ, of ~1 ps. This implies that with 95% confidence the measured mean of the distribution is within $\pm$ 0.2 ps of the true mean and that values of δd, which are the difference of two such measurements, should be accurate to $\pm$ 0.3 ps [7]. Since the clock-data timing calibration is based on a single δd_c measurement, the zero of the Δt axis has an uncertainty of $\pm$ 0.3 ps. However the relative uncertainty of Δt corresponding to different δd_L measurements is much smaller and is governed by the precision of the adjustable delay line. The jitter of the time measurement corresponding to the latch path increases as the metastability point is approached, with the last points on the delay curves shown in Fig. 7 having a σ of 2.5-3.0 ps.

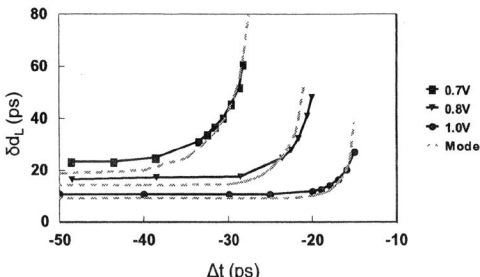

Fig. 7. Measured latch delay, δd_L, vs. clock-data delay, Δt, at different values of VDD.

We believe that this increased jitter arises primarily from the interaction of the clock-data timing jitter with the true latch delay vs. clock-data delay characteristic. Since this characteristic is nonlinear, becoming ever steeper as the metastability point is approached, the measured jitter distribution broadens and becomes skewed in the direction of the metastability point. This behavior has been reported and studied in detail in earlier technologies [4] and was observed qualitatively in our measurements. This effect will tend to make the measured δd_L values increasingly greater than the actual values as metastability is approached. In Fig. 7 the hardware data are compared to the model predictions from circuit simulations using a nominal FET model and full parasitic extraction. The agreement between the model and hardware is reasonable considering the model has not been adjusted for process deviations from nominal. Our data do indicate that the experimental δd_L values increase less rapidly than the model predicts as the metastability point is approached, while the skew effect would tend to make them increase more rapidly than model predictions. The estimated set up time, T_s, of the latch is about 20 ps at VDD = 1.0V, increasing to about 40 ps at VDD = 0.7V.

In Fig. 8, δd_L values, together with the number of valid counts per second with output equal to "1", are plotted against Δt for a different (slower) chip. If there were no jitter or other noise in the system the transition in count rate at the metastability point would be a perfect step function of zero width. In the presence of Gaussian distributed clock-data jitter one would expect the width of the transition to reflect the width of the jitter distribution with the metastability point at the value of Δt where the number of counts per second is reduced to half. The full width of the metastability transition (>99% of the transition) is 1 ps at VDD = 1.0V, increasing to 2 ps at VDD = 0.7V. Assuming that the non-zero width of the transition is due entirely to jitter in clock-data timing, this implies a maximum jitter (at least $\pm 3\sigma$) of about 1 ps and 2 ps at 1.0V and 0.7V respectively. This very precisely controlled clock-data timing which is facilitated by having both clock and data edges originate from the same pulse generator transition is key to our latch characterization technique.

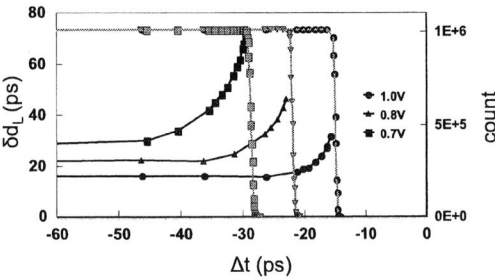

Fig. 8. Latch delay and number of counts of captured events vs. clock-data delay at different values of VDD.

Statistical and systematic variations in the characteristics of FETs in gates unique to the output clock and data paths as indicated by dots in Fig. 3 may introduce a small error in the measurement. The systematic differences are minimized by making the physical layout for such gates identical. Circuit simulations show that the standard deviation for Δt from random statistical variations is < 0.4 ps. In PDSOI technology, the circuit delay may vary by a few percent depending on its switching history [8]. In implementing this test structure in PDSOI, the circuit configuration is such that data, clock and latch paths all switch at the nearly same time interval from the previous switching event, thus eliminating any potential history effect induced errors. We have further verified the robustness of our measurement scheme by demonstrating that the experimental results do not change as the I/O VDD is varied.

EXTENSIONS OF THE TECHNIQUE

In this work we have focused on a level sensitive latch and, in particular, the operation in which a falling D edge arrives in the vicinity of a falling C edge for that level sensitive latch. All aspects of both rising and falling D edges in the vicinity of both rising and falling C edges, for both level sensitive and edge triggered latches can be made with a straightforward modifications of the test circuit in Fig. 3. However, care must be taken to preserve the full common mode rejection features of our approach that enable sub-ps precision in the measurements.

Fig. 9 shows the modified circuit to study events with both rising and falling D edges in the vicinity of a falling C edge for a single level sensitive latch. Table 2 provides the digital values of inputs a1, a2, and a3 for selecting different paths through the circuit to enable the timing measurements. First consider the case of a3 = "1" corresponding to rows 2, 3, and 4 of Table 2. For complementary data and clock input pulses this circuit functions in a way very similar to that of the circuit in Fig. 3, with waveforms analogous to those shown in Fig. 4. The latch operation in which a falling D edge arrives in the vicinity of a falling C edge can again be analyzed as previously described. Next consider the case of a3 = "0" corresponding to rows 5, 6, and 7 of Table 2. Again the circuit is driven with complementary

TABLE 2
Settings a1, a2, and a3 for selecting the output signal

a1	a2	a3	Path
0	0	1	latch
0	1	1	data
1	0	1	clock
0	0	0	latch
0	1	0	data
1	1	0	clock

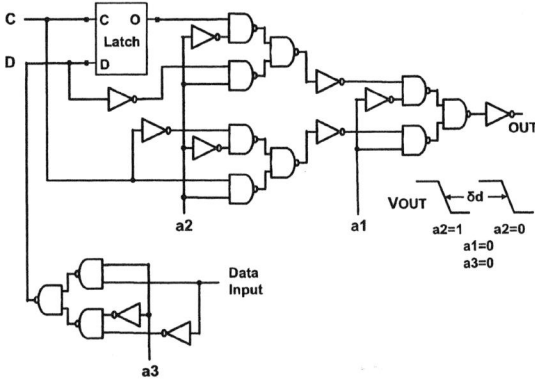

Fig. 9. Extension of the circuit shown in Fig. 3 for measuring both rising and falling D edges. The waveforms near the output correspond to measuring the latch delay for a rising data edge in the vicinity of a falling clock edge.

clock and data pulses generated in the manner shown in Fig. 4. In this case, however, the data pulse is inverted by the circuit preceding the D input of the latch so that now the rising D edge arrives in the vicinity of the falling C edge. All of the output pulses as selected by a1 and a2 settings are now complementary pulses of approximately the same width, just as they were all true pulses of approximately the same width in the first case. The timing information can again be measured as the small timing differences between the falling leading edges of the different output pulses. Notice that in this case when the clock-data delay is measured (a2 = "1" and a1 toggled between "1" and "0") a correction corresponding to one inverter delay must be made to account for the additional inverter present in the data path compared to the clock path. Operated as described this circuit configuration preserves all the common mode rejection features inherent in the original design.

SUMMARY

The latch metastability test structure described here is implemented at the 65-nm-technology node in PDSOI technology. Latch delay is measured as a function of calibrated clock-data delay for model-to-hardware correlation. Highly differential circuit design and measurement techniques enable the measurements to be made with sub-ps precision. The width in time of the metastability transition as determined from the output error count as the latch is tuned through this region shows that the jitter in the clock-data timing has a standard deviation of less than 0.2 ps at VDD = 1.0V.

ACKNOWLEDGEMENT

We gratefully acknowledge Carl Anderson for suggesting this experiment, Thomas Roewer and Mark B. Ritter for ideas on circuit design, Keith Jenkins for advice on measurements, and the support of many individuals in the IBM Systems and Technology Group in the implementation of the test structures.

REFERENCES

[1] Mark B. Ketchen and Manjul Bhushan, "Product representative "at speed" test structures for CMOS characterization" IBM J of R&D, vol. 50, 2006, pp. 451-468.

[2] Harry J. M. Veendrick, "The behavior of flip-flops used as synchronizers and prediction of their failure rate", IEEE J. of Solid-State Circuits, vol. SC-15, pp. 169-176, 1980.

[3] Clemenz L. Portmann and Teresa H.Y. Meng, "Metastability in CMOS library elements in reduced supply and technology scaled applications", IEEE J. of Solid-State Circuits, vol. 30, pp. 39-46, 1995.

[4] Branka Medved Rogina and Bozidar Vojnovic, "Metastability evaluation method by propogation delay distribution measurement", The Fourth Asian Test Symposium, pp. 40-44, 1995.

[5] Lee-Sup Kim and Robert W. Dutton, "Metastability of CMOS Latch/Flip-flop", IEEE Journal of Solid-State Circuits, vol. 25, pp.942-950, 1990.

[6] Carlos Tokunaga, David Blaauw and Trevor Mudge, "True random number generator with a metastability-based quality control", Proceedings 2007 IEEE International Solid-State Circuit Conference, pp. 404-405.

[7] Statistical Methods for Six Sigma in R&D Manufacturing (2003), Anand M. Joglekar, Wiley Interscience.

[8] S.K. H. Fung, N. Zamdmer, P. J. Oldiges, J. Sleight, A. Mocuta, M. Sherony, S-H. Lo, R. Joshi, C.T. Chuang, I. Yang, S. Crowder, T.C. Chen, F. Assaderaghi, and G. Shahidi, "Controlling floating-body effects for 0.13μm and 0.10μm SOI CMOS," 2000 IEDM Technical Digest, pp. 231-232.

SESSION 7

MOS Modeling and Characterisation

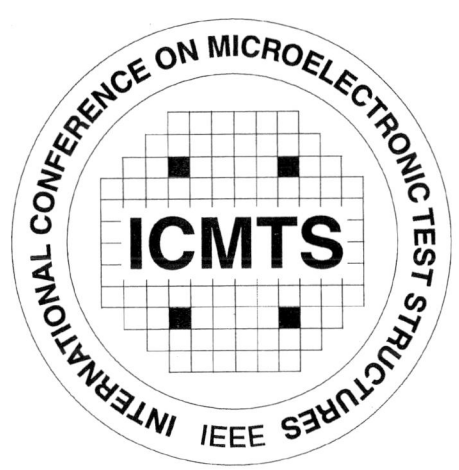

March 26, 13:40–15:00

Co-Chairs: Kevin McCarthy, *University College Cork, Ireland*
Bill Verzi, *Agilent Technologies, USA*

2008 IEEE Conference on Microelectronic Test Structures, March 24-27, Edinburgh, UK

7.1

Characterization of MOSFETs Intrinsic Performance using In-Wafer Advanced Kelvin-Contact Device Structure for High Performance CMOS LSIs

Rihito Kuroda[1], Akinobu Teramoto[2], Takanori Komuro[3], Weitao Cheng[2], Syunichi Watabe[1], Ching Foa Tye[1], Shigetoshi Sugawa[1] and Tadahiro Ohmi[2]

[1] Graduate School of Engineering, Tohoku University
Aza-Aoba 6-6-10, Aramaki, Aoba, Sendai 980-8579
[2] New Industry Creation Hatchery Center, Tohoku University
[3] Agilent Technologies International Japan Ltd.

ABSTRACT

In this work, a new MOSFETs characterization method utilizing in-wafer Kelvin contact device structure is developed. The developed method can eliminate the parasitic series resistances in MOSFETs and allows us to characterize the short channel transistor intrinsic current-voltage characteristics as well as the quantitative effects of the parasitic series resistance to the device performance, very stably and accurately. The developed analysis is useful for the characterization and parameter extractions of fabricated MOSFETs for the device/process development and optimization of ultra-thin gate insulator short channel CMOS LSIs for higher performance.

INTRODUCTION

As the device scaling has reached the sub-hundred nanometer regime, its outcome in current drivability has become less fruitful than that simply estimated by the scaling rule. This standstill originates from the several reasons such as gate leakage current (I_G), immaturity of the gate insulator/silicon interface, the source/drain series resistance and capacitance parasitic effect, mobility degradation, velocity saturation limit and so on [1-3]. Then, to breakthrough these conflicts, it is primarily important to be able to quantitatively characterize and differentiate the effect of each physical mechanism that degrades or limits the intrinsic device performance and to set the effective and quantitative targets of research and development. However, it has become more difficult to do so since more and more physical mechanisms coexist in complex in the simply measured MOSFET electrical characteristics in the scaled down devices. Among the various origins that degrade the device performance, series resistances in MOSFETs such as metal wires, silicon-metal or silicon-silicide contacts and source/drain diffusion region degrade the intrinsic transistor performance more significantly than others in more scaled down devices. It is because that these series resistance increase their values with device size scaling due to the narrower wires, smaller contact area and shallower junctions. There are many ongoing researches to increase the intrinsic device performance by increasing the effective carrier mobility of

MOSFETs [4-7]. However, even though the carrier mobility is increased, total current drivability of MOSFETs cannot be increased unless series resistance is sufficiently decreased concurrently [6]. In other words, it is difficult to even characterize the effect of the technologies to increase carrier mobility at the development stage if large series resistance exists.

Then, this work provides a useful method using the in-wafer advanced Kelvin-contact (AKC) device structure. Using the developed method, we are able to quantitatively analyze the MOSFETs intrinsic electrical characteristics without parasitic series resistances and the effects of the series resistance to the device characteristics.

Advanced Kelvin-Contact (AKC) Method

Current-Voltage characteristics of MOSFETs are affected by the parasitic resistance schematically described in Figs.1 (a-b), where Figs.1 (a) and (b) show the physical positions of the series resistance and equivalent circuit model of MOSFETs, respectively.

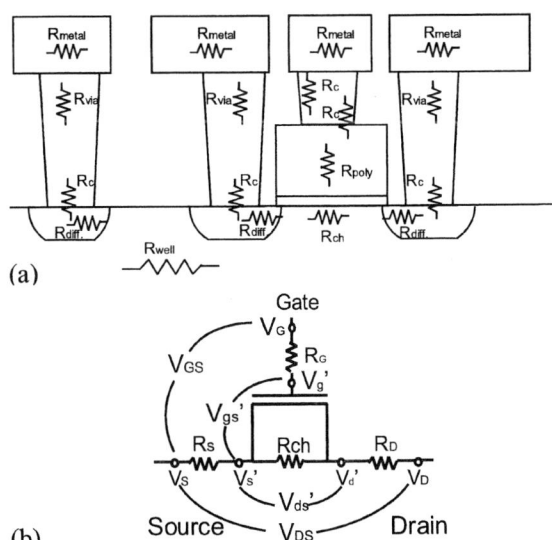

Fig. 1 (a) Schematic view of the channel and parasitic resistance in MOSFET and (b) equivalent circuit. In (b), internal and external or input voltages to the MOSFET are indicated.

978-1-4244-1801-5/08/$25.00 ©2008 IEEE

The following equations that are derived from the long channel MOSFET drain current (I_D)- gate voltage (V_G) equations show the negative feedback effect in I_D-V_G characteristics due to the source and drain terminals series resistance (R_S and R_D).

$$g_{meff} = g_{mi}\left(1 - I_{DS} \cdot (R_S + R_D)/V_{DS}\right), \text{ in linear region} \qquad (1)$$

$$g_{meff} = g_{mi}\left(1 - I_{DS} \cdot R_S / (V_{GS} - V_{th})\right), \text{ in saturation region} \qquad (2)$$

where, g_{mi} and g_{meff} are intrinsic and effective trans-conductance, respectively. These equations indicate that the R_S and R_D degrade the device I_D-V_G characteristics more significantly as transistor intrinsic current drivability increases.

Then, to characterize the MOSFET intrinsic I-V characteristics, the AKC device structure is developed as shown in Figs.2 (a-b), where Figs.2 (a) and (b) show the layout and the schematic image of the source and drain contacts in AKC device structure, respectively.

In this structure, two contacts array are aligned for both source and drain active regions, namely current and potential lines indicated by S_C, D_C and S_P, D_P, in Fig. 2(b) respectively [8]. Similarly for gate and well terminals, two metal wires are connected at the gate electrode and the well active regions, respectively. Using the two contacts for each terminal as Kelvin-contacts [9-10] that connect each other very closely to the intrinsic device, we can eliminate the series resistance and measure the quasi-intrinsic device I-V characteristics.

Devices used in this work are fully-depleted (FD) SOI CMOSFETs with dual poly-silicon gate electrodes. Gate insulator is thermally oxidized 2.6 nm thick SiO_2. Thickness of the SOI and BOX layers are 40 and 100 nm, respectively. SOI layer impurity concentration is 6×10^{16} cm^{-3}. The contact electrode and metal wires are formed by sputtered Al-Si (Si: 0.5 %). Resistivity of n$^+$ and p$^+$ diffusion regions, contact resistance and sheet resistance of metal layer are summarized in table 1.

Table 1
The resistivity of various regions and contacts in the fabricated device.

region	resistivity
n$^+$ silicon region	9.12×10^{-4} Ω-cm
p$^+$ silicon region	1.07×10^{-3} Ω-cm
contact to n$^+$ silicon region	2.79×10^{-7} Ω-cm^2
contact to p$^+$ silicon region	1.62×10^{-7} Ω-cm^2
Al-Si metal layer	1.04×10^{-1} Ω/square

Results and Discussion

Fig.3 shows the measured I-V characteristics of a nMOSFET with W_{eff}/L_{eff}=20.0/0.30 μm in linear region by the conventional measurement, that is, only potential lines indicated by S_p, D_p, G_p and W_p in Fig. 2(a) are used. The conventional measurement is carried out in the CASCADE Microtech's probing system SUMMIT1200 with Agilent 4156C parameter analyzer's signal monitoring units (SMU). Thus in this case, Kelvin contact for each terminal is at the measurement probes.

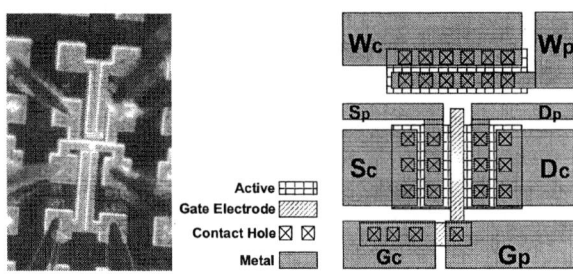

(a)

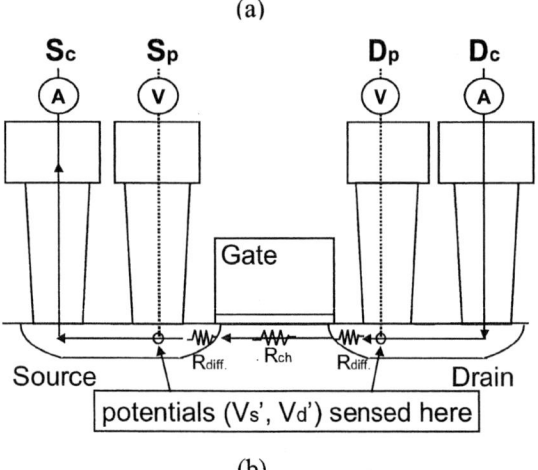

(b)

Fig. 2 (a) Microscopic picture and layout view of the developed advanced Kelvin-contact (AKC) MOSFET structure and (b) schematic view of the source and drain Kelvin-contacts in AKC structure. Using this structure, we can eliminate the parasitic resistance of probes, metal wires, silicon-metal contacts up to the Kelvin contact point at the n$^+$ diffusion under the S_P and D_P contacts.

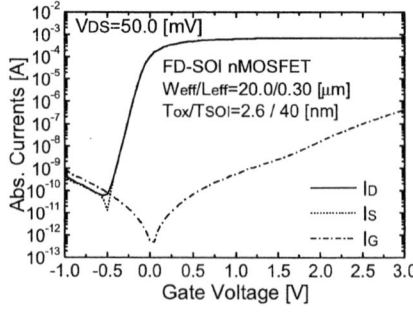

Fig. 3 I_D-V_G characteristics of a nMOSFET measured by the conventional measurement.

156

During the conventional measurement, voltage meters are connected to the lines indicated by S_C, D_C, G_C and W_C in Fig. 2 (a) to sense the voltage drop up to the connection point to the S_p, D_p, G_p and W_p lines. Fig. 4 (a) shows the input and internal source and drain voltages (V_S and V_D, and V_s' and V_d') and calculated internal source to drain voltage (V_{ds}') as a function of drain current during the measurement in Fig. 3. It is obvious that V_s' and V_d' deviate from V_S and V_D and V_{ds}' decreases as I_D increases. Fig. 4 (b) shows the series source and drain resistance (R_S and R_D) calculated from the result shown in Fig. 4 (a) by the following equation.

$$R_{S,D} = \left| \left(V_{S,D} - V_{s,d}' \right) / I_D \right| \qquad (3)$$

R_S and R_D are stably 6.04 Ω for various drain current indicating that measurement works accurately. Also, the source and drain series resistance from the probing pad to the metal-n$^+$ silicon contacts are 6.00 Ω according to the calculation from Table 1. Then, the result of R_S and R_D shown in Fig. 4(b) are in good agreement with the calculation of series resistance from the contact and the metal wire resistance.

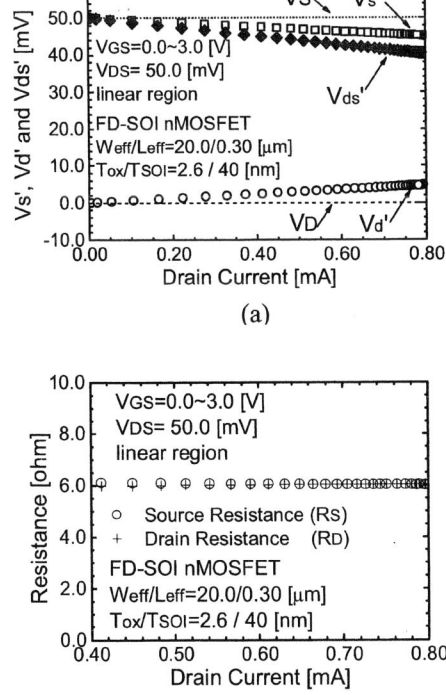

(a)

(b)

Fig. 4 (a) Internal source and drain voltage (Vs' and Vd') and calculated internal source to drain voltage (Vds') as function of drain current measured in Fig.3. (b) Series source and drain resistance (R$_S$ and R$_D$) calculated from Fig.4 (a).

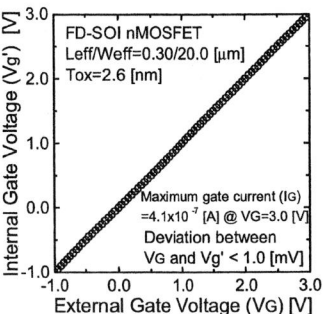

Fig.5 Internal gate voltage (Vg') as function of the input gate voltage (V$_G$).

Fig.5 shows the internal gate voltage (V_g) as a function of input gate voltage (V_G). The deviation is negligibly small because gate current (I_G) is small in this device geometry. However, when the gate insulator is ultra-thin with large leakage current, internal gate voltage drop due to the gate current would negatively impact the current drivability due to the gate over drive voltage reduction.

Figs.6 and 7 show the I_D-V_G and g_m-V_G characteristics of FD-SOI nMOSFET measured by various Kelvin-contacts combinations when V_{DS}=50.0 mV for Fig. 6 and when V_{DS}=2.0V for Fig. 7, respectively. Here, for the result labeled by "All-Kelvin", AKC is applied to gate, source and drain terminals. For "Source NOT Kelvin", AKC is applied to gate and drain terminals but not to source terminal. For "Drain NOT Kelvin", AKC is applied to gate and source terminals but not to drain terminal. For "All NOT Kelvin" (conventional), AKC is not applied to any of the terminal.

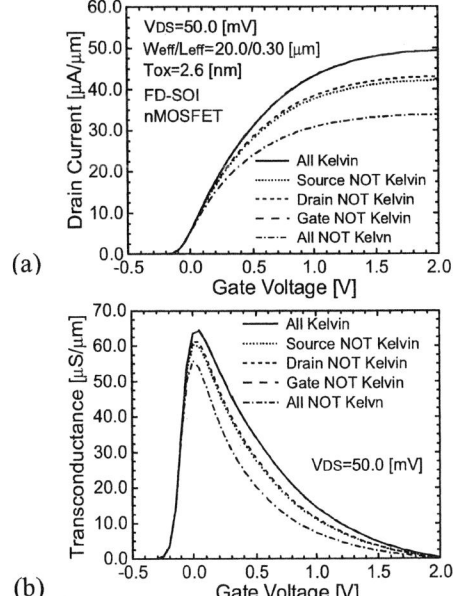

(a)

(b)

Fig. 6 (a) I$_D$-V$_G$ and (b) g$_m$-V$_G$ characteristics of FD-SOI nMOSFET measured by various Kelvin-contacts combinations when V$_{DS}$=50.0 mV

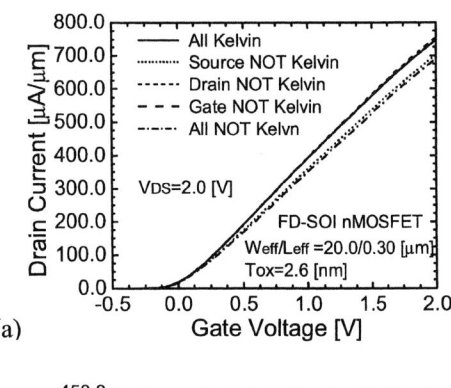

(a)

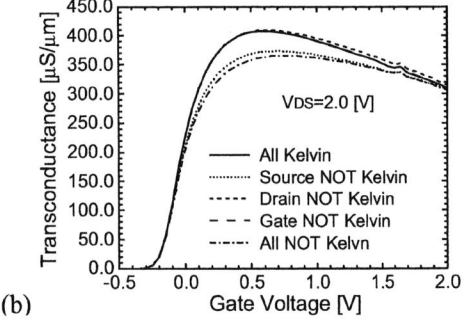

(b)

Fig. 7 (a) I_D-V_G and (b) g_m-V_G characteristics of FD-SOI nMOSFET measured by various Kelvin-contacts combinations when V_{DS}=2.0V

The results clearly show that R_S and R_D degrade the intrinsic transistor current drivability as indicated by equations (1-2). That is, R_S and R_D affect the I-V characteristic similarly in the linear region, while R_S effect more significantly than R_D in the saturation region. Fig. 8 shows the I_D-V_D characteristics for n- and pMOSFET measured by conventional measurement and the AKC method applied to all of the terminals, respectively. The measured devices intrinsically exhibit very high current drivability, which are degraded by the series resistance.

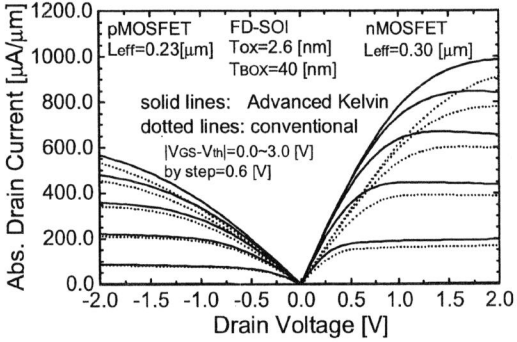

Fig. 8 I_D-V_D characteristics of FD-SOI n- and p-MOSFET measured by conventional and the AKC method, respectively.

Then, the effect of the series resistance to the current drivability is quantitatively analyzed for different channel length as shown in Figs.9(a-b), where the improvement in measured I_D using the AKC Method from the conventional measurement is plotted as a function of L_{eff} for linear region and saturation region, respectively. The results clearly show that intrinsic current drivability is degraded more significantly for shorter L_{eff}, such as 33 % in linear region of nMOSFET with L_{eff}=0.30 μm. These results indicate that use of the AKC method increases its significance for shorter channel devices for the characterization of the MOSFETs, such as the characterization of intrinsic and actual current drivability, parameter extractions and so on. Figs.10(a-b) show the I_D of nMOSFET with L_{eff}=0.30 μm as a function of the length of gate to contact holes measured by conventional and the AKC method in linear and saturation regions, respectively. In (b), saturation region, for the conventional measurement, I_D tends to saturate. It is due to the effect of the contact resistance that becomes more significant than the series resistance in gate to contact silicon region at the gate to contact hole length of 1.10μm. However in the AKC method, the drain current without contact resistance increases for shorter gate to contact length. This result quantitatively shows us the importance of the conjunction of the contact resistance and the gate to contact holes length reductions to increase current drivability of MOSFETs.

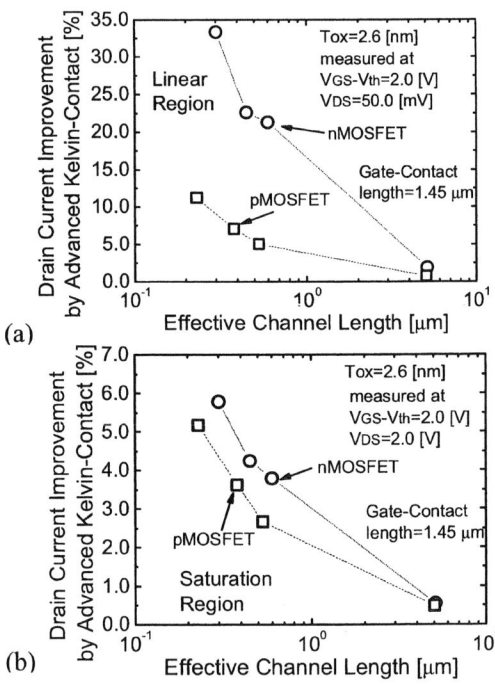

(a)

(b)

Fig. 9 Drain current improvement from the conventional measurement to the AKC method for n- and pMOSFETs as function of effective channel length for (a) linear region and (b) saturation region.

158

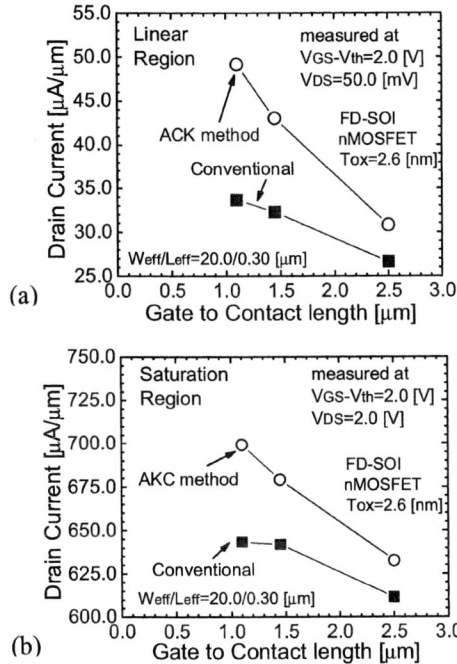

(a)

(b)

Fig. 10 Drain current of a nMOSFET as function of the length of the gate to source and drain contact holes measured by the conventional and the AKC method for (a) linear region and (b) saturation region, respectively.

Furthermore, Fig. 11 shows the results of the set of the I_D-V_G measurement and pad probing repeated for 100 times by conventional and AKC method. In the conventional measurement, measurement stability and accuracy are dominated by the conditions of the probing to the pads and sometimes very high resistance appears at the probing points. On the contrary, for the AKC method, series resistances including the probing points do not affect the measurement and the stability and the accuracy are highly increased as indicated by the very small RMS error for 100 times repeated measurements.

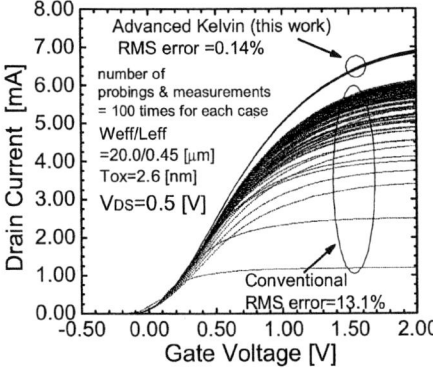

Fig. 11 I_D-V_G characteristics of 100 times of probings and measurements by conventional and AKC method. Measurement stability in AKC method is shown to be very high.

Conclusions

The AKC method allows us to characterize the short channel transistor intrinsic I-V characteristics as well as the quantitative effects of the parasitic series resistance to the device performance, very stably and accurately. The developed analysis is very useful for the detailed characterization of fabricated MOSFETs to evaluate the device performance and process quality for the effective research target determination and developments as well as parameter extractions for physics based compact MOSFET models. Consequently it is very useful for the device/process development and optimization of ultra-thin gate insulator short channel CMOS LSIs for higher performance.

Acknowledgement

This work was supported partly by Japanese Ministry of Education, Culture, Sports, Science and Technology under Grant-in-Aid for Specially Promoted Research (project No. 18002004), also partly by Japan Society for the Promotion of Science Research Fellowship Program (No. 19˙1356). Authors would gratefully acknowledge Agilent Technologies International for their support.

References

[1] P. R. Karlsson, K. O. Jeppson, "Extraction of seires-reistance-independent MOS transistor model parameters," IEEE ED. Lett., vol.13, pp.581-583, 1992.
[2] G. Timp, A. Agarwal, F. H. Baumann, T. Boone, M. Buonanno, R. Cirelli et al., "Low leakage, ultra-thin gate oxides for extremely high performance sub-100nm nMOSFET's," IEDM Tech. Dig., pp.930-933, 1997.
[3] S. Mathew, L. K. Bera, N. Balasubramanian, M. S. Joo and B. J. Cho, "Channel mobility degradation and charge trapping in high-k/metal gate NMOSFETs," Thin Solid Films, vol,462-463, pp.11-14, 2004.
[4] A. Teramoto, T. Hamada, M. Yamamoto, P. Gaubert, H. Akahori, K. Nii, et al., "Very high carrier mobility for high-performance CMOS on a Si(110) surface," IEEE T-ED, vol.54, pp.1438-1445, 2007.
[5] T. Teruka, S. Nakaharai, Y. Moriyama, N. Sugiyama, and S. Takagi, "Selectively-formed high mobility SiGe-on-Insulator pMOSFETs with Ge-rich strained surface channels using local condensation technique," 2004 VLSI Symposium Tech. Dig., pp.198-199, 2004.
[6] T. Ohmi, A. Teramoto, R. Kuroda and N. Miyamoto, "Revolutional progress in silicon technologies exhibiting very high speed performance over a 50GHz clock rate," IEEE T-ED, vol.54, pp.1471-1477, 2007.
[7] W. Cheng, A. Teramoto, M. Hirayama, S. Sugawa and T. Ohmi, "Impact of improved high-performance Si(110)-oriented metal-oxide-semiconductor field-effect transistors using accumulation-mode fully depleted silicon-on-insulator devices," JJAP. vol.45, 3110-3116, 2006.
[8] N. Kasai, H. Mori, T. Matsuki, I. Yamamoto, and K. Koyama, "Separation of intrinsic and parasitic MOSFET parameters using a multiple built-in Kelvin test structure," in Proc. IEEE 1997 ICMTS, pp.198-202, 1997.
[9] F. Werner, "A Method of Measuring Earth Resistivity," Bulletin of the Bureau of Standards, Vol. 12, pp. 469-478, 1915.
[10] Y. Shimizu, M. Nakamura, T. Matsuoka, and K. Taniguchi, "Test Structure for Precise Statistical Characteristics Measurement of MOSFETs," in Proc. IEEE 2002 ICMTS, pp.49-54, 2002.

New Y-Function-Based Methodology for Accurate Extraction of Electrical Parameters on Nano-Scaled MOSFETs

Dominique Fleury*†, Antoine Cros*, Hugues Brut* and Gérard Ghibaudo†

*STMicroelectronics, 850 rue Jean Monnet, F-38926 Crolles, France
†IMEP, Minatec, 3 Parvis Louis Neel, 38016 Grenoble, France
Email: dominique.fleury@st.com, Telephone: +33 (0) 438 923 314

Abstract—We developed a new Y-function-based extraction methodology to overcome the difficulties encountered by applying the conventional techniques. Our method relies on a robust recursive algorithm which requires a limited number of input parameters on which the results have a weak dependence, and so an increased reliability. The obtained results are in line with the previous methods, but show an improved accuracy. Finally, parameter extraction performed through this technique has provided accurate and reliable results over a large range of MOSFET architectures.

Index Terms—Y-function, compact model, CMOS, low field drain current, electrical parameters, MOSFET

I. INTRODUCTION AND STATE OF THE ART

The extraction of electrical parameters allows understanding the physics of MOSFETs, and so is a main concern for circuit design and device engineering. Although many techniques have been developed during the past decades [1]–[8], only a few of them remain reliable for modern CMOS technologies in which aggressive downscaling and numerous technological enhancements have been applied. The shift-and-Ratio (S&R) and "total resistance"-based techniques [5], [7], [8] assume a constant channel mobility (μ) between the different gate lengths. Consequently they fail for analysing submicron MOS-FETs since previous studies disclose that halos [9], [10], strain [11] or neutral defects [12] change the mobility as a function of the channel length (L).

Y-function-based techniques [1]–[4] relay on the low field drain current model (BSIM3v3-like [13]) and does not require any assumption about the $\mu(L)$ behavior. This method has been first introduced by Ghibaudo *et al.* [1], then completed by including a quadratic mobility attenuation factor (θ_2) [2], [14] to consider the strong impact of the surface roughness on the electrical transport in thin gate oxide MOSFETs.

Nevertheless, the θ_2-effect (*i.e.* the surface roughness effect) is often present on modern devices. It causes non-linearity of the Y-function and makes the extraction technique require several input parameters to define the extraction range where θ_2 can be neglected without impacting the results. This problem has already been noticed by Tanaka *et al.* who tried introducing new parameter-sensitive functions in the extraction procedure [3], then a corrective factor (Φ) to suppress the θ_2-effect [4] and linearize the Y-function for performing a straitforward extraction (as in [1]). However the results obtained by this way are very sensitive to the choice of Φ, which may be sometimes doubtful especially if the measurement is noisy.

Globally, none of these techniques enable to perform reliable extractions for all kinds of nano-scaled MOSFETs without being impacted by some uncertainties resulting from the choice of several input parameters. Thus, we propose a new Y-function-based technique relying on a recursive algorithm and gathering all requirements to overcome the drawbacks of the other techniques. It provides robustness and accuracy over all model parameters and has shown coherent results on several technologies.

II. IMPROVED LOW FIELD $I_D(V_G)$ MODEL

Our electrical parameter extraction is performed on $I_D(V_G)$ characteristic, in inversion regime ($V_G > V_{th}$, where V_{th} is the threshold voltage) and for low lateral electric field (usually $V_D = 25$mV). It relies on the well known drain-current relationship (1) [1], [2], [14] which is a BSIM3v3-like model [13].

$$I_D = \frac{\beta V_D V_{Gt}}{1 + \theta_1 V_{Gt}} \tag{1}$$

In (1), V_{Gt} is the gate drive voltage ($V_{Gt} = V_G - V_{th} - V_D/2$), β is the current factor of the transistor and θ_1 is a linear mobility attenuation factor. The latter represents the mobility limitation caused by phonon scattering [15] and includes the series resistance effect (R_{SD}): $\theta_1 = \theta_{1,0} + R_{SD}\beta$ [1]. As detailed in [16], the θ_2-effect becomes significant for thin gate oxide devices, in strong inversion, when the perpendicular electric field is high enough to confine the carriers close to the oxide-channel interface. This effect cannot be well modeled by a simple linearly-dependent mobility attenuation approach, which was established for thick gate-oxide devices ($t_{ox} = 22$nm) when this phenomenon was still insignificant [1]. As a consequence the θ_2-effect is no more negligible on modern devices (t_{ox} is generally thinner than 20Å) and requires to upgrade the original model (1) by replacing the mobility attenuation with a second-order function of V_{Gt} [14]. Furthermore Rais *et al.* have noticed that the θ_2-effect is triggered for higher gate biases than V_{th}, since it becomes significant only for high electric fields [17]. As a consequence a complete model must consider the θ_2-effect to fit with the modern MOSFETs behavior, but also converge towards (1) to remain fully compatible with the previous technologies where θ_2 is negligible. The general form of a such model can be written as (2), where α_1 and β' have be to determined as a function of θ_1, $\theta_2 = \alpha_2$ and β. In (2), a polynomial function of

V_{Gt} as been set to model the mobility attenuation and the late trigger of the θ_2-effect has been considered through shifting V_{th} by ΔV_{th} in the quadratic term in V_{Gt}.

$$I_D = \frac{\beta' V_D V_{Gt}}{1 + \alpha_1 V_{Gt} + \alpha_2 (V_{Gt} - \Delta V_{th})^2}, \quad \alpha_2 = \theta_2 \quad (2)$$

Obviously, (2) must be fully compatible with the previous technologies for which $I_D(V_G)$-curves are still well modeled using (1). The first consequence is that (2) must match with the well-known linear $I_D(V_G)$-relationship when V_{Gt} is close to 0 (3). The latter gives $\beta' = \beta(1 + \theta_2 \Delta V_{th}^2)$ which means that introducing θ_2 and ΔV_{th} in the model acts as if the current factor was virtually increased.

$$I_D = \beta V_D V_{Gt} + o(V_{Gt}) \quad (3)$$

Finally, (2) should be equal to (1) when the θ_2-effect is weak enough to be neglected (4); it comes immediately $\alpha_1 = \theta_1$ (knowing $\beta'(\theta_2 = 0) = \beta$).

$$\lim_{\theta_2 \to 0}(2) = (1) \Leftrightarrow \frac{\beta V_D V_{Gt}}{1 + \alpha_1 V_{Gt}} = \frac{\beta V_D V_{Gt}}{1 + \theta_1 V_{Gt}} \quad (4)$$

The complete model can thus be written as (5), where β' and α_1 have been expressed as a function of θ_1, θ_2 and β.

$$I_D = \frac{\beta V_D V_{Gt}(1 + \theta_2 \Delta V_{th}^2)}{1 + \theta_1 V_{Gt} + \theta_2 (V_{Gt} - \Delta V_{th})^2} \quad (5)$$

For simplifying the extraction procedure, it is very useful to reshape (5) as (6) by dividing all terms by $(1 + \theta_2 \Delta V_{th}^2)$ and performing the variable substitution described in (7). Finally, the simplified equation of the complete model (6) makes the Y-function extraction easy and enables to calculate the very same parameters as in [1], while taking into account the characteristics of submicron MOSFETs.

$$I_D = \frac{\beta V_D V_{Gt}}{1 + \Theta_1 V_{Gt} + \Theta_2 V_{Gt}^2} \quad (6)$$

$$\Theta_1 = \frac{\theta_1 - 2\theta_2 \Delta V_{th}}{1 + \theta_2 \Delta V_{th}^2}, \quad \Theta_2 = \frac{\theta_2}{1 + \theta_2 \Delta V_{th}^2} \quad (7)$$

III. EXTRACTION METHODOLOGY

The Y-function was introduced by Ghibaudo et al. to provide an easy and reliable way to extract V_{th} and β on MOSFETs having a weak θ_2-effect [1] (8). In (8), g_m is the current transconductance (by definition, $g_m = \partial_{V_G} I_D$) and I_D refers to the current model without θ_2-effect (1).

$$Y \stackrel{def}{=} \frac{I_D}{\sqrt{g_m}} \bigg|_{\theta_2 = 0} = \sqrt{\beta V_D} V_{Gt} \quad (8)$$

Unfortunately, the Y-function calculated from the complete model (including the θ_2-effect) takes a non-linear form (9) which complicates the extraction procedure described in [1].

$$Y = \sqrt{\frac{\beta V_D}{1 - \Theta_2 V_{Gt}^2}} V_{Gt} \quad (9)$$

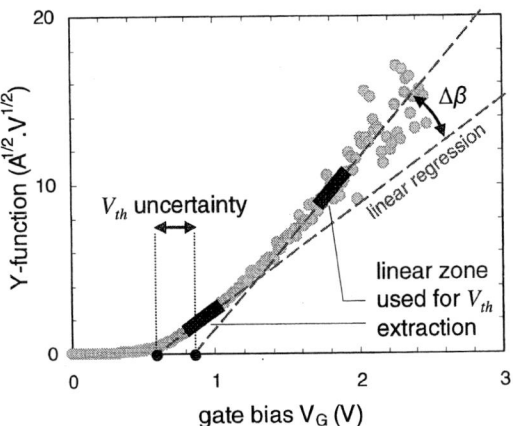

Fig. 1. Uncertain V_{th} and β extractions caused by the choice of the bias range where the θ_2-effect should be insignificant.

Indeed, this non-linearity causes strong uncertainties on V_{th} and β extraction as it relies, to perform the linear regression, on a bias range where the θ_2-effect should be insignificant [4]. Hence the extraction accuracy depends on several input parameters defining this bias range (cf. Fig.1). To solve this problem we propose a new extraction scheme relaying on the 3 following steps:

1) extraction of V_{th} and β using a recursive algorithm,
2) extraction of Θ_1 and Θ_2 from the $\Theta_{eff}(V_{Gt})$-function,
3) extraction of ΔV_{th} and calculation of θ_1 and θ_2.

A. Recursive extraction of the threshold voltage

The threshold voltage (V_{th}) is a mandatory parameter which governs the entire extraction accuracy and coherence. Its determination requires a special care to ensure a high precision in all further calculations. If an iterative technique has already been developed by Mourrain et al. to linearize the Y-function [2], the convergence of this algorithm seems very difficult to control and, in many cases, lead to a strong inaccuracy over the final results. We propose a new recursive algorithm based on the ξ-function ($\xi = 1/Y^2$) to determine V_{th} with a high accuracy. We assume that the exact value of V_{th} is unknown and introduce a rough estimator V_{th}^* which differs from V_{th} by an error $\varepsilon = V_{th}^* - V_{th}$. Thus, $V_{Gt} = V_{Gt}^* + \varepsilon$ (where $V_{Gt}^* = V_G - V_{th}^* - V_D/2$) and the ξ-function can be written as a function of ε (10).

$$\xi \stackrel{def}{=} \frac{1}{Y^2} = \frac{1}{\beta V_D} \left\{ \frac{1}{(V_{Gt}^* + \varepsilon)^2} - \Theta_2 \right\} \quad (10)$$

If V_{th}^* is close enough to V_{th} (i.e. if the relative error $\varepsilon/V_{Gt}^* \to 0$), $1/(V_{Gt}^* + \varepsilon)^2$ can be approximated using (11).

$$\frac{1}{\left(1 + \frac{\varepsilon}{V_{Gt}^*}\right)^2} \underset{\varepsilon/V_{Gt}^* \to 0}{\simeq} 1 - 2\frac{\varepsilon}{V_{Gt}^*} + o\left(\frac{\varepsilon}{V_{Gt}^*}\right) \quad (11)$$

Then (11) can be used to express the ξ-function as a polynomial function of $1/V_{Gt}^*$ (12), for determining β and ε using

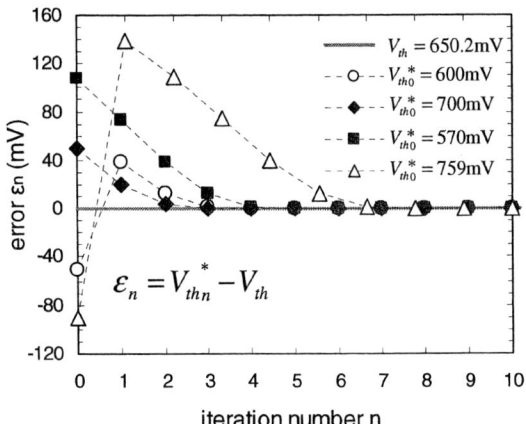

Fig. 2. Evolution of ε_n as a function of iteration number n. This algorithm show a fast convergence when V_{th0}^* is chosen within $-80\text{mV} < \varepsilon_0 < 100\text{mV}$. After 7 steps $\varepsilon_{N>7}$ reaches the calculation precision.

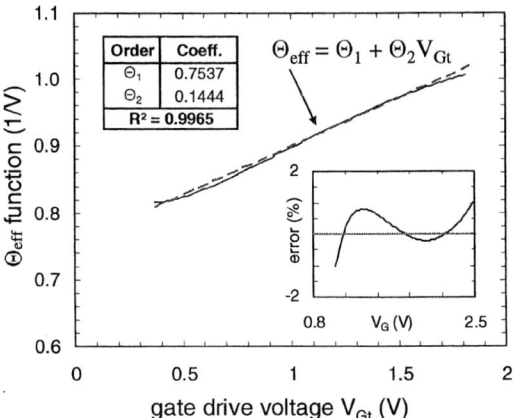

Fig. 3. Extraction of Θ_1 and Θ_2 from the Θ_{eff}-function on a 40nm-length MOSFETs of CMOS 45nm technology. The linear regression well reproduce the trend of data: $R^2 \simeq 0.9999$.

a pth-degree polynomial regression in $1/V_{Gt}^*$ ($p = 0, 2, 3$).

$$\xi \simeq \frac{1}{\beta V_D} \left\{ \frac{1}{V_{Gt}^{*2}} - \frac{2\varepsilon}{V_{Gt}^{*3}} - \Theta_2 \right\} \quad (12)$$

Moreover, the polynomial form of ξ (12) makes up a consistent system which can be used in a recursive algorithm. It requires an input value V_{th0}^* and corrects V_{thn}^* by ε_n for each iteration $n \geq 1$, to finally converges towards V_{th} (13).

$$V_{thn+1}^* = V_{thn}^* + \varepsilon_n, \quad V_{thN}^* \underset{N \to \infty}{\to} V_{th} \quad (13)$$

If the initial value V_{th0}^* is close enough to V_{th}, ε/V_{Gt}^* becomes small enough to ensure the validity of the assumption made in (11), and thus makes the algorithm converge towards V_{th}. By running this algorithm for several values of V_{th0}^*, it has been demonstrated that the error must typically satisfy $-80\text{mV} < \varepsilon_0 < 100\text{mV}$ to enable a good convergence. Fig.2 shows the evolution of V_{thn}^* along 10 iterations for 4 several initial values chosen within this condition. The convergence to V_{th} is reached with an insignificant error after a few steps ($N \simeq 7$), validating the robustness of our algorithm. The best way to find an initial value always close to V_{th} is to measure the extrapolated threshold voltage ($V_{th,ext}$) which provides an acceptable systematical ε_0. $V_{th,ext}$ is equal to the gate bias linearly extrapolated from the $I_D(V_G)$ inflexion point (maximum of transconductance) and can be expressed as (14), where the subscript M refers to the gate bias for which g_m reaches its maximum value [18]. Thus $V_{th,ext} - V_D/2$ should be a close (and slightly underestimated) value of V_{th} which always enables the algorithm to converge.

$$V_{th,ext} = V_{G,M} - \frac{I_D(V_{Gt,M})}{\max(g_m)} \quad (14)$$

B. Accurate extraction of Θ_1 and Θ_2

Once V_{th} and β are extracted with accuracy, they can be used to perform the extraction of Θ_1 and Θ_2. Indeed the Θ_{eff}-function makes a useful expression to extract these parameters

from a linear regression in V_{Gt} (15).

$$\Theta_{eff} = \left(\frac{\beta V_D}{I_D} - \frac{1}{V_{Gt}} \right) = \Theta_1 + \Theta_2 V_{Gt} \quad (15)$$

As shown on Fig.3, the polynomial regression well reproduces the behavior of the Θ_{eff}-function with a $R^2 \sim 1$ and a relative fit error lower than 1%.

C. Extraction of ΔV_{th} and calculation of θ_1 and θ_2

As mentioned before, the Θ_1-parameter does not model only phonon limited mobility but also includes a component from the θ_2-effect ($\Theta_1 = (\Theta_1' - 2\Delta V_{th}\Theta_2)$, where $\Theta_1' = \theta_1/(1 + \theta_2\Delta V_{th}^2)$). Thus, θ_1 needs to be extracted from Θ_1 for keeping a parameter tightly linked to the device physics and enable to performed coherent comparisons with the reference technique [1]. If Θ_2 is voluntary set to zero in (5), the model is shifted from the measurements, even at low gate biases, where the Θ_2 value should not have any impact (because the θ_2-effect is triggered only at high electric fields). The transconductance shift seen on Fig.4 rightly results from the $-2\Delta V_{th}\Theta_2$ term included into Θ_1, and ΔV_{th} can be easily determined by solving that a first-order mobility attenuation model (in Θ_1') must fit with the measurements (g_m^{meas}) for low gate biases (17) (for instance $V_{Gt} \sim V_{Gt,M}$)

$$g_m^{(1)} \overset{def}{=} \frac{\partial I_D^{(1)}}{\partial V_{Gt}} = \frac{\beta V_D}{[1 + \underbrace{(\Theta_1 + 2\Theta_2\Delta V_{th})}_{=\Theta_1'} V_{Gt}]^2} \quad (16)$$

$$g_m^{(1)} = g_m^{meas} \Leftrightarrow \Delta V_{th} = \frac{\sqrt{\beta V_D/g_m^{meas}} - 1 - \Theta_1 V_{Gt}}{2\Theta_2 V_{Gt}} \quad (17)$$

Once ΔV_{th} is known, θ_2 and θ_1 can be calculated using (7) and there is no more reason to keep the Θ_i parameters into the calculation. Finally a flowchart describing the whole extraction procedure can be found on Fig.5.

162

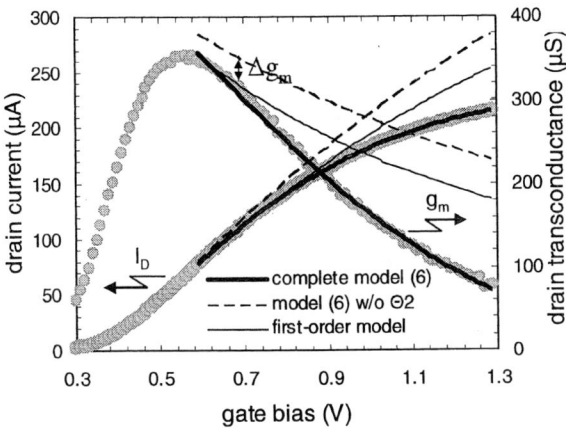

Fig. 4. Comparison of $I_D(V_G)$- and $g_m(V_G)$-curves fitting (bold line) with the complete model (6) including Θ_1 and Θ_2, (dotted line) with the complete model where Θ_2 has been suppressed (keeping Θ_1 intact), (thin line) with a first-order model where Θ_2 has been suppressed and Θ_1 replaced by Θ_1'. 100nm-length nMOS from the 65nm CMOS technology ($t_{ox} \simeq 13$Å).

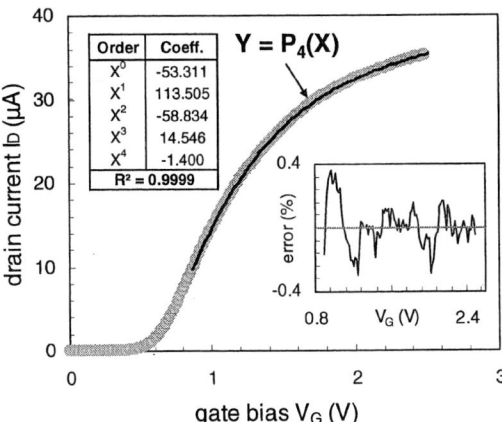

Fig. 6. 4th-order polynomial regression performed on the $I_D(V_G)$-curve of a 40nm length MOSFET from the 45nm technology. Data are very well fitted ($R^2 \approx 1$) and the relative error $\varepsilon = (P_4 - I_D)/I_D$ shown in insert does not exceed $\pm 0.4\%$.

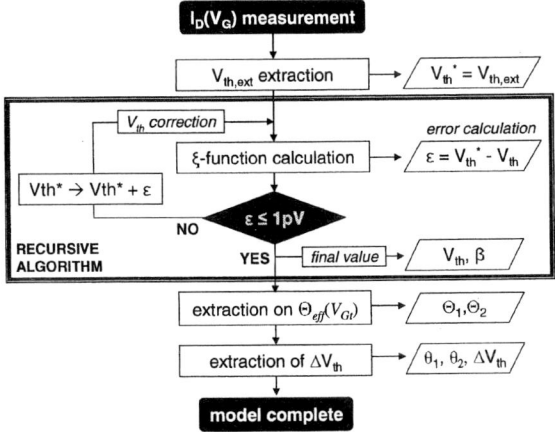

Fig. 5. Flow chart of the whole extraction methodology.

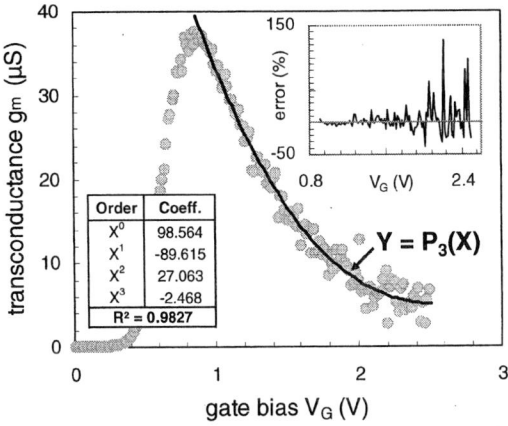

Fig. 7. 3th-order polynomial regression performed on the $g_m(V_G)$-curve of a 40nm length MOS transistor from the 45nm-technology. Despite a high noise amplitude generated by the first order derivation, measurement is very well fitted ($R^2 = 0.9827$) and the relative error is distributed around 0%.

IV. APPLICATION AND RESULTS

The downscaling and the use of new materials (for instance high-κ) lead to increase the measurement noise amplitude [19] and make the extraction harder to be performed with accuracy. To improve the signal-to-noise ratio a solution may consist in increasing the integration time on the measurement equipments (HP4156 Semiconductor Analyzer in this case). Unfortunately, this solution would increase the measurement time, which is usually minimized to satisfy the industrial requirements and allow statistical analysis to be performed. Furthermore the limited number of points contained in one curve can be a serious source of inaccuracy for V_{th} and β, depending on the V_G-step used for the measurements.

A. Converting the measurements into analytical functions

If $I_D(V_G)$-curves seem to have a good signal-to-noise ratio (SNR) even on small devices measured with a fast method, g_m

discloses a poor SNR which can alter the extraction results. Furthermore, Discrete Fourier Filtering (DFT) and most of classical filtering techniques does not get rid of the uncertainties resulting from the measurement step. Hence, fitting $I_D(V_G)$- and $g_m(V_G)$-curves with 4 and 3-order polynomial equations seems the best way to suppress the measurement noise and transform discrete data into continuous analytical functions. The best results are obtained when the regression is applied using the least square fitting technique from the maximum of the transconductance ($V_{G,M}$) to the end of the curve. Fig.6 and Fig.7 show fits of $I_D(V_G)$ and $g_m(V_G)$ measurements by those polynomial equations. They show good R^2 values very close to 1 and an error (ε) centered around 0% (see inserts)

TABLE I
RESULTS OF THIS EXTRACTION METHODOLOGY PERFORMED ON SEVERAL PMOSFETS FROM THE 45NM-CMOS TECHNOLOGY ($W = 1\mu m$).

L_{mask} (nm)	35	40	50	70	100	500	1000
$-V_{th}$ (mV)	615	650	627	620	588	476	443
$-\Delta V_{th}$ (mV)	377	316	256	355	286	285	261
β (mA/V^2)	2.36	2.18	1.82	1.45	1.07	.259	.130
θ_1 (V^{-1})	.863	.849	.718	.752	.565	.442	.395
θ_2 (V^{-2})	.152	.146	.145	.142	.133	.112	.115
$1 - R^2$ (ppm)	15	26	13	15	18	1.7	1.3

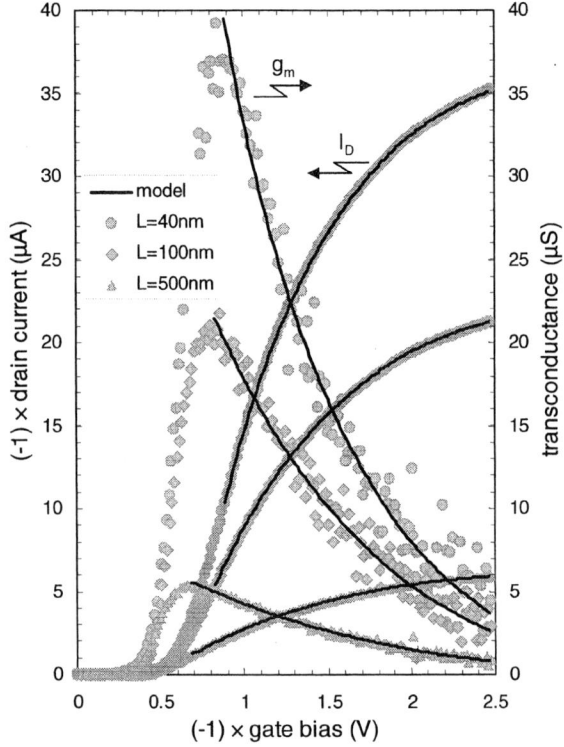

Fig. 8. The model (bold lines) fits $I_D(V_G)$ and $g_m(V_G)$ measurements for several gate lengths (pMOS transistors).

B. Results

Our new methodology has been performed on a set of 45nm-CMOS technology MOSFETs with mask lengths ranging from $L = 35$nm to $1\mu m$. Results summarized in Tab.I seem very coherent (for instance ΔV_{th} remains quite constant over all lengths), and provide very good fits of the measurements ($1 - R^2 \sim 15$ppm), as shown on Fig.8. As explained into [2], [20], the series resistances can be extracted from $\Theta_i(\beta)$ plots. If R_{SD} depends linearly from V_G ($R_{SD} = R_{SD,0} + \lambda V_G$), expressing I_D as a function of its intrinsic value $I_{D,0}$ ($I_D = I_{D,0}(V_D - R_{SD}I_D)/V_D$) allow extracting $\Theta_{1,0}$ and $\Theta_{2,0}$ [20] (18). In (18) the $_0$ subscript refers to the intrinsic (R_{SD}-independent) value of the parameters. The extraction of R_{SD} has been performed on 45nm-technology MOSFETs of length ranging from 35nm to $1\mu m$. The plots show a good alignment

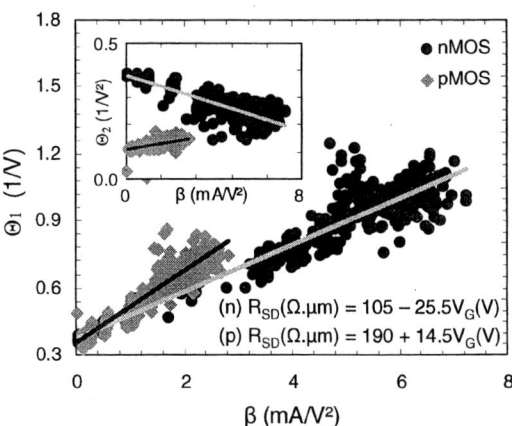

Fig. 9. Extraction of $R_{SD}(V_G)$ from $\Theta_i(\beta)$ scatter plots composed by an extraction performed on 1000 MOSFETs of several length (35nm to $1\mu m$).

(Fig.9), validating a coherent extraction over all lengths.

$$\begin{cases} \Theta_1 = \Theta_{1,0} + R_{SD,0}\beta \\ \Theta_2 = \Theta_{2,0} + \lambda\beta \end{cases} \tag{18}$$

C. Invariability of results towards the range of extraction

Independence of results about the V_G-range of extraction is the main advantage of this technique and a major interest to overcome the problems occurring in the other Y-function-based techniques. As a polynomial replacement is performed on the full inversion range (from $V_{G,M}$ to the higher bound of the measurement), the extraction is weakly concerned by the measurement noise. Thus, as long as the model is representative of the MOSFET behavior, it is not surprising to notice that the results have a weak dependence about the V_G-range boundaries. The lower bound of the V_G-range must be close to $V_{G,M}$, where β and V_{th} can be easily separated from the Θ_2 component. The results show a weak variation when the latter is shifted from $V_{G,M}$ to $V_{G,M} + 300$mV and it is generally set to $V_{G,M} + 100$mV in order to suppress any residual impact from the diffusion current (which is not predicted by our drift current model).

The Fig.10 quantifies the variation of result values when the upper bound is shifted from 1.3V to 2.5V. These variations are very small — below 0.5% for V_{th} and β, about 1% for Θ_1 and Θ_2 — and confirm the robustness of our technique.

D. Benchmarking

To highlight the usefulness and the accuracy gain provided by this methodology, a benchmark comparing these results with those obtained by applying several other Y-function-based techniques has been performed on a same device. Thus the standard Y-function enhanced by θ_2 parameter [1], [2], the iterative Y-function [2] and the $Y-\Phi$ technique [4] are compared with this work. Result uncertainty has been estimated for each technique by slightly changing the input

164

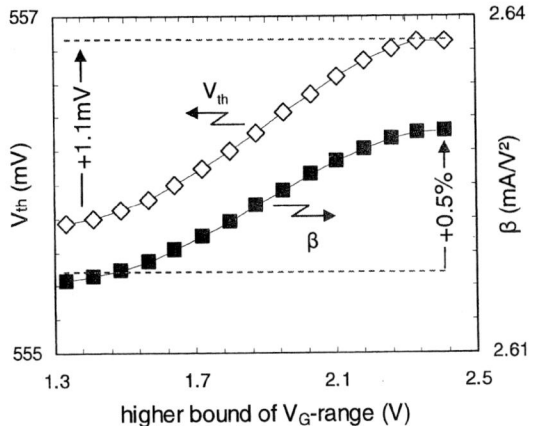

Fig. 10. Evolution of V_{th} and β as a function of the upper bound of the V_G-range of extraction. (1μm-width and 35nm-length pMOS transistors).

TABLE II
COMPARISON BETWEEN DIFFERENT EXTRACTION TECHNIQUES APPLIED ON A PMOS ($L = 35$NM, $W = 1\mu$M) FROM CMOS 45NM TECHNOLOGY.

Technique	this work[a]	Y-function[b]	iterative [2][b]	Y-Φ [4][b,c]
$\lvert V_{th}\rvert$ (mV)	**556 – 557**	564 – 623	503 – 612	570 – 585
β (mA/V^2)	**2.62 – 2.63**	2.54 – 3.36	2.12 – 3.43	2.59 – 2.81
θ_1 (V^{-1})	**.939 – .948**	.810 – 1.46	.960 – 1.69	.777 – .978
θ_2 (V^{-2})	**.146 – .148**	.089 – .212	.089 – .123	.132 – .169

[a]Variations obtained by shifting the upper bound of the extraction range
[b]Variations obtained by shifting the entire extraction range
[c]The Φ-parameter determination is dependent on the extraction range.

parameters or corrective factors values (Tab.II). This methodology returns more accurate results than those obtained from the others, while keeping absolute values truly coherent with our expectations. In addition several extractions has been performed on various technologies (for instance FDSOI with High-κ/metal-gate stack [21]) which all gave very good results and validate the versatility of our technique.

V. CONCLUSION

The high capability of this new extraction technique for performing accurate and reliable extractions on submicron CMOS technologies has been proved. This new methodology overcomes the problem of Y-function non-linearity found on other techniques and offers a straitforward extraction having a low sensitivity about the choice of the input parameters. The high accuracy of this technique have been demonstrated for several key parameters of the MOSFET such as V_{th}, β or R_{SD}. Comparison of the results uncertainty induced by the choice of the input-parameters for this work and several other Y-function-based methods has been investigated. It reveals a weaker dependence and a better results stability for this work. Finally decreasing the points number in the $I_D(V_G)$ measurements should allow to perform automatic extractions with in-line calculations, and thus improving the parametric test with a more reliable iterative technique.

ACKNOWLEDGMENTS

The authors would like to thank the Alliance Crolles2 Advanced Modules and Process Integration teams for providing the devices used in this work.

REFERENCES

[1] G. Ghibaudo, "New method for the extraction of MOSFET parameters," *IEE Electronics Letters*, vol. 24, pp. 543–545, Apr. 1988.
[2] C. Mourrain, B. Cretu, G. Ghibaudo *et al.*, "New method for parameter extraction in deep submicrometer MOSFETs," in *Proc. IEEE Int. Conf. Microelectronic Test Structures (ICMTS'00)*, Monterey, USA, Mar. 2000, pp. 181–186.
[3] T. Tanaka, "Novel extraction method for size-dependent mobility based on BSIM3-like compact model," *Japanese Journal of Applied Physics*, vol. 44, pp. 2424–2427, 2005.
[4] T. Tanaka, "Novel parameter extraction method for low field drain current of nano-scaled MOSFETs," in *Proc. IEEE Int. Conf. Microelectronic Test Structures (ICMTS'07)*, Tokyo, Japan, Mar. 2007, pp. 265–267.
[5] Y. Taur, D. Zicherman, D. Lombardi *et al.*, "A new 'shift and ratio' method for MOSFET channel-length extraction," *IEEE Electron Device Lett.*, vol. 13, pp. 267–269, May 1992.
[6] H. Katto, "Device parameter extraction in the linear region of MOSFET's," *IEEE Trans. Electron Devices*, vol. 18, pp. 408–410, Sep. 1997.
[7] B. Cretu, T. Boutchacha, G. Ghibaudo *et al.*, "New ratio method for effective channel length and threshold voltage extraction in MOS transistors," *IEEE Electron Device Lett.*, vol. 37, pp. 717–719, May 2001.
[8] H. Brut, A. Juge, and G. Ghibaudo, "New approach for the extraction of gate voltage dependent series resistance and channel length reduction in CMOS transistors," in *Proc. IEEE Int. Conf. Microelectronic Test Structures (ICMTS'97)*, Monterey, USA, Mar. 1997, pp. 188–193.
[9] H. van Meer, K. Henson, J.-H. Lyu *et al.*, "Limitation of shift-and-ratio based L_{eff} extraction techniques for MOS transistors with halo or pocket implants," *IEEE Trans. Electron Devices*, vol. 21, pp. 133–136, Mar. 2000.
[10] K. Romanjek, F. Andrieu, T. Ernst *et al.*, "Characterization of the effective mobility by split C(V) technique in sub 0.1μm Si and SiGe PMOSFETs," *Solid State Electronics*, vol. 49, pp. 721–726, 2005.
[11] F. Andrieu, T. Ernst, C. Ravit *et al.*, "In-depth characterization of the hole mobility in 50-nm process-induced strained MOSFETs," *IEEE Electron Device Lett.*, vol. 26, pp. 755–757, Oct. 2005.
[12] A. Cros, K. Romanjek, D. Fleury *et al.*, "Unexpected mobility degradation for very short devices : A new challenge for CMOS scaling," in *Proc. IEEE IEDM Tech. Dig.*, San Francisco, USA, Dec. 2006, pp. 663–666.
[13] Y. Cheng, M.-C. Jeng, S. Liu *et al.*, "A physical and scalable I-V model in BSIM3v3 for analog/digital circuit simulation," *IEEE Trans. Electron Devices*, vol. 44, pp. 277–287, Feb. 1997.
[14] T.-C. Ong, P. K. Ko, and C. Hu, "50-a gate-oxide MOSFET's at 77 k," *IEEE Trans. Electron Devices*, vol. 34, pp. 2129–2135, Oct. 1987.
[15] G. Reichert and T. Ouisse, "Relationship between empirical and theoritical mobility models in silicon inversion layers," *IEEE Trans. Electron Devices*, vol. 43, pp. 1394–1398, Sep. 1996.
[16] S.Takagi, "On the universality of inversion layer mobility in si MOSFET's: Part I—effects of substrate impurity concentration," *IEEE Trans. Electron Devices*, vol. 41, pp. 2357–2362, Dec. 1994.
[17] K. Rais, "On the high electric field mobility behavior in Si MOSFET's from room to liquid helium temperature," *Physica Status Solidi (a)*, vol. 145, pp. 217 – 221, 1994.
[18] G. Ghibaudo, "Analytical modeling of the MOS transistor," *Physics States Solid (a)*, vol. 113, p. 223, 1989.
[19] J. Jomaah and G. Ghibaudo, "Low frequency noise in si-based MOS devices," in *Proc. 18th Int. Conf. Noise and Fluctuations (ICNF'05)*, Salamanca, Spain, Sep. 2005, pp. 181–186.
[20] A. Cros, S. Harrison, R. Cerutti *et al.*, "New extraction method for gate bias dependent series resistance in nanometric double gate transistors," in *Proc. IEEE Int. Conf. Microelectronic Test Structures (ICMTS'05)*, Leuven, Belgium, Apr. 2005, pp. 69–74.
[21] C. Fenouillet-Beranger, S. Denorme, B. Icard *et al.*, "Fully-depleted SOI technology using high-K and single-metal gate for 32nm node LSTP applications featuring 0.179μm^2 6T-SRAM bitcell," in *Proc. IEEE IEDM Tech. Dig.*, Washington, USA, Dec. 2007, pp. 267–270.

2008 IEEE Conference on Microelectronic Test Structures, March 24-27, Edinburgh, UK

7.3

A Novel Biasing Technique for Addressable Parametric Arrays

Brad Smith, Uma Annamalai†, Alexandre Arriordaz, Venkat Kolagunta, Jeff Schmidt and
Mehul Shroff

Freescale Semiconductor, †University of Arkansas

3501 Ed Bluestein Blvd, Mailstop K-10, Austin, TX 78721, USA

ABSTRACT

Addressable arrays that use switches to isolate the devices being tested are limited in size and utility by the parasitic leakage caused by those switches. A new biasing technique that removes the drain-source bias from these switches has been studied to address this problem. Simulations performed in a 90 nm low-power technology predicted more than a two-decade drop in parasitic leakage of the array. Experiment data performed on a 90 nm technology confirmed this improvement.

INTRODUCTION

Addressable parametric arrays are very useful for generating large volumes of data in a reasonable area. Being an array, all the devices under test (DUTs) share the same bond pads used to connect to the electrical tester. This gives them a distinct area advantage over discrete devices.

To form an array, a method is needed for selecting which device under test (DUT) is to be tested. MOSFET DUTs provide the option of using the DUTs themselves as switches. The drain and/or source nodes of the DUTs can be tied together and the gates of the DUTs not being tested can be biased to an "off" state [1,2]. This gives the cleanest MOSFET measurements, but a defective device could induce measurement issues with multiple other MOSFETs that are sharing a column (or even the whole array).

This work addresses the type of array architecture in which pass gates are used to select which DUT's nodes are to be connected to the testing bus [3-7]. Adding pass gates to the DUTs' nodes adds parasitic leakage that limits the current that can be measured

from the DUTs themselves. This often means that leakage current may not be measured accurately in an array configuration. Even if the devices used in the pass gates have much lower leakage than the DUT's (e.g., using low-leakage, thick-oxide devices in the pass gates and thin-oxide devices as DUTs), ganging tens or hundreds of them onto an I/O bus fast negates this difference. CMOS pass gates contain two devices, each of which needs to be relatively wide in order to minimize its parasitic resistance, again multiplying the pass gate's impact on the bus' parasitic leakage.

ARRAY DESIGN

The generic array configuration shown in Fig. 1 shows the style of addressable array that was used throughout this work. The parametric tester provides an address and the address decode circuitry selects one of the DUTs. That DUT's selection circuitry enables CMOS pass gates that connect the DUT's nodes to a bus that goes to the bond pads and eventually to the tester. An "m by n" configuration is shown in Fig. 1, but an "1 by n" configuration was also used, in which there is only one column and the DUT I/O bus is connected directly to the bond pads without first going through the mux.

A basic unit cell of an array element is shown in Fig. 2. The dashed lines are meant to indicate that all the pass gates switch on/off together (i.e., all "on" or all "off"). Note that having separate Gate Force and Gate Sense lines are not necessarily required, but are a convenient way to measure and correct for any drop in gate voltage at the DUT's gate, should high gate current be a worry for a particular DUT. The drain

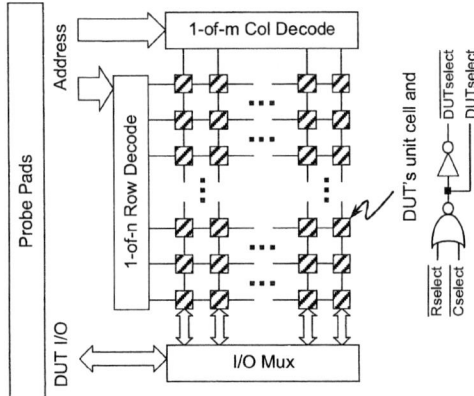

Fig. 1: A generic array configuration

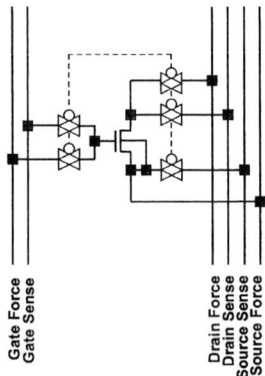

Fig. 2: A basic unit cell showing a DUT and its associated pass gates.

978-1-4244-1801-5/08/$25.00 ©2008 IEEE

166

and source nodes each have a Force and a Sense line. The Source Force is connected directly to the DUT because any switching device will add non-negligible resistance, which would adversely affect the DUT's measurements.

Selected DUT

That the pass gates will have some non-zero "on" resistance is unavoidable. Making them larger will minimize the resistance, but add to the parasitic leakage current. Therefore, care must be taken in their design to optimize both resistance and leakage. This resistance means that for nodes that carry significant current (especially the drain), a higher voltage will have to be forced so the voltage at the DUT's node is the desired value. For example, if a DUT's drain current is 1 mA and the pass gate's effective resistance is 1 kΩ, the pass gate will drop 1 V across it. In order to place 1 V on the DUT's drain, the Drain Force node will have to be set to 2 V. The Sense lines can be used to dynamically set the proper Force values.

Unselected DUTs

When the pass gates of the DUTs not being measured are turned "off", the DUTs' internal nodes are disconnected from the tester that is driving the DUT being tested. Leakage and other second-order mechanisms will drag the DUT's nodes to some unknown voltage(s). Worst-case, if the drain floats to ground, the full Drain Force voltage will appear across the pass gates that connect the drains to the Drain Force line. As mentioned, the leakage of those pass gates will be indistinguishable from the selected DUT's drain current (as measured on the Drain Force line).

LRDB

In order to create a high-precision array with minimal parasitic leakage, it is important to control the voltages of the unselected DUTs' nodes, especially the drain. If the drain node floats to ground, removing the drain-source bias that's applied to the Drain Force pass gate should significantly reduce the array's parasitic leakage. In addition, forcing this node to be a known voltage should result in more reproducible DUT leakage measurements. To remove the bias across the drain pass gate, a voltage similar to that on the Drain Force line needs to be applied to the unselected DUTs' drains. A new signal, called "Drain Leak", has been introduced in Fig. 3. Drain Leak connects to a DUT's drain node through an extra pass gate that switches opposite of the other pass gates. In other words, it is "off" when the others are "on", and *vice-versa*. The dashed lines are intended to indicate this. In this configuration, the drain nodes of the unselected DUTs are set to the Drain Leak voltage. This biasing scheme is referred to as Leakage Reduction DUT Biasing (LRDB). With the drain-source bias removed from the Drain Force pass gates, the unselected DUTs contribute significantly lower parasitic current to the

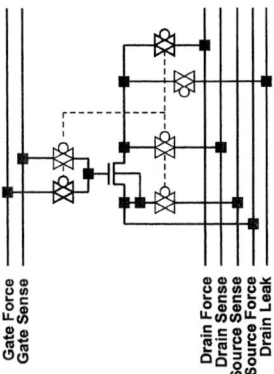

Fig. 3: A unit cell similar to Fig. 2 with an extra pass gate to tie the DUT's drain node to the Drain Leak line when the DUT is unselected (referred to as LRDB).

Drain Force bus, providing higher resolution measurements of the device being measured.

Adding the Drain Leak line addresses the problem of drain leakage but may cause another: If the one or more of the unselected DUTs turns on, Drain Leak could be forced to flow excessive current. There are two ways to avoid this. First, the gates of the DUTs not being tested could be tied to their sources, as shown in Fig. 4. This is the style of cell that was used throughout this work. The extra pass gate makes sure that the unselected DUTs are biased "off", but does not cover the case of a defect that shorts the DUT's drain and source nodes. As before, the dashed lines are meant to indicate which pass gates switch together. The second way to avoid excessive Drain Leak currents is to place an extra pass gate in series with the DUT's drain node, as shown in Fig. 5. This pass gate would switch on/off along with the other normal pass gates – opposite of the LRDB pass gate. Thus, an unselected DUT is completely disconnected from the Drain Leak line, preventing high current even if the DUT is shorted to ground. This obviously adds resistance in series with the DUT, but such a resistance

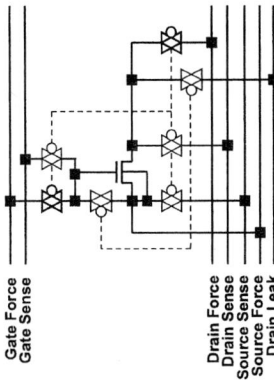

Fig. 4: An LRDB unit cell with an extra pass gate to tie the DUT's gate node to its source node when the DUT is unselected.

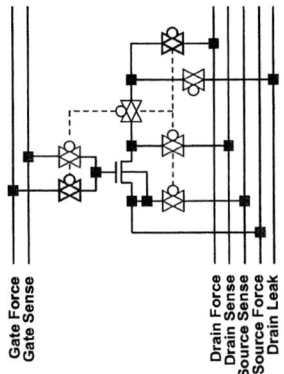

Fig. 5: An LRDB unit cell with an extra pass gate connected to the DUT's drain node to prevent current flow when the DUT is unselected.

can be tolerated on the drain node. The Drain Sense line tells the tester exactly what the DUT's real drain voltage is.

This technique has been demonstrated using NMOS devices, but there is no reason it can't be used for PMOS and other devices as well.

SIMULATION RESULTS

SPICE simulations were performed on an array with 128 DUTs connected together in a one-column configuration. The simulations were done using 3.3 V devices as switches in a 90 nm low-power CMOS technology. No address decode circuitry or other periphery was included, as this was intended to directly compare the unit cell from Fig. 2 without LRDB to the unit cell from Fig. 5 with LRDB. All DUTs were unselected, leaving the Drain Force current to be nothing but the array's parasitic leakage current. The results are listed in Table I, normalized per DUT. The LRDB unit cell showed a leakage current two orders of magnitude lower than the unit cell without LRDB.

Next, full arrays of a 1x32 configuration were simulated (schematics identical to those later measured on Si). The Drain Force and Gate Force voltages were held constant while Drain Leak was swept to measure the effect of LRDB on Drain Force current. Fig. 6 shows that the array's parasitic leakage level started at around 0.1 pA for Drain Force = 0 V and dropped quickly as the Drain Force voltage was increased. Above a Drain Leakage voltage of around 0.2 V, its

Table I: Simulated parasitic array leakage in a 1x128 array with and without LRDB.

	Drain Force Current (Gate Force=0 V)
Unit cell from Fig. 2, without LRDB	4 pA/DUT
Unit cell from Fig. 5, with LRDB	0.04 pA/DUT

effect disappears, leaving an array leakage of around 1 pA – a two-decade drop. However, the DUT's own leakage current (as measured in discrete or array form) was comparable to that of the array itself, so the effect of LRDB was nearly invisible. The simulation was repeated using -0.1 V on Gate Force to induce a lower-leakage DUT (Fig. 7). The array leakage was the same, but the effect on the DUT was much more visible.

It should be noted that simulations in this situation need to be validated with Si data because the key node in the circuit (the DUT's drain) is floating. The simulator is likely assuming that the node is effectively grounded, but that may or may not be true on Si. Models are typically not calibrated for this sort of design, since floating nodes are generally a bad design practice. Nevertheless, the simulations do indicate that LRDB can help significantly in the worst-case scenario, where the DUT's drain is grounded.

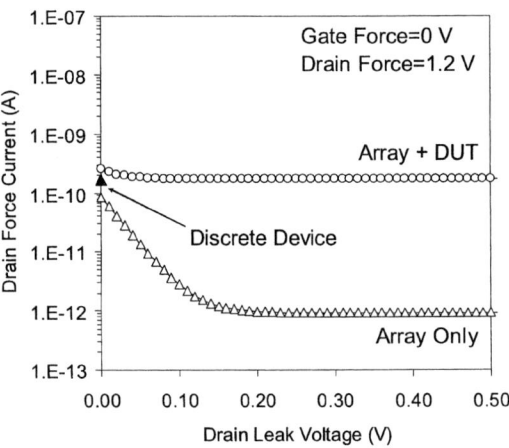

Fig. 6: Simulated Drain Force current in a 1x32 array as a function of Drain Leak voltage.
Drain Force=1.2 V and Gate Force=0 V.

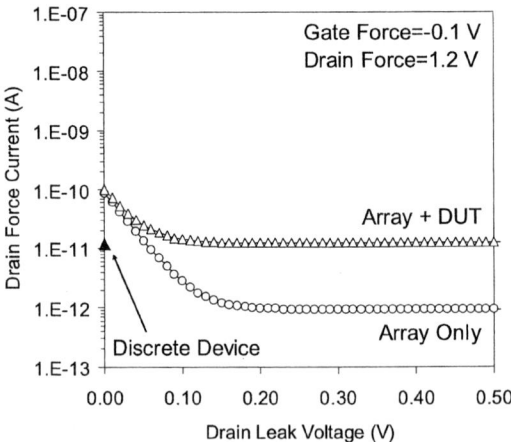

Fig. 7: Simulated Drain Force current in a 1x32 array as a function of Drain Leak voltage.
Drain Force=1.2 V and Gate Force=-0.1 V.

(a)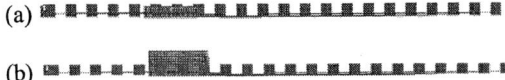

(b)

Fig. 8: Layout of (a) a 1x32 array, and (b) a 4x32 array. Both are in a 22-pad test module.

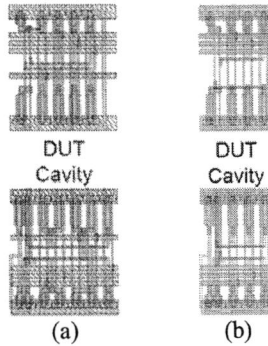

DUT   DUT
Cavity   Cavity

(a)   (b)

Fig. 9: Layout of the unit cell (a) with and (b) without LRDB.

EXPERIMENTAL RESULTS

Addressable arrays were fabricated in a 90 nm low-power CMOS technology using 3.3 V thick-oxide pass gates. Both a 1x32 configuration and a 4x32 configuration were tested. The results from each showed the same trends so unless otherwise noted, the results reported here are from the 1x32 configuration. The layouts of the arrays are shown in Fig. 8. The size of the unit cell was defined by the pass gates and was fairly compact (Fig. 9). The DUT cavity could easily be expanded for larger DUTs if needed, but it was left at its minimum size for this work. As can be seen, the addition of two pass gates to enable the LRDB technique did not cause any area problems for the array because the bond pads controlled the horizontal size of the test module.

Pass Gate Leakage

To test the basic premise of LRDB, an individual pass gate was biased in a similar manner to how it would be used in an array: One side (Drain Force) was held at 2 V while the other side (Drain Leak) was ramped from 0 V to 2 V. The well voltages of the pass gate devices were held constant at 0 V and 3.3 V, as would be the case in the array. As can be seen in Fig. 10, the leakage current of the pass gate dropped exponentially with the drain-source bias applied to it, with measurements and simulation in reasonable agreement. With only a few hundred millivolts applied to the other side of the pass gate, the leakage current dropped to an immeasurably small value. This suggests that the Drain Leak voltage doesn't have to follow the Drain Force voltage precisely. Rather, as

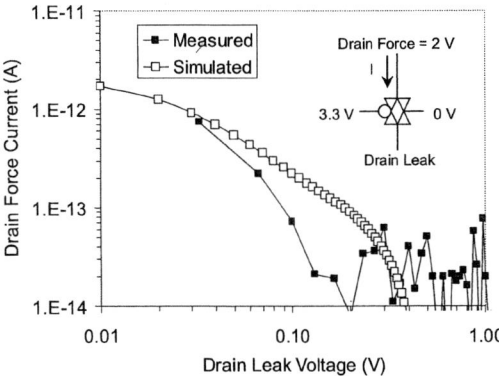

Fig. 10: Measured and simulated leakage of a single 3.3 V pass gate biased as it would be in an array using LRDB.

long as Drain Leak is above about 400 mV, the desired reduction in array leakage is realized.

Parasitic Array Leakage

The leakage currents of addressable arrays in a 1x32 configuration, with and without LRDB, were measured as a function of the Drain Leak voltage. Drain Force was held at 1.2 V and Gate Force was either 0 V (standard measurement) or -0.1 V (provides a lower-leakage DUT). The Drain Force current was measured both with no DUTs selected and with a single DUT selected. The former measures the leakage of the array by itself; the latter shows how that affects the measured DUT leakage. The "array-only" leakages in Fig.s 11 and 12 (Drain Force current with no DUTs selected) showed a response just like that seen with a pass gate alone, confirming that the leakage reduction in the unselected DUTs' pass gates is visible in a full array. This is effectively a sweep from the worst-case situation (the DUT's drain is grounded) to the best-case situation (the DUT's drain is some higher voltage). At Drain Leak = 0 V, the Drain Force current was a few nA, matching what was measured on a similar array without LRDB (at Gate Force = 0 V). When Drain Leak was increased past 0.2 V, the array's parasitic leakage dropped below a measurable level. That two-decade drop in background leakage made the low-leakage DUT's current clearly visible (Fig. 11). In Fig. 12, the DUT's leakage was high enough that the impact of the LRDB was fairly subtle. The leakage measured on an identical discrete DUT (*i.e.*, no array) is also included on the plot for comparison and matches what was measured in the array with Drain Leak > 0.2 V. The improvement of LRDB, therefore, enabled the measurement of a low DUT leakage current by significantly lowering the array's parasitic current.

Thus, forcing a Drain Leak voltage – even one as low as 0.2 V – produced a two-decade drop in the array's parasitic leakage, enabling measurement of lower DUT currents.

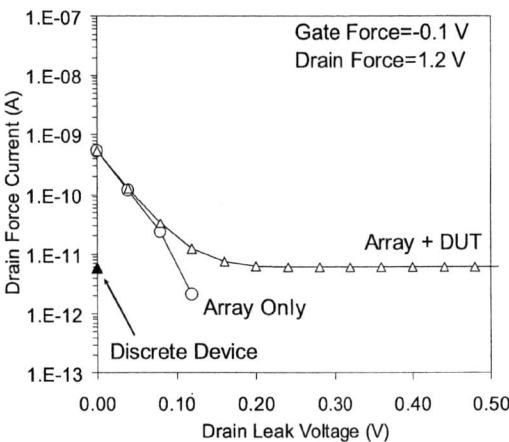

Fig. 11: Drain Force current in a 1x32 array as a function of Drain Leak voltage. Drain Force=1.2 V and Gate Force=-0.1 V.

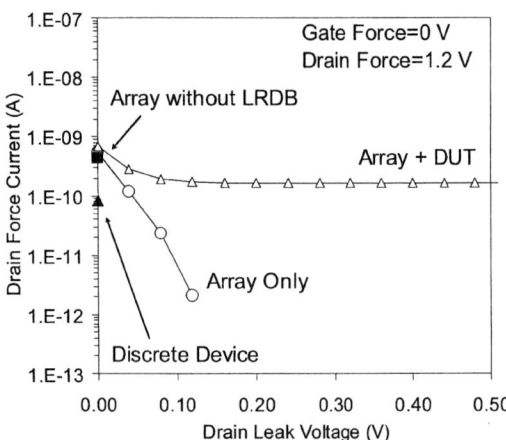

Fig. 12: Drain Force current in a 1x32 array as a function of Drain Leak voltage. Drain Force=1.2 V and Gate Force=0 V.

Device Measurements

NMOS devices of W=1 μm and L=0.10μm were measured in arrays with a 1x32 configuration, with and without LRDB. All DUTs in the array were identical to enable studies of the array itself. Drain Force was held at 1.2 V. The measurements were repeated with Drain Leak at 0 V and at 1.2 V.

The resulting ID-VG curves are shown in Fig. 13. The measurements from the array without LRDB showed a leakage floor of a few nA, agreeing with the previous data. For low currents, the non-LRDB measurement deviated from the values obtained from an identical discrete device, indicating influence from the array. With LRDB and forcing Drain Leak to 0 V, the low-current limitation was again a few nA. Raising Drain Leak to 1.2 V recovered the measurement, matching what was measured on an identical discrete device.

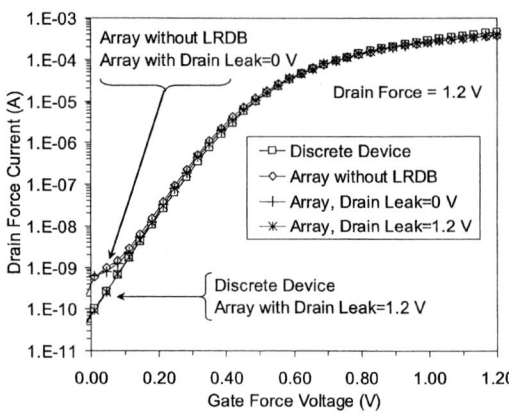

Fig. 13: NMOS characteristics as measured in discrete form (*i.e.*, device by itself, connected directly to bond pads), in an array without LRDB and in an array with LRDB and different voltages on Drain Leak.

The difference was not as dramatic as in Fig. 11, but it clearly demonstrated the viability of the LRDB biasing scheme.

CONCLUSION

Parametric addressable arrays were fabricated and simulated using a low-power 90 nm CMOS technology. The addition of a Drain Leak line, connected to every DUT's drain through an extra pass gate, was shown to cause a significant drop in the array's parasitic leakage current, enabling the array to measure much lower DUT currents. A drop of roughly two orders of magnitude in parasitic leakage was both simulated and measured experimentally. This technique allows either higher-resolution current measurements and/or many more DUTs to be placed into an addressable array while maintaining acceptably low parasitic leakages.

REFERENCES

[1] A. Mizumura, T. Ohishi, N. Yokoyama, M. Nonaka, S. Tanaka, and H. Ammo, "A study of 90nm MOSFET subthreshold hump characteristics using newly developed MOSFET array test structure", *Proc. IEEE Int'l Conf. on Micro. Test Structures*, Apr 2005.

[2] L. Portmann, C. Lallement and F. Krummenacher, "A High Density Integrated Test Matrix of MOS Transistors for Matching Study", *Proc. IEEE Int'l Conf. on Micro. Test Structures*, Mar 1998.

[3] N. Izumi, H. Ozaki, Y. Nakagawa, N. Kasai and T. Arikado, "Evaluation Of Transistor Property Variations Within Chips On 300-mm Wafers Using a New MOSFET Array Test Structure", *IEEE Trans. Semi. Mfg.*, vol. 17, no. 3, pp. 248-254, Aug 2004.

[4] K.Doong, *et al*, "Field-Configurable Test Structure Array (FC-TSA): Enabling design for monitor, model and manufacturability", *Proc. IEEE Int'l Conf. on Micro. Test Structures*, Mar 2006.

[5] K. Gettings, V. Gettings and D. Boning, "Test Circuit for Study of CMOS Process Variation by Measurement of Analog Characteristics", *Proc. IEEE Int'l Conf. on Micro. Test Structures*, Mar 2007.

[6] R. Lefferts and C. Jakubiec, "An Integrated Test Chip for the Complete Characterization and Monitoring of a 0.25μm CMOS Technology That Fits Into Five Scribe Line Structures 150μm by 5,000μm", *Proc. IEEE Int'l Conf. on Micro. Test Structures*, Mar 2003.

[7] K. Agarwal, *et al*, "A Test Structure for Characterizing Local Device Mismatches", *Symp. On VLSI Circuits*, 2006.

SESSION 8
RF

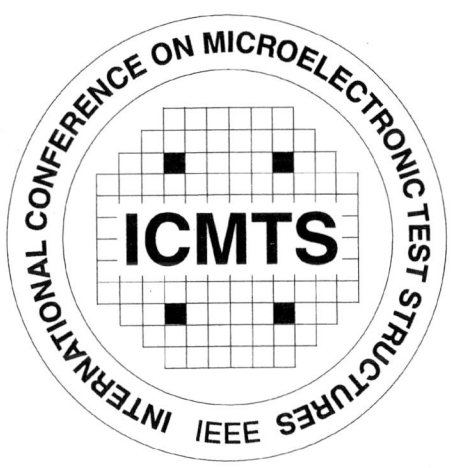

March 26, 15:30–16:50

Co-Chairs: Franz Sischka, *Agilent Technologies, Germany*
Ulrich Schaper, *Infineon, Germany*

2008 IEEE Conference on Microelectronic Test Structures, March 24-27, Edinburgh, UK

8.1

2.6-GHz RF Inductive Power Delivery for Contactless On-Wafer Characterization

Jonathan Tompson, Adam Dolin** and Peter Kinget

Department of Electrical Engineering,
Columbia University, New York, NY, USA

**now with Anadigics, Warren, NJ, USA

ABSTRACT

This paper presents the critical components of a contactless IC testing infrastructure to power on-chip characterization circuits. This includes an inductively-coupled, contactless power delivery system implemented on a 90nm CMOS technology using 150μm x 150μm on-chip and off-chip spiral inductors, low loss rectifiers and on-chip voltage regulators to create a constant and repeatable 1-V, 8.5-mW DC source. We present the measured process variation of ring oscillator test circuits in contrast to the on-chip voltage source variation to demonstrate the feasibility of process variation analysis using this system.

INTRODUCTION

A number of contactless testing solutions have been recently presented in literature utilizing different coupling mediums, such as RF coupling through micro-antenna structures [1,2] and optical coupling through CMOS photo-detectors [3,4]. The aim of a contactless approach is to provide access to on-wafer test structures in the early stages of the fabrication cycle, without damage to the wafer and without the need for ESD circuits. The delivery of DC power to the test structure is the most significant challenge in developing a contactless testing macro.

Power delivery though RF micro antennas, when realistically sized, is difficult due to low levels of coupling, thus requiring traditional mechanical probing connections to inject DC power when RF coupled power is insufficient [1], or alternatively, large on-chip antenna devices [2]. Optical coupling techniques have shown considerable promise in power delivery with small die space requirements [4], but require the design and characterization of specialized photodiodes typically not readily available in standard CMOS technologies.

The magnetic coupling of an off-chip spiral inductor positioned above an on-chip spiral inductor can achieve relatively high levels of coupling, enabling efficient contactless power delivery. At GHz frequencies the size of the spirals can be made comparable to traditional pad structures and the associated ESD circuits in standard CMOS technologies. Furthermore, the architecture and results presented here demonstrate that precise voltage rectification and regulation can be realized on-chip to present an accurate DC power supply. The generation of a repeatable, accurate DC power supply is the most important challenge in demonstrating a feasible contactless testing system, and is the focus of this paper.

SYSTEM ARCHITECTURE

To demonstrate a proof of principle, a compact and robust architecture was chosen as illustrated in Fig. 1; the on-chip power delivery macro consists of a spiral inductor, a CMOS rectifier, a linear regulator with OTA and power transistor, and a CMOS voltage reference; a ring oscillator serves as the device under test and load circuit; for testing an off-chip spiral inductor mounted on a micro-positioner is placed over the macro and driven by a standard RF signal source.

The inductors were simulated with an EM simulator, EMX, to obtain broadband frequency domain s-parameter data. An equivalent circuit, as shown in Fig. 2, was extracted using Matlab optimization routines. Since changes to the transformer's physical structure result in direct changes to the values of the lumped circuit elements, this design flow makes circuit optimization more intuitive. For instance; the

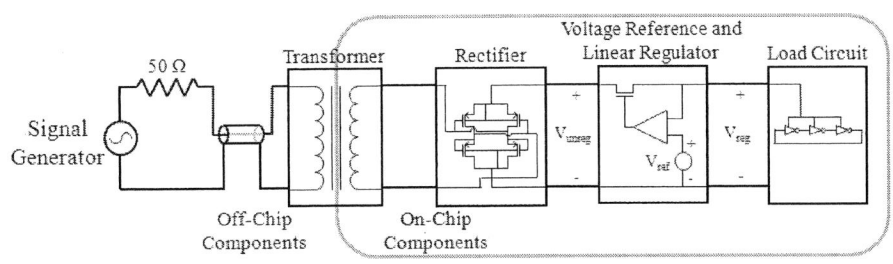

Fig. 1. Diagram of the Proposed Contactless IC Testing Architecture

978-1-4244-1801-5/08/$25.00 ©2008 IEEE

175

numerous degrees of freedom in transformer topology can be optimized to find a structure with desirable low series resistances and high coupling coefficient at a given frequency, even though important second order effects make precise hand calculations too difficult. The dimensions of the inductor were then optimized for maximum power transfer efficiency (PTE) through circuit simulation of the equivalent transformer circuit in combination with the RF rectifier (described later). We obtained the best PTE with a 3-turn 150μm x 150μm off-chip inductor and a 5-turn 150μm x 150μm on-chip inductor realized in the top metal layer. Minimum metal spacing of the spiral traces was used to result in desirable high Q and low loss inductors. The spacing between the on-chip and off-chip inductor was assumed to be less than 50μm.

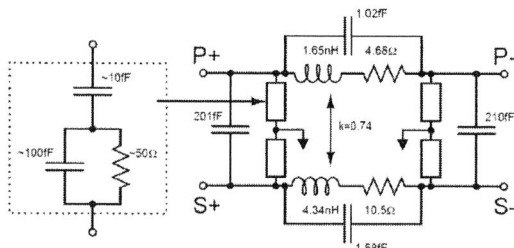

Fig. 2. Transformer Equivalent Circuit Model.

On-chip ring oscillators are commonly used to characterize device performance and process variation across and between wafers. We used a digitally programmable, variable load, 57-stage ring oscillator as a circuit under test to load the power delivery macro.

The RF rectifier block, shown in Fig. 3, uses a combination of low threshold MOS diodes (M3 and M4) and low threshold pass-transistors (M1 and M2) to implement efficient full-wave rectification of the RF signal in the used 90nm CMOS technology. PMOS devices were used to increase isolation from the substrate and careful consideration was given to the resonance conditions between the transformer structure and large rectifier capacitances. To tune the rectifier input impedance and to decrease effective series resistance, multiple copies of the rectifier structure were connected in parallel to decrease current density and thus series voltage drop. For RF frequencies, a combination of pass transistors and diode devices provides a reasonable tradeoff between low series resistance (typical of diode rectifiers) and low V_T drop (typical of pass transistor rectifies). Rectifier topologies based on pass transistors only were not feasible at RF since they require special gate drivers to prevent shoot through currents and the associated power losses.

A compact on-chip linear regulator was used to ensure robustness and simplicity, where the OTA is a single stage design biased in weak-inversion to minimize power consumption. A low power, on-chip bandgap

reference based on [5], was designed for use in this project.

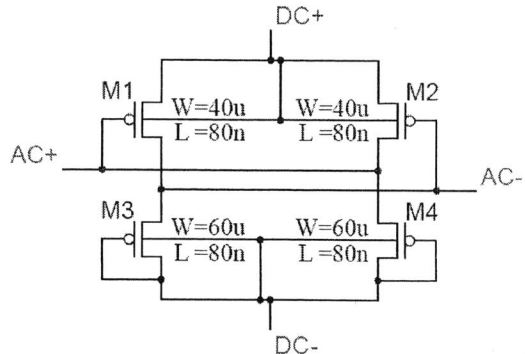

Fig. 3. 2.6GHz RF Rectifier Schematic.

IC IMPLEMENTATION

The designs were fabricated in a standard 90nm CMOS technology and tested for functionality. Fig. 4 shows the die photograph and the layout of the power delivery macros and the ring oscillators serving as circuit under test. Note that the power supply circuitry is of comparable size as the 150μm x 150μm inductor devices. For future designs these devices could be placed inside or underneath the inductors further reducing die space using the design techniques demonstrated in [6]. Alternatively they can be placed in space reserved for wafer dicing to preserve metal layer resources near the DUT. To test circuits early on in the fabrication cycle, layouts using only one layer of metal can be considered for future designs.

To verify this proof-of-principle design we added the top and bottom DC bondpads to observe internal DC voltages and digitally control the ring oscillator load during testing, while the left hand side bondpads are for RF output. Ultimately no DC bondpads will be needed and the RF output can be coupled back through the transformer; also, control signals can be modulated on the RF input signal as is standard practice in other inductive link designs [8,9].

The system is duplicated into two separate test cases; the *spiral only test case* and the *transformer test case*. In the *spiral only test case* the primary inductor of the transformer is to be fabricated on an external substrate. The secondary inductor is realized on chip in the top metal layer. The proper functionality of the *spiral only test case* is the ultimate goal of this project. Also available is an output buffer and output probing pads to bring the ring oscillator output signal to a spectrum analyzer for the measurements of the ring oscillator process variation.

In the *transformer test case* the secondary inductor is realized on-chip in the top two metal layers while the primary is laid out in the two metal layers below such that the entirety of the transformer is on-chip. The purpose of this case was to demonstrate system functionality without the need of fabricating external probing inductors. The ring oscillator output signal is not externally available for this case.

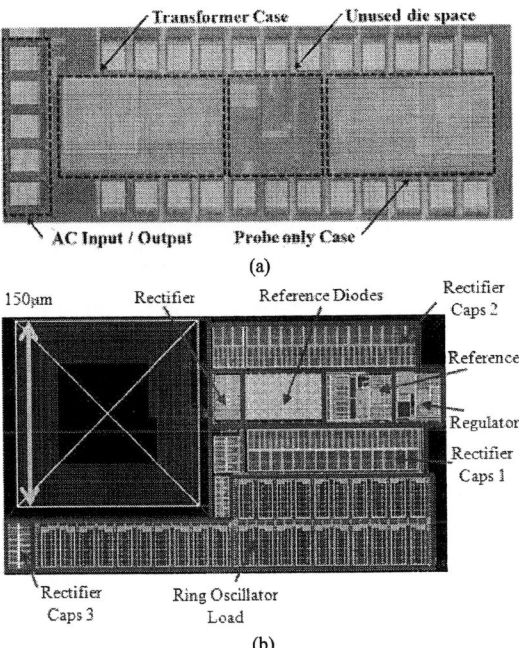

(a)

(b)

Fig. 4. (a) Die Photograph; (b) Layout of Probe Only Test Case.

As of the time of this writing, prototypes for the external thin film, probing inductors to test the *spiral only test case* are being fabricated with photolithography in our in-house clean room facilities. This process includes chrome-gold spirals, with polyimide insulator deposited on thin glass substrates. These devices are to be mounted on PCBs and placed in close proximity to the surface of the DUT using micro-positioners. However, due to photolithography complications, the fabrication process details are not yet finalized and the testing results using this system will be presented in a future work.

To still demonstrate external magnetic coupling, but without fully optimized external probing inductors, an external probe inductor structure was manufactured on a PCB. However, only a single turn, 127μm inductor device with transmission line feed and matching circuitry could be fabricated due to the layout constraints on a standard 5mil PCB process. Such inductor geometry leads to substantially poorer coupling as with optimized thin-film probing inductors, given that the magnetic field strength from a low turn device is significantly reduced.

MEASUREMENTS AND RESULTS

In a first set of tests we measured the *transformer test case* to demonstrate the feasibility of reliably delivering DC power to a test circuit using the presented test macro. In Fig. 5 a 16-dBm RF sinusoid[1] between 2.6GHz and 2.7GHz is delivered to the on-chip primary inductor. This results in sufficiently large *Vunreg* so that the on-chip regulator can provide a stable 1-V power supply to drive the ring-oscillator load which represents 8.51mW of loading. We have further measured that a 14-dBm RF signal to the transformer is sufficient to drive a 4.0mW ring-oscillator load. Additionally, we have found an excellent *Vreg* insensitivity to variations in loading conditions, RF input power, and RF input frequency. In Fig. 5, 100mV variations in *Vunreg* due to variations in received power as a result of changes in input frequency from the optimal 2.69GHz, result in less than 1mV changes to *Vreg*. Typically, we have measured 10mV/V sensitivity of *Vreg* to changes in *Vunreg*.

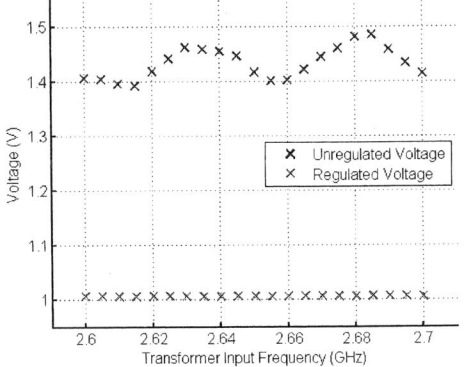

Fig. 5. Measured Regulated Voltage versus Input Frequency.

Fig. 6 illustrates the measured statistical distribution of the regulated and the unregulated voltages across 52 TT dies from the same wafer. A 13-dBm 2.69-GHz RF sinusoid was applied to the transformer test case with 4mW of DC loading. The average unregulated voltage was approx. 1.3V with a σ of 72.5mV, due to impedance mismatches, device variations, variable loading conditions and other effects. The regulated voltage had a mean of 1.0V with a σ of only 14.5mV. Additional tests with an input power at 12dBm and 11dBm which are below the nominal power of 13dBm are summarized in Table. 1. Note that an 11-dBm input power is marginally sufficient for the tested loading conditions and would not be used in practice.

---

[1] During our initial testing we had access to a 16dBm RF generator which became unavailable later; subsequent tests had to be done with maximum RF power of 13dBm.

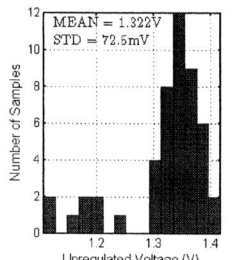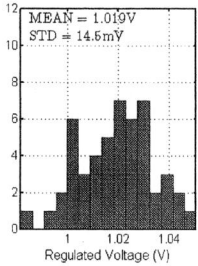

Fig. 6. Measured Statistical Distribution of Vunreg and Vreg across 52 samples.

Input Power (dBm)	$\mu(V_{unreg})$ (V)	$\sigma(V_{unreg})$ (mV)	$\mu(V_{reg})$ (V)	$\sigma(V_{reg})$ (mV)
11	0.970	31.4	0.954	30.6
12	1.111	58.3	1.010	18.5
13	1.322	72.5	1.019	14.5

Table 1. Measured Statistical Variation of Transformer Test Case Voltages for Varying RF Input Power; 13dBm is the Nominal RF Input Power

To demonstrate a practical application for this contactless power delivery macro, we investigated the possibility of monitoring the frequency variation of the 57-stage ring oscillator load during a second set of tests. Note that the ring oscillator inverter devices were sized with non-minimum length, and a large number of stages was chosen such that the frequency was approximately 500MHz; low enough to prevent coupling with the 2.69 GHz RF power carrier. However, larger device sizes and larger number of stages reduce the variability of the ring oscillator output frequency.

Fig. 7 illustrates the statistical variation of the 57-stage ring oscillator on the 52 dies tested, when the unregulated supply, V_{unreg}, of the probe only test case is driven by an external 1.35V DC power supply. An average frequency of 432.6MHz with a σ of 13.4MHz was found. The mean and sigma of the regulated supply voltage was also measured and found to be 1.019V and 14.5mV respectively. The supply pushing of the ring oscillator has been measured to be less than 510 MHz/V; the σ in the output frequency due to supply variations is then 7.4MHz which is significantly smaller than the total measured ring oscillator frequency σ. In addition, even when the unregulated voltage is significantly lower and higher than 1.35V nominal, the variation in regulated voltage in contrast to the variation in ring oscillator frequency is still statistically significant. A summary of these results is shown in Table 2.

As shown in [7], for future work it will be possible to design ring oscillators with fewer stages, followed by dividers, which will exhibit more pronounced frequency variations. Additionally designs highlighting different process variation effects can be used.

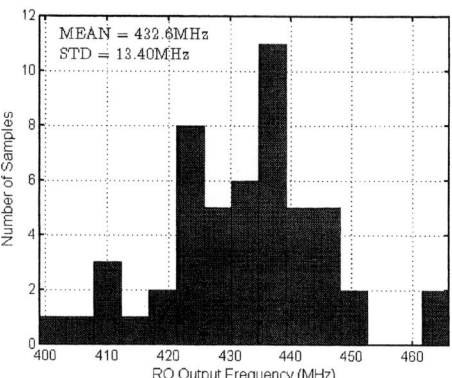

Fig. 7. Measured Statistical Distribution of the Ring Oscillator Frequency Across 52 Samples.

V_{unreg} (V)	$\mu(V_{reg})$ (V)	$\sigma(V_{reg})$ (mV)	$\mu(freq)$ (MHz)	$\sigma(freq)$ (MHz)
1.1	0.993	8.84	417.2	10.37
1.35	1.019	15.0	432.6	13.40
1.6	1.021	14.9	433.9	13.77

Table 2. Measured Statistical Variation of Probe Only Test Case Regulated Voltage and Ring Oscillator Frequency.

In a third set of tests we brought the single turn PCB inductor (described above) into close proximity to the on-chip inductor of the *spiral only test case*. Due to the low number of turns and large size of this inductor, it was only able to couple a small amount of power into the on-chip devices. We found that applying our maximum RF power of 13dBm to the probing inductor resulted in only 735mV unregulated voltage at the output of the on-chip rectifier. This is insufficient to adequately provide enough voltage for the regulator to output a regulated 1V V_{DD}. We anticipate far better results with thin-film external probing inductors which can be designed with the correct size and number of turns.

We used the PCB inductor to perform a preliminary evaluation of the probe misalignment on the system performance. As shown in Fig. 8, it was found that within +/- 50μm lateral misalignment from the optimal probe position, the unregulated voltage drop was no less than 350mV. Within +/- 5μm displacement, which is a reasonable alignment tolerance for most micro-positioner devices, an unregulated voltage drop of only 5mV was found suggesting good insensitivity to probe misalignment. Also note that our earlier test have demonstrated that the regulated output voltage is very accurate over a large range of unregulated voltage values which further adds to the system robustness.

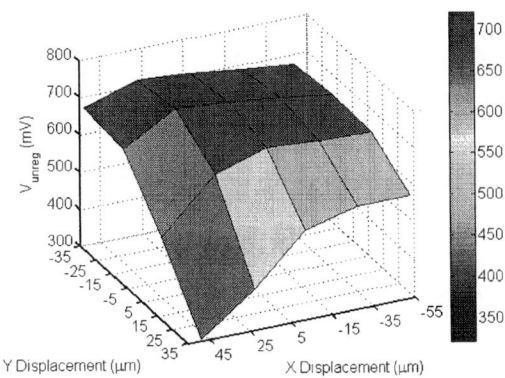

Fig. 8. Measured effect of PCB Probe Misalignment on Unregulated Voltage

SUMMARY

We present a system to power integrated test circuits without making physical contact for use in wafer characterization testing. The system is capable of supplying 8.5mW of regulated DC power from 16dBm AC power, to digital switching loads. The on-chip regulated 1V supply repeatability was shown to be sufficient for variation analysis of ring oscillators. Using an external PCB probe device promising insensitivity to lateral probe misalignment of less than 5mV drop within +/- 5μm was recorded.

Future extensions to this work will include supply macros capable of transferring digital data through modulation of the power carrier and the load impedance as has been shown feasible in literature [8,9], with minimum impact on power transfer efficiency. To save space, components can be placed inside or beneath the integrated inductors [6]. To enable evaluation of circuits early on in the fabrication cycle, layouts can be developed that require only one layer of metal interconnect.

ACKNOWLEDGMENTS

The authors thank J.O. Plouchart of IBM for technical discussions; IBM for financial support through an IBM Faculty Award; Integrand Software for the use of the EMX simulator; Prof. John Kymissis and his students for help with thin-film inductor design and fabrication; and UMC for test chip fabrication.

REFERENCES

[1] C. Sellathamby, M. Reja, L. Fu, B. Bai, E. Reid, S. Slupsky et. al. "Non-contact wafer probe using wireless probe cards," Proceedings of the IEEE International Test Conference, Nov. 2005, pp. 6.

[2] B. Moore, M. Margala, C. Backhouse. Design of wireless on-wafer submicron characterization system. IEEE Transactions of the Very Large Scale Integration (VLSI) Systems, Vol. 13, No. 2, Feb. 2005, pp. 169-180.

[3] S. Sayil, D. Kerns Jr., S. Kerns. Comparison of contactless measurement and testing techniques to a new all-silicon optical test and characterization method. IEEE Transactions on Instrumentation and Measurement, Vol. 54, No. 5, Oct. 2005, pp. 2082-2089.

[4] M. Babazadeh, J. Estabil, B. Borot, G. Johnson, N. Pakdaman, W. Doedel, J. Vickers, et. al. First look at across-chip performance variation using non-contact performance-based metrology. IEEE Advanced Semiconductor Manufacturing Conference, Vol. 17, 2006, pp. 278-283.

[5] H. Banba, H. Shiga, A. Umezawa, T. Miyaba, T. Tanzawa, S. Atsumi, K. Sakui. A CMOS bandgap reference circuit with sub-1-V operation. IEEE Journal of Solid-State Circuits, Vol. 34, No. 5, May 1999, pp. 670-674.

[6] F. Zhang, P. Kinget. Design of components and circuits underneath integrated inductors. IEEE Journal of Solid-State Circuits, Vol. 41, No. 10, Oct. 2006, pp. 2265-2271.

[7] M. Bhushan, M. Ketchen, S. Polonsky, A. Gattiker. Ring oscillator based technique for measuring variability statistics. IEEE International Conference on Microelectronic Test Structures, Mar. 2006, pp. 87-92.

[8] J.F. Gervais, J. Coulombe, F. Mounaim, M. Sawan. Bidirectional high data rate transmission interface for inductively powered devices. IEEE Canadian Conference on Electrical and Computer Engineering, Vol. 1, May 2003, pp. 167-170.

[9] Y. Hu, J.F. Gervais, M. Sawan. High power efficiency inductive link with full-duplex data communications. IEEE International Conference on Electronics, Circuits and Systems, Vol. 1, Sept. 2002, pp. 359-362.

2008 IEEE Conference on Microelectronic Test Structures, March 24-27, Edinburgh, UK

8.2

Advanced Test Structure Design for Dielectric Characterisation of Novel High-K Materials

John A. O'Sullivan*, Wenbin Chen**, Kevin G. McCarthy* and Gabriel M. Crean**

*Dept. of Electrical and Electronic Engineering, University College Cork, Ireland
**Tyndall National Institute, Ireland

Abstract—The extremely high level of integration currently required within the wireless industry poses significant challenges at all technology steps from materials through processing to packaging. Capacitor design is a very important element of this technology integration challenge. Recently there has been strong interest in the use of PMNT ($Pb(Mg, Nb)TiO_3$) as a thin-film layer in semiconductor processes [1]. The high dielectric constant of PMNT makes it an attractive material for the fabrication of MIM (Metal/Insulator/Metal) decoupling capacitors. Before PMNT can be used in conjunction with a modern Si process for capacitor design the PMNT layer must be accurately characterised. This paper addresses the issue of thin-film characterisation through wafer-probe measurements and electromagnetic simulation (EM) of coplanar waveguides. Based on the results obtained a design methodology for optimum test structure layout is presented.

I. INTRODUCTION

Growth in the microelectronics industry has resulted in high demand for high dielectric constant ferroelectric thin films. High dielectric constant material thin films are now commonly used in microwave devices. The characteristics of the high dielectric constant material, such as dielectric constant and the dielectric loss tangent directly determine the quality of the devices and system.

PMNT is a promising ferroelectric material due to its high dielectric constant and other material properties [1]. PMNT thin films have been manufactured using various techniques, such as Solgel, pulsed laser deposition, metalorganic chemical vapour deposition and rf sputtering deposition techniques.

Due to the benefits of high dielectric constant, PMNT becomes an attractive material for the fabrication of MIM (Metal/Insulator/Metal) decoupling capacitors. Before PMNT can be used in conjunction with a modern Si process for capacitor design the PMNT layer must be accurately characterised. Coplanar-waveguide (CPW) lines can be utilized to measure scattering parameters (S-parameters), from which the permittivity of PMNT thin films can be calculated. Furthermore, the Quasi-TEM assumption for the analysis of microwave passive devices has been made for a long time with high accuracy, and the assumption of its validity has been experimentally proven up to 110 GHz with 5% error margin [2]. In this paper a design methodology for optimum test structure layout is presented based on the coplanar-waveguide structures.

In section II, a measurement analysis of an Impedance Standard Substrate (ISS) is presented. EM simulations are also carried out on the ISS substrate and the results obtained are

used to analyse a basic PMNT based structure in section III. Section IV presents a cross-section of the final PMNT based structure, along with a mask layout for the coplanar-waveguide lines.

II. ANALYSIS AND MEASUREMENT OF IMPEDANCE STANDARD SUBSTRATE

One of the main difficulties in verifying a particular wafer probe measurement method is the lack of an available 'known' standard that can be used for reference purposes. This problem becomes increasingly acute as device size decreases. However, one standard that may be used for measurement verification is an impedance standard substrate (ISS). Generally ISS standards are used for calibration purposes. However, the verification lines on the ISS substrates can be used as a method to test a dielectric constant extraction technique and also to test the accuracy of EM simulations. Some of the standards used for on-wafer calibrations are illustrated in figure 1.

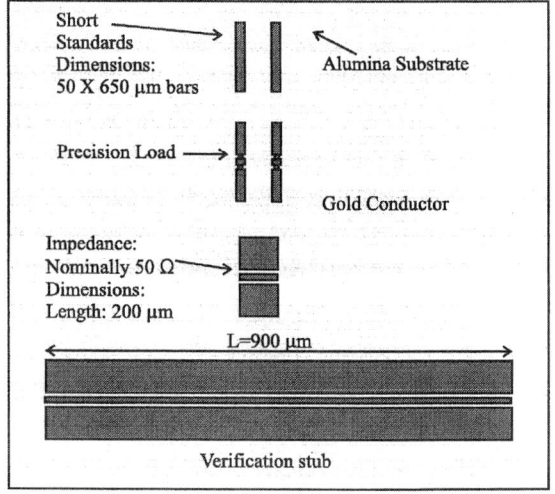

Fig. 1. Short, load and thru calibration standards on an ISS.

The ISS standard used in this work has an alumina substrate with a dielectric constant of $\epsilon_r = 9.9$. The fact that the value of ϵ_r is known allows for the verification of various dielectric parameter extraction techniques. The verification stub illustrated in figure 1 is essentially a coplanar waveguide. S-parameters measurements have been carried out on this

978-1-4244-1801-5/08/$25.00 ©2008 IEEE

180

coplanar line. The measured s-parameter data can be converted to a chain (or ABCD) matrix [3]. The measured characteristic impedance can then be extracted from the ABCD matrix using equation 1.

$$Z_0 = 50\sqrt{\frac{B}{C}} \qquad (1)$$

Figure 2 illustrates the measured and simulated characteristic impedance from $1 - 10$ GHz. As can be seen there is good correspondence between the results. The line was designed to have a characteristic impedance of 50Ω [4]. Many of the methods used to extract the properties of a particular substrate are based on knowing either the dielectric constant of the substrate or the characteristic impedance of the transmission line [5], [6]. Therefore, the measured characteristic impedance can be used in conjunction with the equations available in the relevant literature to extract a value for the dielectric constant of the ISS substrate. The mode of propagation through a coplanar waveguide is quasi-TEM (transverse electromagnetic). A set of equations, based on the work of Heinrich, yields accurate results for the extraction of the dielectric constant [7], [8]. The methods are based on obtaining ϵ_r for a TEM mode of propagation and subsequently using this to find ϵ_r for a quasi-TEM situation. ϵ_r for the TEM situation is found by applying equations 2 through to 5.

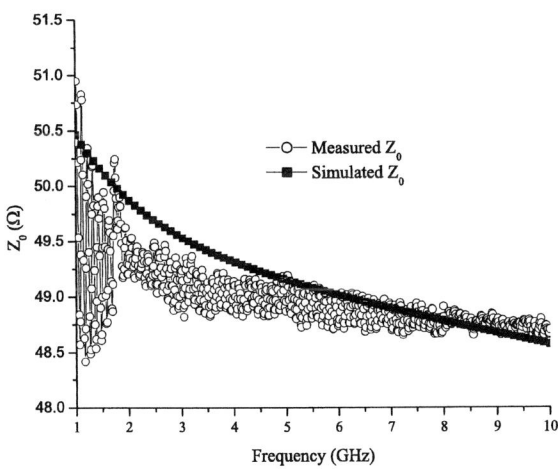

Fig. 2. Measured and simulated characteristic impedance of 900μm verification stub of ISS based coplanar waveguide.

$$K = \frac{(S_{11}^2 - S_{21}^2 + 1)}{2S_{11}} \qquad (2)$$

$$R = K \pm \sqrt{K^2 - 1} \qquad (3)$$

$$T = \frac{S_{11} + S_{21} - R}{1 - (S_{11} + S_{21})R} \qquad (4)$$

$$\varepsilon_{r-TEM} = \frac{j\ln(T)}{2\pi\left(\frac{d}{\lambda}\right)}\frac{1-R}{1+R} \qquad (5)$$

$$\varepsilon_r = \frac{\frac{\varepsilon_{rTEM}}{2\varepsilon_0 c_0 Z_0} - F_{up}}{F_{low}} \qquad (6)$$

Z_0 is obtained from equation 1.

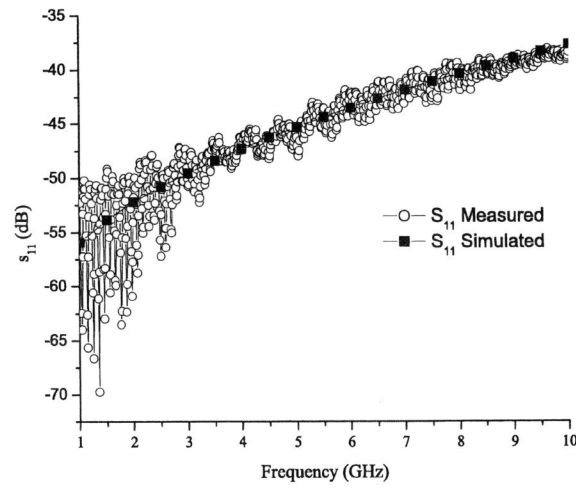

Fig. 3. Measured and simulated S_{11} versus frequency.

K, R and T are based entirely on the measured s-parameters. ϵ_{r-TEM} (for the quasi-TEM situation)is then extracted using equation 6. Z_0 is obtained from equation 1. Complete elliptical integrals of the first kind are used for the determination of the F_{up} and F_{low} parameters [7]. The measured characteristic impedance is 49.1Ω at 5GHz. Using this value, in equation 6 yields $\epsilon_r = 10.07$. Therefore, using equation 6 ϵ_r is calculated to within an accuracy of 2.0%. These equations show that the value of ϵ_r can be extracted from coplanar waveguide measurements as long as the value of the measured characteristic impedance is relatively constant.

III. ELECTROMAGNETIC DESIGN OF OPTIMUM THIN-FILM STRUCTURES

Figure 3 illustrates the measured and simulated input reflection coefficient for the 900μm ISS coplanar waveguide. The simulated results have been obtained from the EM simulator Momentum [9]. The excellent correspondence between the results demonstrates the validity of the simulation techniques applied.

These simulation techniques have been applied to more advanced substrates that incorporate a high-k thin-film layer. For ISS based coplanar lines that are fabricated on a continuous piece of alumina (Substrate height = 625 μm $\pm$ 25 μm) characterisation of the dielectric properties is straightforward. The situation is different for thin-film layers fabricated on a silicon substrate. Multi-layered thin-film structures are more difficult and time consuming to analyse accurately. The final

181

fabricated PMNT structure will be presented in section IV, but initial EM simulations are carried out on the less complex cross-section of figure 4. Consider the cross-section shown in figure 4. In this case the PMNT layer, the S_iO_2 layers and the Si substrate all influence the characteristics of the coplanar line and so the parameter extraction procedure becomes more complicated. In order to extract the dielectric constant of the PMNT layer to the greatest possible degree of accuracy an optimum coplanar test structure is required. Such a structure can be designed using EM simulation techniques.

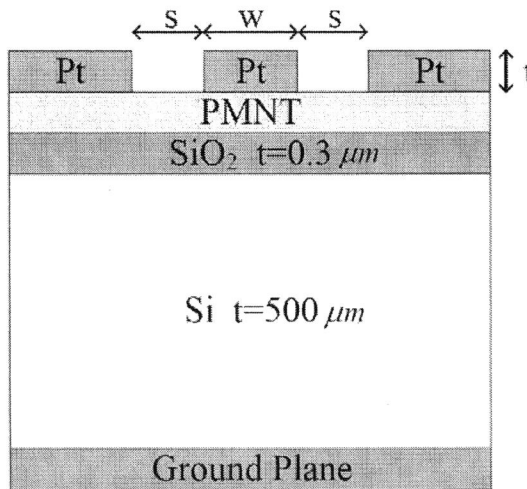

Fig. 4.   Cross-section of structure used for EM simulations.

Detailed EM simulations have been carried out in order to determine an optimum test structure for thin-film layer characterisation. The results of these simulations are summarised in figure 5 through to figure 8. Figure 5 illustrates the variation of the propagation constant with the variation of the spacing between the central conductor and the adjacent ground planes. The smaller the spacing between the central conductor and the adjacent ground planes the greater the influence of the thin-film PMNT layer on the propagation constant. This is further illustrated in figure 6. This shows the change in the propagation constant of a coplanar waveguide for a change in the dielectric constant of the PMNT layer from 200 to 300. As can be seen the PMNT layer has a much greater influence on the propagation constant for the $1\mu m$ spacing. Therefore, the spacing between the central conductor and the ground planes for optimum thin-film coplanar characterisation structures should be minimised.

Figure 7 illustrates the variation in the propagation constant for various central conductor widths for a change in the dielectric constant from 100 to 200. As can be seen a narrow conductor is most suitable to the requirements of this investigation.

The next design variable to be analysed is the thickness of the PMNT layer. Figure 8 illustrates the change in the propagation constant for various thicknesses of PMNT for

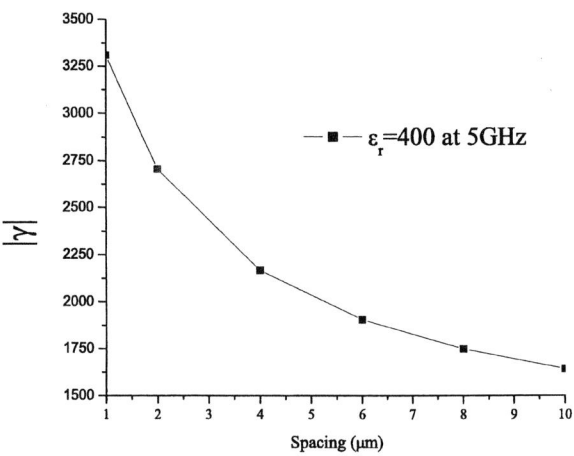

Fig. 5.   Variation of propagation constant with conductor to ground plane spacing.

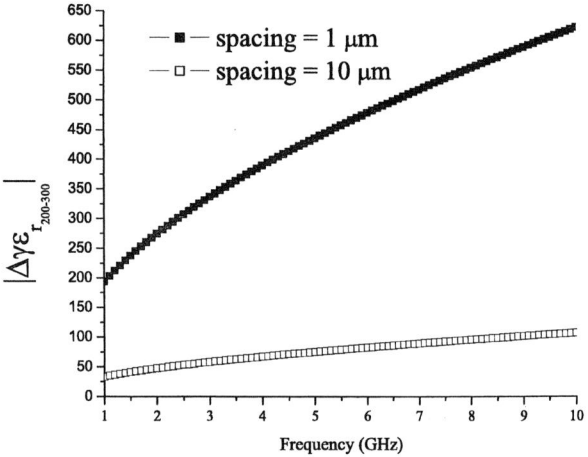

Fig. 6.   EM analysis of spacing variation of PMNT structures.

a change in the dielectric constant of the PMNT layer. As anticipated maximum change in the propagation constant is observed for the maximum simulated PMNT thickness ($t = 1\mu m$). The dimensions available to designers will vary from process to process. However, the results presented here serve as an important guideline for the design of optimum test structures. The coplanar waveguide should have 1) a minimum conductor to ground plane spacing, 2) a minimised conductor width and 3) maximum PMNT layer thickness.

IV. FURTHER SIMULATION FOR COMPLEX THIN-FILM STRUCTURE

In order to realize PMNT thin-films in a silicon environment a structure, such as that illustrated in figure 9, is required. For processing reasons, a platinum metal layer is required between

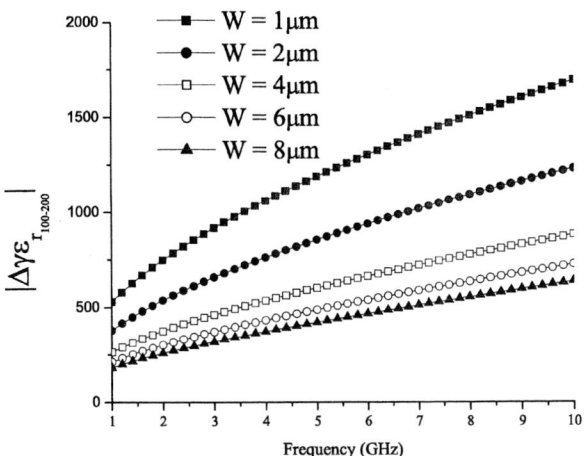

Fig. 7. EM analysis of conductor width variation of PMNT structures.

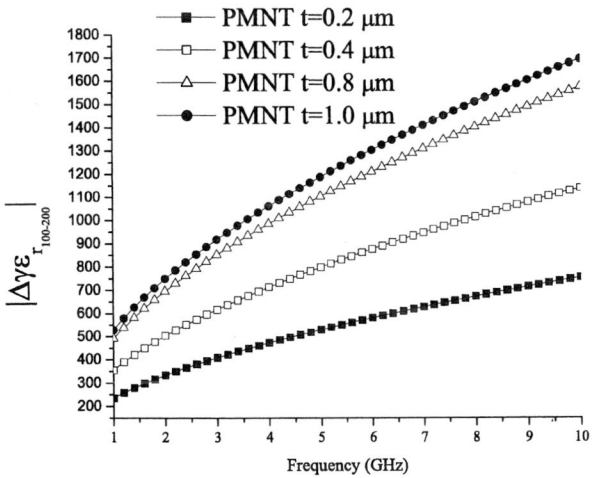

Fig. 8. EM analysis of PMNT thickness variation of PMNT structures.

the PMNT layer and the T_iO_2 layer. In this case, the PMNT layer, the extra platinum layer, the T_iO_2 layer, the S_iO_2 layer and the Si substrate all influence the overall characteristics of the coplanar line and so the parameter extraction procedure becomes much more complicated. The situation is even more complex for thin-film layers because of the second conductor layer.

The design methodology of section III has been applied to this structure. A number of different coplanar-waveguide structures have been designed, as illustrated in the mask layout of figure 10. The upper right corner of figure 10 shows a set of 6 coplanar-waveguide structures. For these 6 structures the center conductor width is 1 μm. The conductor to ground plane spacing is different for each of the 6 structures and varies form 0.5 μm to 10 μm. The upper left corner of figure 10 contains a similar set of structures except in this case the center conductor

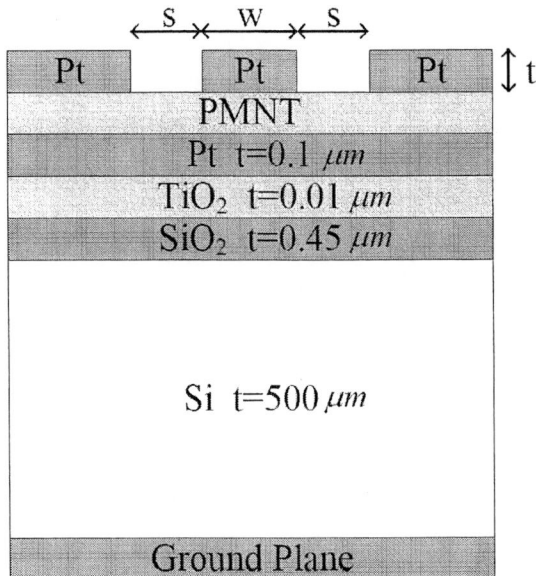

Fig. 9. Cross section of the final PMNT based structure.

width is 10 μm. The remaining structures on the mask of figure 10 are the same as those just described except for variations in the length of the coplanar-waveguide lines.

Figures 11 and 12 illustrate some Momentum results for a 500 μm line with a center conductor width of 1 μm and a center conductor to ground plane spacing of 1 μm for the structure of figure 9. Figure 11 illustrates the propagation constant, while figure 12 shows the S_{11} and S_{21} values of the line.

V. CONCLUSION

In this paper, an advanced test structure design methodology for dielectric characterisation of novel high-K materials has been presented. The dielectric constant of an alumina substrate has been successfully extracted from s-parameter measurements of a coplanar waveguide. More importantly EM simulation methods have been verified and subsequently implemented in the design of optimum test structures for the characterisation of thin-film high-k dielectric layers.

A set of guidelines that show the optimum dimensions that should be utilised in the design of coplanar dielectric characterisation structures has been presented. The proposed design methodology for optimum test structure layout has been implemented on the Momentum electromagnetic simulator based on the coplanar-waveguide method and quasi-TEM assumption. Mask design for the coplanar-waveguide has been finished based on the proposed design methodology.

ACKNOWLEDGMENT

This work is supported in part by the European Union Sixth Framework Programme through STRP Project, 033103 (CAMELIA).

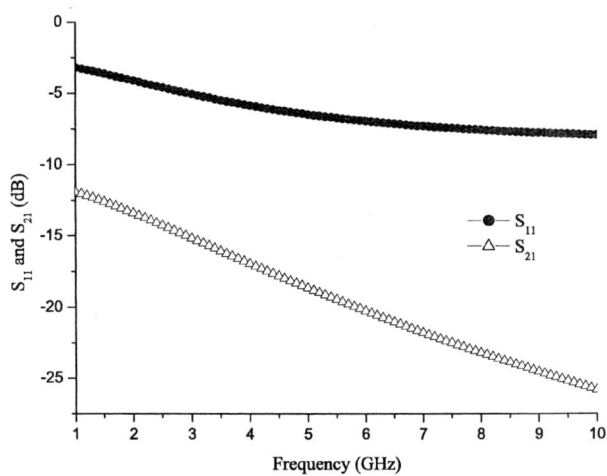

Fig. 12. EM analysis of s-paramters for the final PMNT based structure.

REFERENCES

[1] A. Fundora, E. Martinez and J. M. Siqueiros, "A study of electrical and optical fatigue in PMNT films on titanium nitride substrates," *Journal of Physics*, 7 December 2006.

[2] M. Y. Frankel, R. H. Voelker and J. N. Hilfiker, "Coplanar Transmission Lines on Thin Substrates for High Speed Low Loss Propagation," *IEEE Trans. Microwave Theory Tech.*, vol. 42, no. 3, pp. 396, 1994.

[3] G. Gonzalez, "Microwave Transistor Amplifiers Analysis and Design," Prentice Hall, New Jersey, 2nd edition, 1997, p. 62.

[4] Cascade Microtech Application Note, "On Wafer Vector Network Analyzer Calibrations and Measurements," 1997, Pub. Name PYRPROBE-0597.

[5] F. V. Hanna, "Finite boundary corrections to coplanar stripline analysis," *Electronics Letters*, vol. 16, no. 15, pp. 604-606, 17 July 1980.

[6] M. D. Janezic and D. F. Williams, "Permittivity characterization from transmission-line measurement," *IEEE MTT-S 1997 International Microwave Symposium Digest*, vol. 3, pp. 1343-1346, 8-13 June 1997.

[7] W. Heinrich, "Quasi-TEM description of MMIC coplanar lines including conductor-loss effects," *IEEE Trans on MTT*, vol. 4, no.1, pp. 45-52, January 1993.

[8] N. Berger, K. Biller, H. O. Ruoss, F. M. Landstorfer, "Broadband non-destructive determination of complex permittivity with coplanar waveguide fixture," *Electronics Letters*, vol. 39, no.20, pp. 1449-1451, October 2003.

[9] "Momentum," Agilent Technologies User Manual, [online], Available: *http://eesof.tm.agilent.com/docs/adsdoc2004A/pdf/mom.pdf*, September 2004.

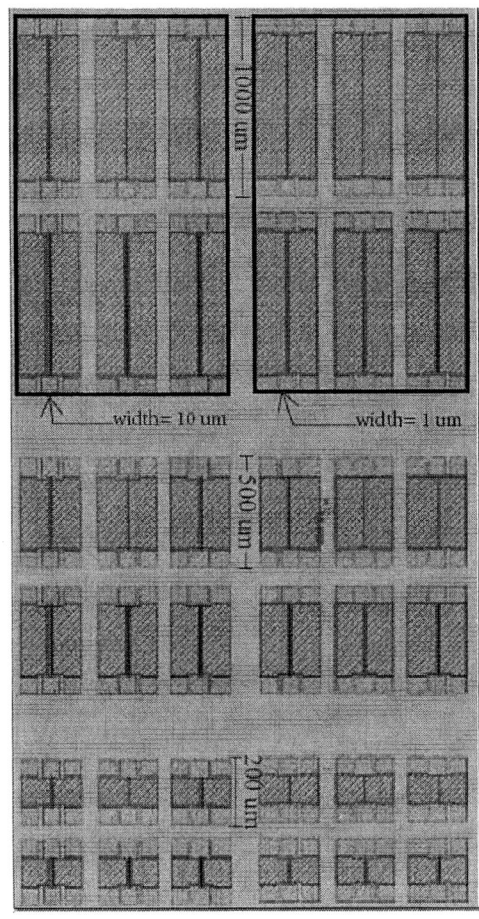

Fig. 10. Mask design for CPW's.

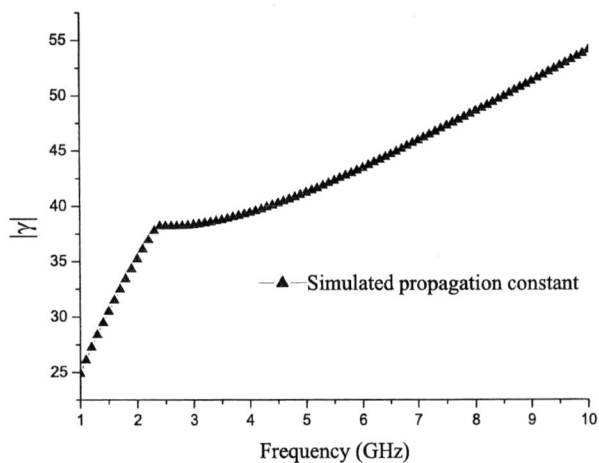

Fig. 11. EM analysis of propagation constant of final PMNT based structure.

Characterization of T-Shaped Terminal Impedances of Differential Short Stubs in Advanced CMOS Technology

Chiaki Inui and Minoru Fujishima

School of Frontier Science, The University of Tokyo
Kashiwa, Chiba, Japan

ABSTRACT

For short stubs in advanced CMOS technology, small terminal impedances are achieved by employing differential transmission lines and making virtual ground. However, no quantitative evaluation method for the terminal impedance of a differential short stub has been reported. To characterize the terminal impedance accurately, we propose the use of a T-shaped terminal impedance model of differential short stubs, where terminal impedance is evaluated by applying differential-mode and common-mode signals. In this paper, we describe the T-shaped terminal impedance model and the characterization procedure for terminal impedance. From measurement data, it is shown that the T-shaped terminal impedance of differential short stubs is successfully evaluated.

INTRODUCTION

In a recent millimeter-wave CMOS design, a large gain in a single-stage amplifier is difficult to achieve [1,2] because MOSFETs in an amplifier should operate at a significant fraction of its cut-off frequency. Since no insertion losses caused by passive circuits, such as matching circuits, can be compensated by MOSFETs in a millimeter-wave band, the optimization of passive circuits is inevitable to suppress insertion loss. Here, short stubs are often used in matching circuits [3] and filters. Noted, however, that the parasitic impedance at a grounded terminal degrades the quality factor of what as a resonator and increases insertion loss, as shown in Fig. 1. To improve the performance of a short stub, not only a reduction in its parasitic impedance but also its accurate characterization is required. In this paper, we propose a new characterization method for T-shaped terminal impedance for differential short stubs, which are utilized for reducing differential terminal impedance. By applying the new method, it will be shown that differential and common terminal impedances can be successfully characterized.

CHARACTERIZATION OF TERMINAL IMPEDANCE OF SHORT STUB

In on-chip short stubs, the terminal of a transmission line with a quarter wavelength is connected to a ground metal. Here, in a layout based on advanced CMOS technology, ground metal is constructed using a mesh structure to satisfy density rules. However, since the impedance of mesh metals is not negligible, the terminal of the short stub has an impedance Z_R, as shown in Fig. 2. Consequently, perfect ground cannot be achieved at the on-chip stub terminal. This Z_R causes a discrepancy between the measurement and simulation results, and increases insertion loss. Using an input impedance Z_{in} of a transmission line terminated with arbitrary load impedance [4], the terminal impedance Z_R is derived as

$$Z_R = \frac{Z_0 \left(Z_0 \sinh \gamma \ell - Z_{in} \cosh \gamma \ell \right)}{Z_{in} \sinh \gamma \ell - Z_0 \cosh \gamma \ell}, \qquad (1)$$

where Z_0, γ and ℓ are the characteristic impedance, a propagation constant and the length of the transmission line, respectively. On the other hand, the terminal impedance of a differential short stub can be described by a T-shaped equivalent circuit, as shown in Fig. 3, which is divided into the Z_d of the terminal impedance in the differential mode, and the Z_c of that in the common mode. For the evaluation of differential short stubs, when a pair of stubs is divided as shown in Fig. 4 considering the 180° phase difference between two ports, the voltage reflection coefficient in the differential mode Γ_{diff} is described as

$$\Gamma_{diff} = \frac{\left(S_{11} - S_{12} \right) - \left(-S_{22} + S_{21} \right)}{2}, \qquad (2)$$

where S_{ij}'s are elements of a scattering matrix. Since Γ_{diff} gives the input impedance Z_{in_diff}, the terminal impedance Z_d is obtained using Eq. (1). When the input signal is given by a common mode, and a pair of stubs is divided, as shown in Fig. 5, considering no phase difference between two ports, the voltage reflection coefficient in the common mode Γ_{com} is described as

$$\Gamma_{com} = \frac{\left(S_{11} + S_{12} \right) + \left(S_{22} + S_{21} \right)}{2}. \qquad (3)$$

Γ_{com} gives the terminal impedance $Z_d + 2Z_c$ in Fig. 5. Finally, we can obtain Z_d and Z_c from the measured two-port S parameter. The calculation procedure is summarized as a flowchart in Fig. 6.

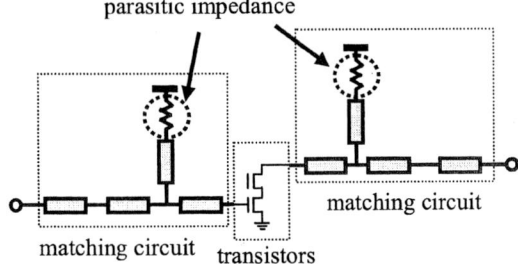

Fig. 1. Parasitic impedance at grounded terminal in matching circuits.

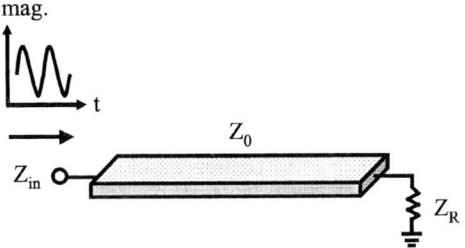

Fig. 2. Short stub with terminal impedance Z_R.

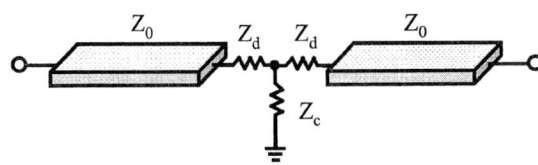

Fig. 3. T-shaped terminal impedance in differential short stubs.

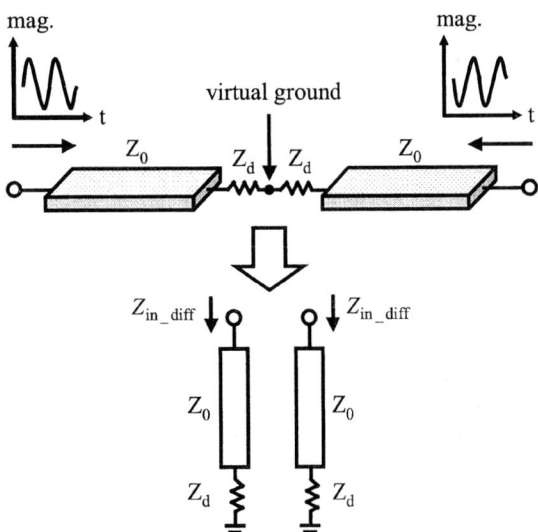

Fig. 4. Terminal impedances of short stubs in differential mode.

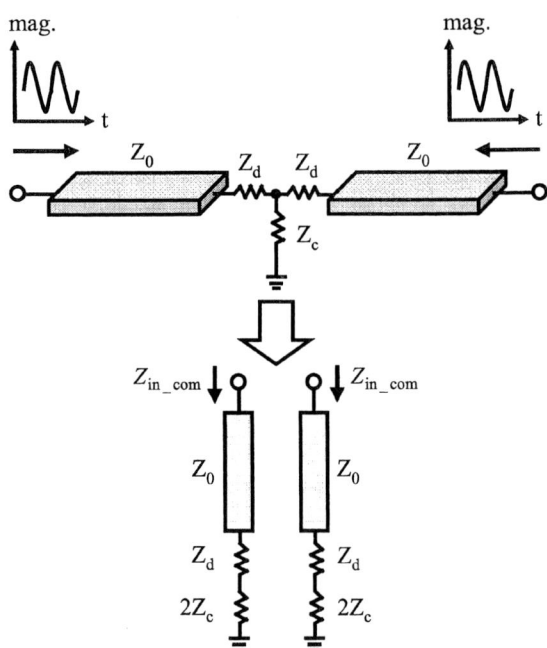

Fig. 5. Terminal impedances of short stubs in common mode.

measurement data : S_{11}, S_{12}, S_{21}, S_{22}

$$S_{\text{diff}} = \frac{S_{11} + S_{22} - S_{12} - S_{21}}{2}, \quad S_{\text{com}} = \frac{S_{11} + S_{22} + S_{12} + S_{21}}{2}$$

$$Z_{\text{in\_diff}} = 50 \times \frac{1 + S_{\text{diff}}}{1 - S_{\text{diff}}}, \quad Z_{\text{in\_com}} = 50 \times \frac{1 + S_{\text{com}}}{1 - S_{\text{com}}}$$

$$Z_c = \frac{Z_0 \left(Z_0 \sinh \gamma \ell - Z_{\text{in\_diff}} \cosh \gamma \ell \right)}{Z_{\text{in\_diff}} \sinh \gamma \ell - Z_0 \cosh \gamma \ell},$$

$$Z_d + 2Z_c = \frac{Z_0 \left(Z_0 \sinh \gamma \ell - Z_{\text{in\_com}} \cosh \gamma \ell \right)}{Z_{\text{in\_com}} \sinh \gamma \ell - Z_0 \cosh \gamma \ell}$$

Z_d and Z_c are obtained.

Fig. 6. Flowchart for obtaining Z_d and Z_c from two-port S-parameter.

186

MEASUREMENT RESULTS

The TEG of differential short stubs is fabricated using a six-metal 90nm CMOS process. The structure of the transmission line is based on a slow-wave transmission line as shown in Fig. 7 [5]. Figure 8 shows a chip photograph. Figures 8(a) and 9 show a differential short stub with a parallel-plate capacitor (1.4pF) that can be externally biased through a DC pad. Figures 8(b) and 10 show a differential short stub with parallel-plate (1.0pF) and comb (10.0pF) capacitors. A simple straight transmission line is also evaluated for use as a differential short stub, as shown in Fig. 8(c). Two-port S-parameters are measured with a vector network analyzer (37397D/Anritsu). Measurement data for analysis are obtained after de-embedding. For the evaluation of the devices in Figs. 8(a) and (b), firstly, short stubs in a single-end structure are evaluated using S11, and secondly, the one in differential structure is evaluated by using S11, S12, S21, and S22. For the evaluation of the device in Fig. 8(c), short stubs in a differential structure are evaluated by applying the flowchart in Fig. 6 under the condition that there is no parasitic impedance between the two stubs. Figures 11(a) and (b) show a smith chart of the terminal impedance in the single-end structure. Figures 11 (c) and (d) show terminal impedances with a differential short stub with a parallel-plate capacitor (1.4pF), (e) and (f) show terminal impedances with a differential short stub with parallel-plate (1.0pF) and comb (10.0pF) capacitors. Figures 11 (g) and (h) show terminal impedances for a straight transmission line. Figures 11(a), (c) and (d) show that the main terminal impedance appeared in the single-end structure is not from Z_d but from Z_c. In Figs. 11(d) and (f), a lower terminal impedance is achieved using a bypass capacitor of 11.0pF than using that of 1.4pF. In Fig. 11(g), since both the resistance and reactance of terminal impedance in a differential structure are within $\pm 1\Omega$ at 0.04-110GHz, it is found that short stubs are terminated almost ideally. In the case of a simple transmission line, because the center is not terminated to ground metals, Z_c is almost infinity, as shown in Fig. 11(h).

CONCLUSION

We proposed a new method of characterizing the T-shaped terminal impedances of differential short stubs from the analysis in differential-mode and common-mode signals. By applying the method to the measurement results, it was shown that the resistance and reactance of the terminal impedance in the differential mode are within $\pm 1 \Omega$ at 0.04-110 GHz and that the terminal impedance in the common mode is infinity when a straight transmission line is evaluated as a pair of short stubs in a differential structure. The proposed method is expected to be useful for the improvement of differential short stubs.

ACKNOWLEDGMENT

This study is supported by the Strategic Information and Communication R&D Promotion Program (SCOPE) of the Ministry of Internal Affairs and Communications of Japan. The chip fabrication was facilitated by the Chip Fabrication Program of the VLSI Design and Education Center (VDEC), University of Tokyo, in collaboration with STARC, Fujitsu, Ltd., Matsushita Electric Industrial Company, Ltd., NEC Electronics Corporation, Renesas Technology Corporation, and Toshiba Corporation.

REFERENCES

[1] M. A. Masud, H. Zirath, M. Ferndahl, and H-O. Vickes, "90 nm CMOS MMIC amplifier", *Radio Frequency Integrated Circuits Symposium*, 2004. Digest of Papers. 2004 IEEE 6-8, pp. 201-204, June. 2004.

[2] S. P. Voinigescu, M. Gordon, C. Lee, T. Yao, A. Mangan, and K. Yau, "System-on-chip design beyond 50 GHz", *Proceedings. Fifth International Workshop on System-on-Chip for Real-Time Applications*, pp. 10-13, 2005.

[3] G. Gonzalez, "Microwave transistor amplifiers analysis and design second edition", Printice-Hall, 1997, 1984.

[4] B. C. Wadell, "Transmission line design handbook", Artech House, Norwood, MA 1991.

[5] I. C. H. Lai, H. Tanimoto and M. Fujishima, "Characterization of high Q transmission line structure for advanced CMOS processes," *IEICE Trans. Electron.*, pp. 1872-1879, Dec. 2006.

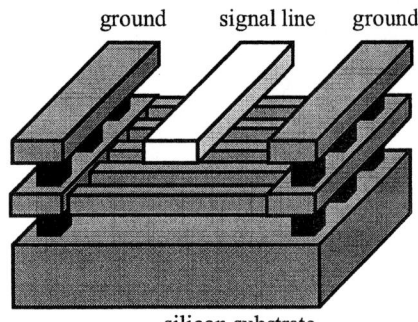

Fig. 7. Structure of slow-wave transmission line (SWTL).

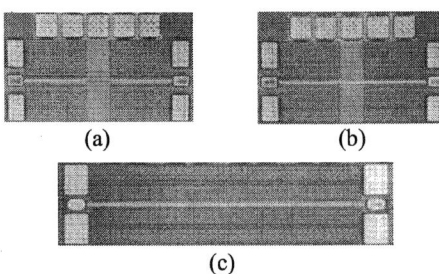

(a)   (b)

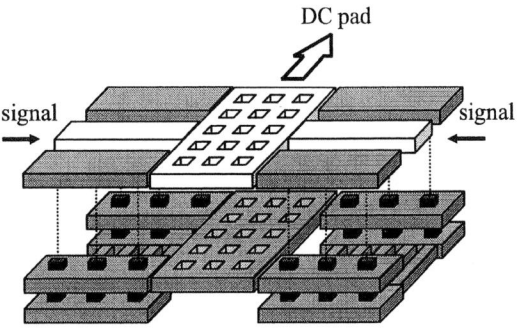

(c)

Fig. 8. Chip photograph of (a) differential short subs with parallel-plate capacitor (1.4pF), (b) differential short stubs with parallel-plate (1.0pF) and comb (10.0pF) capacitors, and (c) straight transmission line.

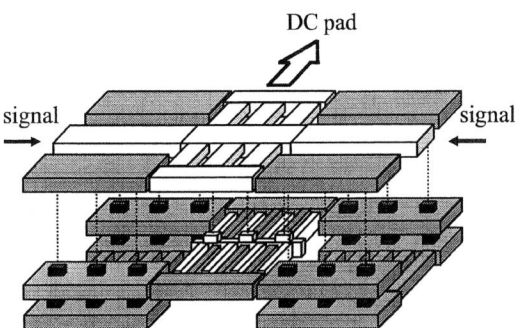

Fig. 9. Structure of differential short subs with parallel-plate capacitor (1.4pF).

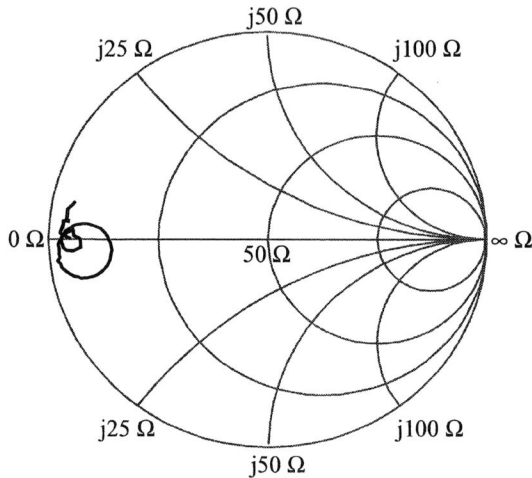

(a) Z_R of terminal impedance with parallel-plate capacitor (1.4pF) in single-end.

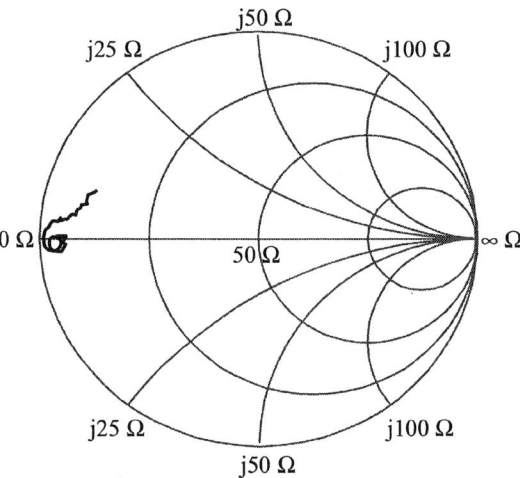

(b) Z_R of terminal impedance with parallel-plate (1.0pF) and comb (10.0pF) capacitors in single-end.

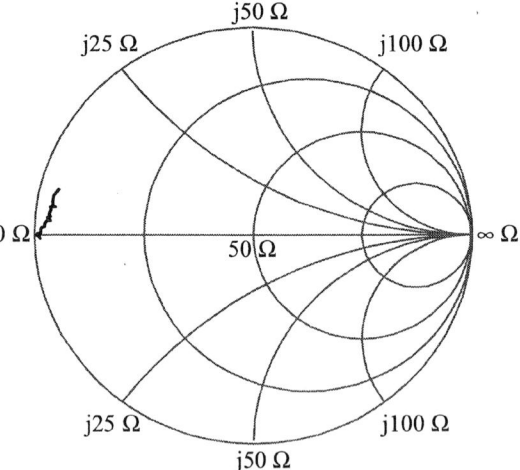

(c) Z_d of terminal impedance with differential short stub with parallel-plate capacitor (1.4pF).

Fig. 10. Structure of differential short stubs with parallel-plate (1.0pF) and comb (10.0pF) capacitors.

188

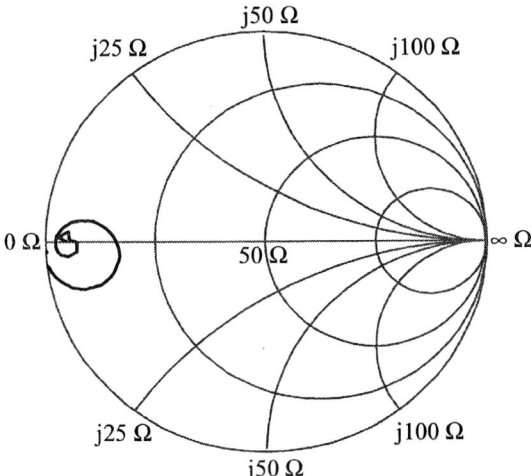

(d) Z_c of terminal impedance with differential short stub with parallel-plate capacitor (1.4pF).

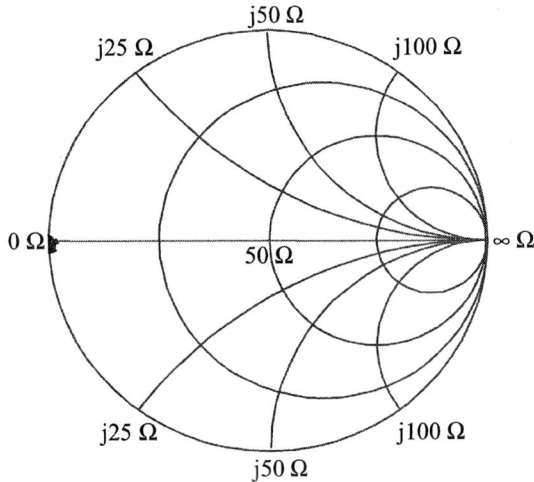

(g) Z_d of terminal impedance for straight transmission line.

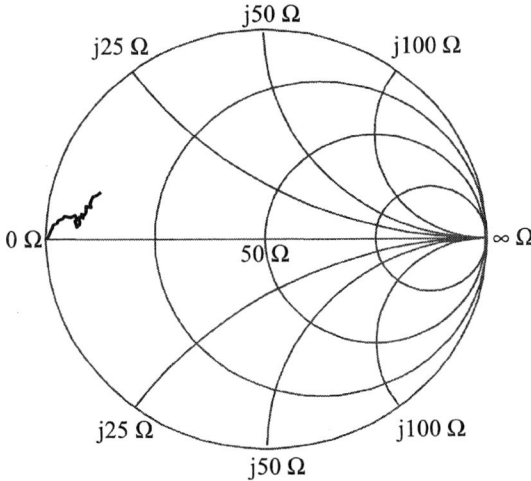

(e) Z_d of terminal impedance with differential short stub with parallel-plate (1.0pF) and comb (10.0pF) capacitors.

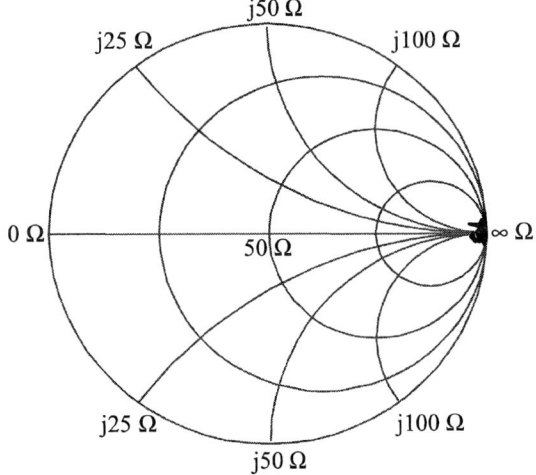

(h) Z_c of terminal impedance for straight transmission line.

Fig. 11. Characterization results of terminal impedances in short stubs at 0.04-110 GHz.

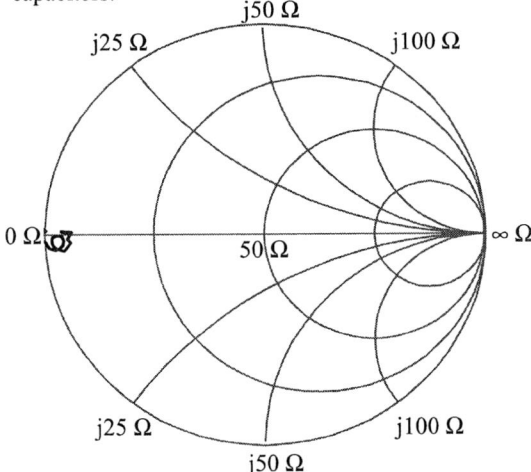

(f) Z_c of terminal impedance with differential short stub with parallel-plate (1.0pF) and comb (10.0pF) capacitors.

2008 IEEE Conference on Microelectronic Test Structures, March 24-27, Edinburgh, UK

8.4

Identifying dielectric and resistive electrode losses in high-density capacitors at radio frequencies

M.P.J. Tiggelman, K. Reimann*, J. Liu*, M. Klee**, W. Keur**, R. Mauczock**, J. Schmitz, and R.J.E. Hueting

University of Twente, MESA+ Institute for Nanotechnology
Department of Semiconductor Components, 7500 AE Enschede, The Netherlands
Telephone: +31 (0)53-4892644, Fax: +31 (0)53-4891034
Email: m.p.j.tiggelman@ewi.utwente.nl

*NXP Semiconductors, 5656 AE, Eindhoven, The Netherlands

**Philips Research, 5656 AE, Eindhoven, The Netherlands

Abstract— A regression-based technique is presented which distinguishes the dielectric loss from the resistive loss of high density planar capacitors in a very wide bandwidth of 0.1–8 GHz. Moreover, the procedure yields useful results if the capacitor deviates from a lumped element model and indicates when the used approximations break down or whether size-dependent loss mechanisms exist.

I. INTRODUCTION

Reconfigurable passive building blocks can reduce system area and costs in handheld applications. Tunable technologies based on micro-electro-mechanical-systems (MEMS) [1], [2], varactor diodes [3], liquid crystals [4] or ferroelectrics [5] support the continuous miniaturization of microwave electronics. We optimize ferroelectric capacitors on silicon as a potential key component for microwave frequency agile applications. The requirements of a low loss tangent $\tan \delta$ (high unloaded quality factor Q) and a high tunability are evident for tunable capacitors at radio frequencies (RF) especially in passive circuits where no active circuitry is present that compensates for dissipation. Fast processing and characterization cycles are desired for the material development of complex ferroelectrics for high density planar capacitors.

We present an easy-to-implement, but still accurate enough, electrical material characterization technique, which distinguishes the dielectric loss from the resistive electrode loss of simple metal-insulator-metal (MIM) test devices. Our technique even provides for a partial compensation of distributed effects and indicates graphically when the used approximations break down or whether size-dependent loss mechanisms exist. In this paper we make use of the geometry, sheet resistance, measured series equivalent capacitance and frequency behavior of more than 2 circular shaped test structures to separate the aforementioned losses at a given frequency. The resistive electrodes, and the inductance overcompensation during a 1-port short-open-load (SOL) calibration have an effect on the measured equivalent capacitance and loss tangent. The loss separation technique is fully exploited in the latter section. It enables the process engineer to gain a swift and better understanding in how to optimize the process flow to improve RF performance.

II. TEST STRUCTURES

We employ circular shaped ground-signal-ground (GSG) capacitor test structures of which only the top electrode is patterned [6] (see Fig. 1).

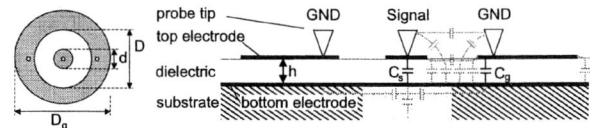

Fig. 1. A top view with 3 probe landing spots and a cross section of the test device. The bottom electrode and the dielectric layer are not patterned. The parasitic capacitances are indicated with striped grey colors.

The top electrodes consist of a Pt-Au stack and the bottom electrode of Pt. The sheet resistance of the top electrodes is $R_{s,t} \approx 70\,\text{m}\Omega/\square$ and the sheet resistance of the bottom electrode is $R_{s,b} \approx 1.4\,\Omega/\square$. A ferroelectric insulator of barium strontium titanate (BST), with a relative permittivity $\varepsilon_r = 170$ and a thickness $h = 110\,\text{nm}$, is sandwiched between the parallel electrodes. The diameter d of the signal path varies from $8\,\mu\text{m}$ to $88\,\mu\text{m}$. The bottom electrode and dielectric layer are both homogeneous and unpatterned. The capacitance of the signal (center) path C_s is in series with the much larger ground capacitor C_g with an outer ring diameter of $D_g = 600\,\mu\text{m}$ and an opening of $D = 100\,\mu\text{m}$. The measured equivalent series capacitance is therefore approximated by C_s.

The parasitic capacitances through the air, the dielectric and silicon are also indicated in Fig. 1. An assessment is made on the capacitance contributions to the test structures. Starting with the capacitance from the signal path to the bottom electrode

$$C_{\text{par,signal path}} = \epsilon_0 \varepsilon_r (\pi d) \left(\frac{2\ln(2)}{\pi} - \text{underetch} \right) \quad (1)$$

we use conformal mapping [7], [8] to get an estimation of the parasitic capacitance, which is lower than 2.1% of C_s, assuming the underetch equals zero. The capacitance from the ground top electrode to the bottom electrode is negligible, because of the very large ground capacitance. The capacitance

978-1-4244-1801-5/08/$25.00 ©2008 IEEE

from the signal path to the ground path through the dielectric is also neglected. The distance between the contact paths is in the μm-range, while the dielectric thickness is in the nm-range, so the electric field is screened by the bottom electrode.

The capacitance between the ground and signal probes and paths is simulated in the planar full-wave electromagnetic program Sonnet [11](more details about the simulations are given in section IV-B). The permittivity of air is set to 1×10^{-4}, then to 1 and the measured equivalent series capacitance is compared. The simulation results show that the additional capacitance is in the low fF-range and is hence negligible. The capacitance from the middle of the bottom electrode downwards into the 675 μm substrate has a very small ratio of the area to the dielectric thickness and is estimated in the fF-range. The capacitance parallel to the bottom electrode through the silicon substrate is determined by performing Sonnet simulations without a bottom electrode. We obtained less than 20 fF for a capacitor with an inner diameter of 40 μm. Since the influence of all parasitic capacitances compared to our high density capacitors differ individually a factor 100 or more we can neglect all these parasites.

III. MODELING

The electrode resistance is modeled to be able to separate the measured dielectric loss from the measured resistive loss. We model the ferroelectric capacitor with an R_s–C_s series equivalent model (see Fig. 2).

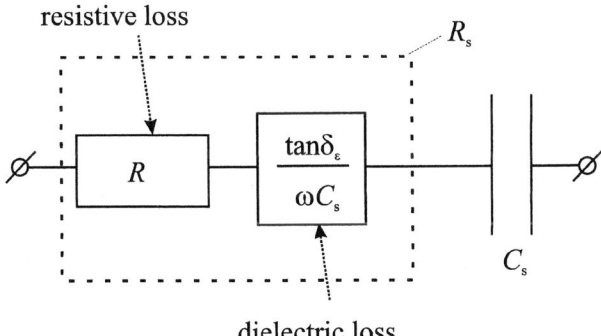

Fig. 2. The measured series R_s–C_s equivalent capacitor model.

with the equivalent series resistance $R_s = \Re(Z_{11})$, the modeled electrode resistance R, the dielectric loss $\tan \delta_\varepsilon$ [9]., and the equivalent series capacitance C_s.

The measured loss tangent then follows

$$\tan \delta = \frac{1}{Q} = \frac{\Re(Z_{11})}{|\Im(Z_{11})|} = \tan \delta_\varepsilon + \omega C_s R \qquad (2)$$

The latter term represents the additional resistive electrode loss at RF. The geometry of the electrodes determines the resistance R [6], [10]. For a lumped element case R is

approximated by the sum of the resistance of the center (signal) path (see Fig. 1 and 3)

$$R_{\text{center}} = \frac{R_{s,t} + R_{s,b}}{8\pi} \qquad (3)$$

and the resistance of the bottom electrode between the signal and ground path

$$R_{\text{ring}} = \frac{R_{s,b}}{2\pi} \ln \left(\frac{D}{d} \right) \qquad (4)$$

and the resistance R_{outer} of the connection to the ground probes. A schematic circuit is depicted in Fig. 3.

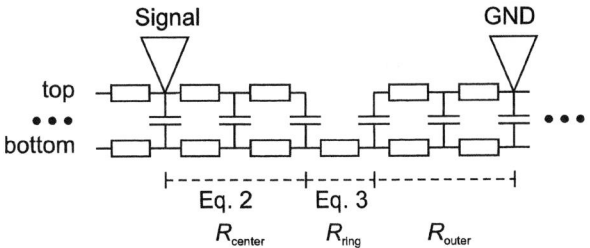

Fig. 3. Any lumped capacitor becomes a distributed R-C chain at RF. The capacitances corresponding to the electrical fields through air and substrate (not drawn) are still negligible small with respect to the large capacitances of the MIM capacitor.

R_{outer} can be neglected in our case because of the larger diameter $D \gg d$. The contact resistance of the electrodes at the probe tip is measured on an unpatterned top electrode with $R_{\text{short,top}} \approx 0.1\,\Omega$ and is hence negligible. We therefore write the resistance

$$R \approx g_R R_{s,b} \quad \text{with} \quad g_R = \frac{1}{2\pi} \left(\frac{1+\eta}{4} + \ln \frac{D}{d} \right) \qquad (5)$$

$$\text{and} \quad \eta = \frac{R_{s,t}}{R_{s,b}}$$

as a product of the sheet resistance $R_{s,b}$ of the bottom electrode and an electrode geometry factor g_R. The sheet resistance ratio η is negligibly small for the typical choice of well conducting top electrodes.

IV. EXPERIMENTAL RESULTS AND DISCUSSION

1-Port S-parameter were measured with an Advantest R3767CG vector network analyzer (VNA) in the frequency range of 300 kHz–8 GHz, and an Infinity GSG probe (probe pitch = 125 μm). All measurements were performed at zero bias with an RF power of $P_{\text{RF}} = -10$ dBm at room temperature using the built-in short-open-load calibration on a CS-5 calibration substrate of GGB Industries. A small error is made in the inductance measurements because of a zero short inductance parameter in the calibration kit. A port offset could be used to account for the finite probe inductance, but was avoided for the sake of simplicity. The data analysis used here is insensitive to the inductance calibration error (see below).

A. Capacitance with frequency

The measured ferroelectric capacitance decreases with frequency due to dielectric relaxation and distributed effects. Relaxation effects cause a delay in ionic polarization decreasing the relative dielectric constant linearly on a logarithmic frequency scale. The decrease is independent of the diameter and is relatively small compared to the much stronger decrease due to distributed effects from a few hundred megahertz onwards (see Fig. 4).

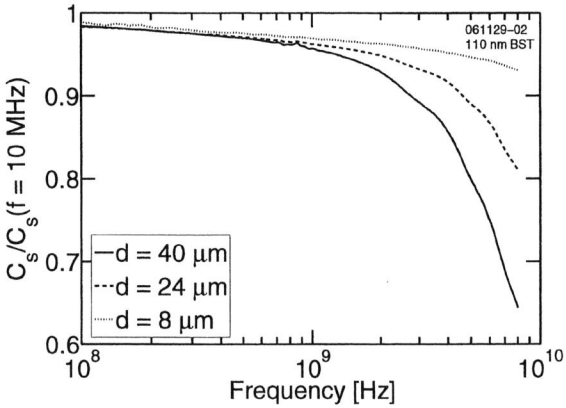

Fig. 4. The normalized capacitance declines above $f = 400\,\text{MHz}$ for the largest capacitor due to the distributed nature of the capacitor.

A lumped capacitor becomes distributed with increasing frequency due to the resistivity of the electrodes (see also Fig. 3). At RF the R–C chain leads to a voltage drop at the edges of the capacitor, which acts as if the area of C_s would be reduced. A reduced R_{center} is accompanied by a reduced C_s (deviation from the linear decline in capacitance in Fig. 4). To confirm this assumption simulations were setup in the planar 3D full-wave electromagnetic simulation program Sonnet [11]. All possible coupling mechanisms are included in the electromagnetic analysis.

B. Capacitor modeling

A simulation in Sonnet starts with defining the mesh and the dimensions of a box. A box consists of 6 grounded and lossless boxwalls in which a test structure is designed. Our complete capacitor stack is visualized in Fig. 5.

The planar dielectric and metal layers are surrounded by grounded boxed walls. The capacitor electrodes are isolated from the grounded sidewalls, similar to the measurements. The ground-signal-ground (GSG) probe pins are simulated in a relatively simple way by three $4\,\mu\text{m} \times 4\,\mu\text{m}$ lossless vias from the top boxwall to the top electrode of the signal and ground path as depicted in Fig. 5 and 6.

The signal via port is placed at the center of the via and both separate ground vias are connected to the top box wall. Our planar circular capacitive test structures are approximated

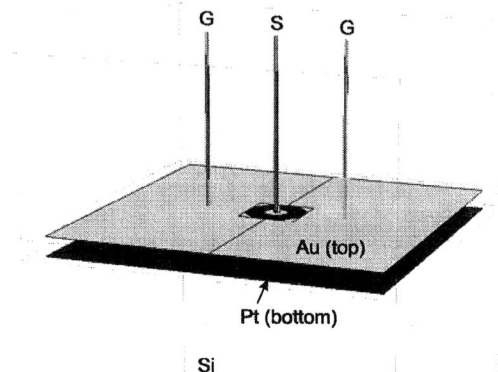

Fig. 5. A 3D image of the planar layers of a MIM capacitor test structure with vias as GSG probe tips inside a Sonnet design box.

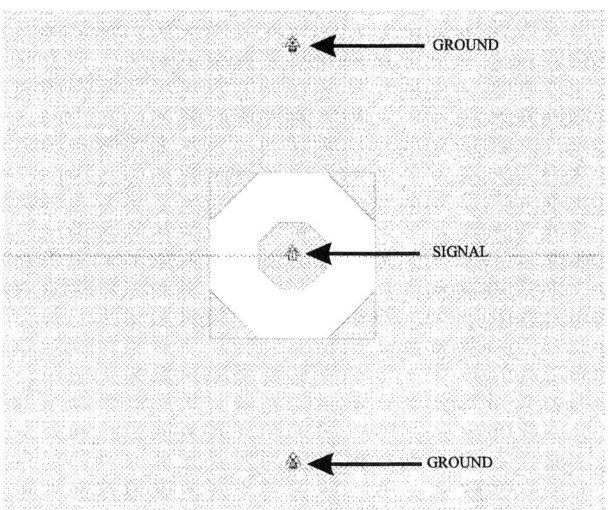

Fig. 6. A zoomed-in top view of a capacitor in Sonnet with an inner signal path diameter of $40\,\mu\text{m}$, three vias act as probe pins of a GSG probe and a $50\,\Omega$ signal port is situated in the center of the DUT.

by a regular octagon.

The MIM capacitors are modeled by 3 planar dielectric layers with material parameter values as given in TABLE I with the relative permittivity ε_r, the relative magnetic permeability μ_r and the dielectric conductivity σ. These values are constant and were chosen to match the measured values at 200 MHz.

Layers	Thickness	ε_r	μ_r	$\tan\delta$	σ
air	$1\,\mu\text{m}$	$1 \cdot 10^{-4}$	$1 \cdot 10^{-4}$	0 %	0 S/m
BST	$0.11\,\mu\text{m}$	170	1	1 %	0 S/m
Si	$675\,\mu\text{m}$	11.9	1	0.04 %	5 S/m

TABLE I

LAYER PROPERTIES OF THE DIELECTRIC LAYERS IN SONNET.

The non-physical parameter values of the air layer

192

minimize the parasitic capacitance and inductance caused by the lossless vias. We therefore do not need to simulate separate calibration or de-embedding structures at the cost of neglecting a part of the device capacitance and inductance. The resistive electrode layers are defined by the measured sheet resistance as mentioned in section II.

1) Modeling distributed effects: The capacitor stack is simulated with lossless and lossy electrodes. The 1-port response of the simulated C_s and $\tan\delta$ at different frequency points are depicted in Fig. 7.

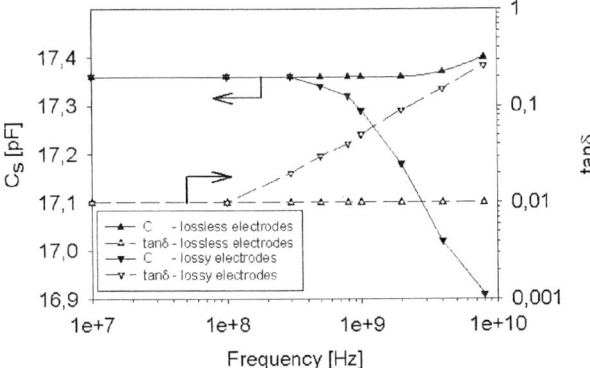

Fig. 7. Planar electromagnetic simulations in Sonnet [11] show the capacitance of a test structure with an inner diameter of 40 μm with lossless and lossy electrodes. No relaxation effects are included in the simulations.

The strong frequency dependent influence of the resistive electrodes on the C_s and $\tan\delta$ is confirmed. The amount of electrode resistance of a high density capacitor determines the extent of the distributed effects, and therefore the resistive electrode loss. This result points out the importance of good conducting electrodes to obtain a low loss tangent at RF.
Furthermore, measurement uncertainty also affects the measured capacitance as described in [10], [12]. The short and load calibration standards are not ideal standards. They also contain parasitic inductances, which add up to the impedance of the standards. This causes inaccuracies in the measurement data if the user-defined calibration kit within the VNA is not adjusted properly, or if no correction factor is taken along in the parameter extraction program. In our case no quantified correction terms by the calibration kit manufacturer are known since our probe dissents from the advised one. An additional inductance caused by an incomplete impedance compensation of the short and load standard affects the measured capacitance with frequency. A manual correction by including an inductance L to the impedance parameters changes the reactive part of the impedance and therefore the capacitance as depicted in Fig. 8.

In our case the VNA assumes zero inductance for the combination of the GSG probe and the short or load measurement where is fact it should be around 10 pH. Adding the missing inductance to the measured impedance of

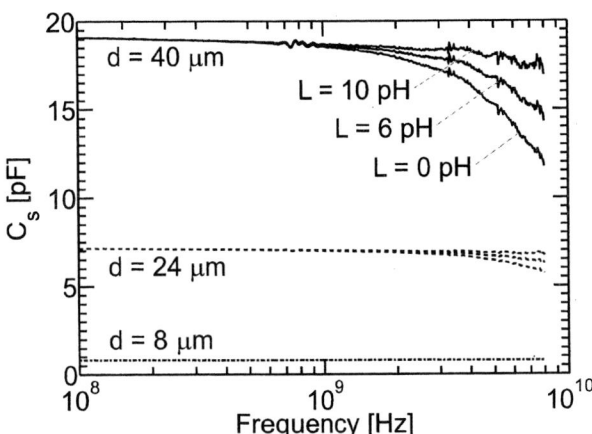

Fig. 8. The effect of an additional inductance of $L = 0$, 6 and 10 pH on the measured equivalent capacitance with frequency for devices under test (DUT)s with an inner diameter of $d = 40$, 24, and 8 μm.

$$Z_{11} = R - \frac{i}{\omega C_{s,L=0}} + i\omega L, \quad \text{with } C_s = \frac{-1}{\omega\Im(Z_{11})} \quad (6)$$

To distinguish the dielectric from the resistive loss a separation technique is proposed in the following section with partial compensation of distributed effects and inductance miss calibration.

C. The loss separation technique

The loss separation technique uses at least three circular shaped test structures with different inner diameters. No additional de-embedding structures are required. Distributed effects are partially compensated by plotting $\tan\delta$ as a function of the *measured* equivalent series capacitance (see Fig. 9). The slope of the curves is equal to $\omega R_{s,b}$, as expressed in (2) and (5). The dielectric loss tangent $\tan\delta_\varepsilon$ is determined by a linear extrapolation to the intercept point at $\tan\delta(C_s g_R)$ in Fig. 9 at $f = 8$ GHz. A deviation from linearity at small radii would suggest an influence of the edges of the capacitors, e.g., by damage during processing caused by reactive ion etching, increasing the loss tangent. A deviation from linearity at large radii suggests strong distributed effects, which decrease the capacitance and increase the loss tangent.

Employing the same visualization of parameters in Sonnet, the distributed effects for physically large capacitors becomes clearly visible as depicted in Fig. 10.

In Fig. 11 the measured $\tan\delta_\varepsilon$ and the resistive loss are separated across the entire frequency span. The (total) loss tangent of test structures with an inner diameter of 40, 24 and 8 μm is shown. A smaller inner diameter decreases the $\tan\delta$ at RF. To calculate the $\tan\delta_\varepsilon$ the measured total loss is extrapolated to $C_s g_R \to 0$ for each frequency using all data from $d = 40$ to 8 μm and $d = 24$ to 8 μm.

A graphical outline of the measurement data of more then two capacitors in Fig. 11 can safely and easily separate the

193

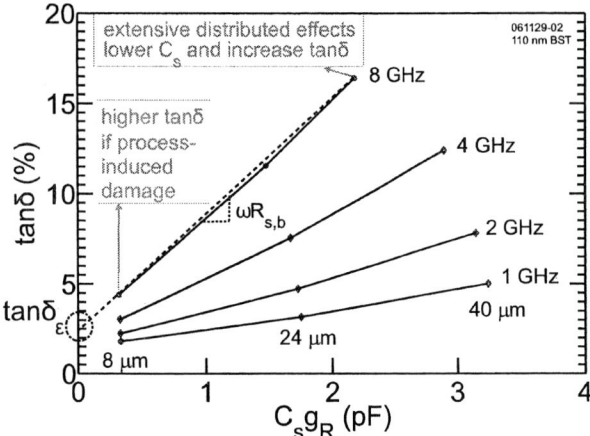

Fig. 9. The measured dielectric and the resistive loss tangent are separated using a linear regression. No inductance compensation is employed after calibration. The Sonnet simulation results at $f = 1\,\text{GHz}$ are indicated by the X-markers and show a close resemblance to the measurement results if $d = 8$ and $24\,\mu\text{m}$ as expected.

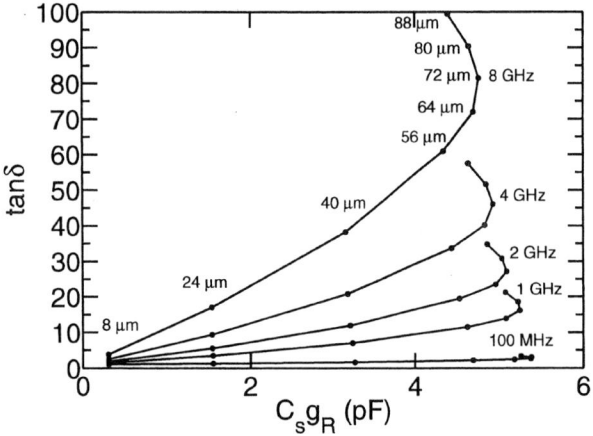

Fig. 10. This Sonnet simulation result shows the influences of distributed effects at multiple frequencies on capacitors with a relatively large inner diameter.

dielectric loss from the resistive electrode loss, as expressed in (2). It is especially suited to measure the frequency dependence of the dielectric loss tangent $\tan\delta_\varepsilon$. The results in Fig. 11 shows a typical ferroelectric behavior of increasing dielectric losses with increasing frequency.

CONCLUSIONS

An easy-to-use regression-based technique for characterizing planar high density capacitors is presented. This technique separates the dielectric losses from the resistive losses at $f = 0.1$–$8\,\text{GHz}$ utilizing three circular test structures. Furthermore, a deviation from the linear regression curve for small radii capacitors would suggest process-induced damage at the edges of the capacitor, and for large radii capacitors the influence of prominent distributed effects. A possible inductance

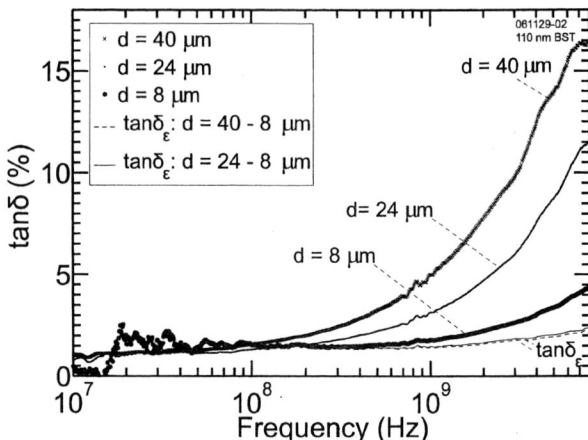

Fig. 11. The measured losses with and without subtraction of the resistive bottom electrode losses.

miss calibration is canceled to a large extent by the analysis technique. The effect of the electrodes on the losses at RF, and the validity of the approximations for the separation are demonstrated.

ACKNOWLEDGEMENTS

This work is partially supported by the European Commission, project NANOSTAR. M.P.J. Tiggelman would like to thank NXP Semiconductors for financing his research.

REFERENCES

[1] S. Chang and S. Sivoththaman, *A Tunable RF MEMS Inductor on Silicon Incorporating an Amorphous Silicon Bimorph in a Low-Temperature Process*, IEEE Electron Device Letters, vol. 27, no. 11, pp. 905–907, November 2006.

[2] Th.G.S.M. Rijks, P.G. Steeneken, J.T.M. van Beek , M.J.E. Ulenaers, A. Jourdain, H.A.C. Tilmans, J. De Coster, and R. Puers, *Microelectromechanical Tunable Capacitors for Reconfigurable RF Architectures*, Journal of Micromechanical Microengineering, vol. 16, pp. 601–611, February 2006.

[3] K. Buisman, L.C.N. de Vreede, L.E. Larson, M. Spirito, A. Akhnoukh, Y. Lin, X.-d. Liu, and L.K. Nanver, *A Monolithic Low-Distortion Low-Loss Silicon-on-Glass Varactor-Tuned Filter with Optimized Biasing*, IEEE Microwave and Wireless Component Letters, vol. 17, no. 1, pp. 58–60, January 2007.

[4] J.A. Yeh, C.A. Chang, C.-C. Cheng, J.-Y. Huang, and S.S.H. Hsu, *Microwave Characteristics of Liquid-Crystal Tunable Capacitors*, IEEE Electron Device Letters, vol. 26, no. 7, pp. 451–453, July 2005.

[5] N. Setter et al., *Ferroelectric thin films: Review of Materials, Properties, and Applications*, Journal of Applied Physics, vol. 100, no. 051606, September 2006.

[6] Z. Ma, A.J. Becker, P. Polakos, H. Huggins, J. Pastalan, H. Wu, K. Watts, Y.H. Wong, and P. Mankiewich, *RF Measurement Technique for Characterizing Thin Dielectric Films*, IEEE Transactions on Microwave Theory and Techniques, vol. 45, no. 8, pp. 1811–1816, August 1998.

[7] S.B. Cohn, *Problems in Strip Transmission Lines*, IEEE Trans. on Microwave Theory and Techniques, vol. 3, Issue 2,pp. 119–126, March 1955.

[8] T.H. Lee, *Planar Microwave Engineering*, ISBN 0521835267, Cambridge University Press, pages 233-237, 2004.

[9] A.K. Tagantsev, V.O. Sherman, K.F. Astafiev, J. Venkatesh, and N. Setter, *Ferroelectric Materials for Microwave Tunable Applications*, Journ. of Electromagnetics, vol. 11, no.1–2, pp. 5–66, September 2003.

[10] P. Rundqvist, A. Vorobiev, S. Gevorgian, and K. Khamchane, *Non-Destructive Microwave Characterization of Ferroelectric Films on Conductive Substrates*, Integrated Ferroelectrics, vol. 60, no. 1, pp. 1–19, February 2004.

[11] (2007) The Sonnet website. [Online]. Available: http://www.sonnetsoftware.com/

[12] P.K. Petrov, N.McN. Alford, and S. Gevorgyan, *Techniques for Microwave Measurements of Ferroelectric Thin Films and Their Associated Error and Limitations,*Meas. Sci. Technol., vol. 16, pp. 583-589, January 2005.

[13] A. Davidson, K. Jones, and E. Strid, *LRM and LRRM Calibrations with Automatic Determination of Load Inductance*, 36th ARFTG Conf. Dig., pp. 57–62, November 1990.

SESSION 9

Interconnect

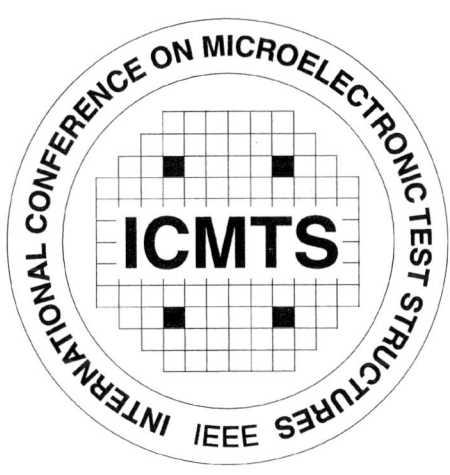

March 27, 09:00–10:20

Co-Chairs: Hi-Deok Lee, *Chungnam National University, Korea*
Larg H. Weiland, *PDF Solutions, USA*

2008 IEEE Conference on Microelectronic Test Structures, March 24-27, Edinburgh, UK

A Study of Cross-Bridge Kelvin Resistor Structures for Reliable Measurement of Low Contact Resistances

N. Stavitski[1], *Student Member, IEEE*, J.H. Klootwijk[2], *Member, IEEE*
H.W. van Zeijl[3], A.Y. Kovalgin[1], R.A.M. Wolters[1,4]

[1] MESA+ Institute for Nanotechnology, Chair of Semiconductor Components
University of Twente, Postbox 217, 7500AE Enschede, The Netherlands
E-mail: n.stavitski@utwente.nl
[2] Philips Research, High Tech Campus 4, 5656 AE Eindhoven, The Netherlands
[3] DIMES, Delft University of Technology, Feldmannweg 17, 2628 CT Delft, The Netherlands
[4] NXP Research Eindhoven, High Tech Campus 4, 5656 AE Eindhoven, The Netherlands

ABSTRACT

The parasitic factors that strongly influence the measurement accuracy of Cross-Bridge Kelvin Resistor (CBKR) structures for low specific contact resistances (ρ_c) have been extensively discussed during last few decades and the minimum of the ρ_c value, which could be accurately extracted, was estimated. We fabricated a set of various metal-to-metal CBKR structures with different geometries, i.e., shapes and dimensions, to confirm this limit experimentally. As a result, a model was developed to account for the actual current flow and a method for reliable ρ_c extraction was created. It was found that in our case of metal-to-metal contacts, the measured CBKR contact resistance was determined by the dimensions of the two-metal stack in the area of contact and sheet resistances of the metals used.

INTRODUCTION

Cross-Bridge Kelvin Resistor (CBKR) structures are the most widely used test structures to characterize metal-semiconductor contacts in the planar devices of VLSI technology [1, 2]. On the other hand, CBKR was found to be very sensitive to lateral current crowding around the contact when the contact window is smaller than the underlying layer. Several simulations and correction methods were introduced in order to account for this current crowding effect [3-6]. However in the low resistance range, the extracted silicide-to-silicon specific contact resistance (ρ_c) values, obtained using CBKR structures, were still orders of magnitude different from the results obtained using other methods [2]. An explanation of this phenomenon is the accuracy problems during the data extraction using CBKR structures in the range of $\sim 10^{-8}\ \Omega \cdot cm^2$ and below [7]. In this case, the lateral current flow around the contacts gives rise to an even higher additional resistance [8, 9]. This effect becomes more pronounced for a lower ρ_c and a higher sheet resistance (R_{sh}) of the underlying layer. Simulations show that for $\rho_c < 10^{-7}\ \Omega \cdot cm^2$ the extracted ρ_c can differ by one or two orders of magnitude from the actual value [6]. Moreover, the trend in the modern technology of high-density integrated circuits is toward lower ρ_c and higher R_{sh} values, due to the

shallower junctions. This will further complicate the interpretation of CBKR measurement results.

Our research is therefore concerned with finding the minimum contact resistance, which can be obtained experimentally using CBKR test structures, and developing a correction model to account for the actual current flow. For that purpose, CBKR structures of different geometries, i.e., dimensions and shapes of the contact area were designed and manufactured. These structures were evaluated for metal-to-metal contacts to assure the case of very low contact resistances.

MEASUREMENT TECHNIQUE AND TEST STRUCTURES DESCRIPTION

A standard four-terminal CBKR test structure is used to determine ρ_c of metal-to-metal contacts (Fig. 1).

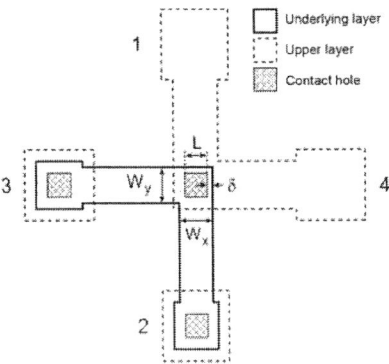

Fig. 1. Four-terminal CBKR structure with geometry parameters definition. In our structures, the contact geometry parameters (δ and L) for both layers are identical, unless mentioned otherwise.

The measurement principle consists of forcing the current (I) between contacts 1 and 2 and measuring the voltage drop (V_{34}) between contacts 3 and 4. The actually measured Kelvin resistance R_k can then be found as

$$R_k = \frac{V_{34}}{I}. \qquad (1)$$

978-1-4244-1801-5/08/$25.00 ©2008 IEEE

In the 1D-Model approach [4], the specific contact resistance can be calculated directly from the contact area A and R_k, assuming that the resistance due to the voltage drop across the actual contact (R_c) equals R_k:

$$\rho_c = R_c A = R_k A. \qquad (2)$$

The 1D-Model does not account for the current flowing in the overlap region (δ) of the underlying layer (Fig. 1), when $\delta > 0$. In that case the so-called 2D-Model should be applied [4]. The analytical model by Schreyer and Saraswat was used in this study as a starting point for the correction. The measured R_k is then a sum of the R_c and the resistance due to the current flow around the contact in the overlap region (R_{geom}) (3). The ρ_c can further be extracted from (4), where R_{sh} is the sheet resistance of the underlying layer. The contact geometry parameters are defined in Fig. 1.

$$R_k = R_c + R_{geom} \qquad (3)$$

$$R_k = \frac{\rho_c}{A} + \frac{4R_{sh}\delta^2}{3W_xW_y}\left[1 + \frac{\delta}{2(W_x - \delta)}\right] \qquad (4)$$

In order to verify the validity of the results obtained, the CBKR structures were designed to cover a wide range of contact sizes (i.e., length L for square contacts and diameter D for round contacts) and δ. Some of the structures were designed with two δ, different for the back and front metal layers: δ_B and δ_F, respectively. To exclude the uncertainty in the definition of δ in the case of round contacts, the metal tap width (V_{tap}, Fig. 2) was varied as well. The details are summarized in Table I. An example of the square and round CBKR structure is presented in Fig. 2.

TABLE I
IMPORTANT GEOMETRY PARAMETERS OF OUR CBKR STRUCTURES

Geometry	$L(D)$ μm	δ μm	δ_F μm	δ_B μm	V_{TAP} μm
Square	1	0.2 – 5	n/a	n/a	n/a
Square	2	0.2 – 5	n/a	n/a	n/a
Square	4.4	0.2 – 5	n/a	n/a	n/a
Square	8.9	0.2 – 5	n/a	n/a	n/a
Square	17.7	0.2 – 5	n/a	n/a	n/a
Square	8.9	n/a	0.2 – 5	0.2 – 5	n/a
Round	5	0.5 – 10	n/a	n/a	1 – 2
Round	10	0.5 – 10	n/a	n/a	1 – 2
Round	20	0.5 – 10	n/a	n/a	1 – 2
Round	10	n/a	1 – 5	1 – 5	2
Round	20	n/a	1 – 5	1 – 5	2

TEST STRUCTURES FABRICATION

The (100) p-type Si wafers with a 1 μm-thick thermal oxide were used to fabricate the test structures for this study. First, a 0.675 or 1.4 μm- thick Al layer was

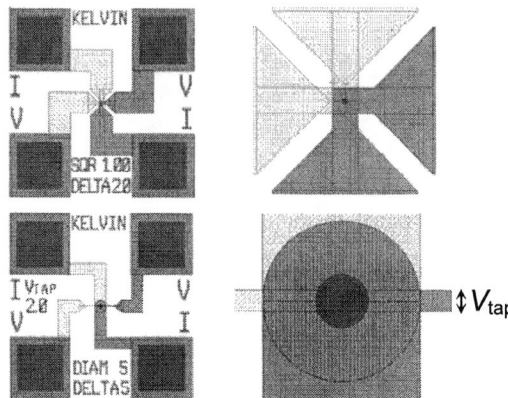

Fig. 2. An example of the newly-designed square and round CBKR structures. The complete structure including the bond pads is on the left- and a blow up of the actual contact is on the right-hand side.

sputtered and patterned using I-line lithography and plasma etching. Then, a 0.8 μm-thick layer of SiO$_2$ was deposited by PECVD and the contact holes were opened. Prior to the second Al deposition, the contacts were in-situ RF-precleaned. The second Al layer of 0.675 or 1.41 μm was sputtered and patterned as the front metallization layer, including the bond pads. Finally, the structures received 20-min annealing at 400 °C in a N$_2$/H$_2$ (10%) mixture.

RESULTS AND DISCUSSION

A. Measured Kelvin resistance R_k for square contacts with symmetric δ for back and front metals

The R_k data as a function of contact size $A = L^2$ and the overlap size δ are given in Fig. 3 and Fig. 4, respectively. It can clearly be seen that R_k increases with increasing δ and decreases with increasing contact size. This is in agreement with the theory (4), demonstrating, that for $\delta > 0$, the lateral current flow gives rise to an additional voltage drop that is included in V_{34}, leading to a higher R_k value.

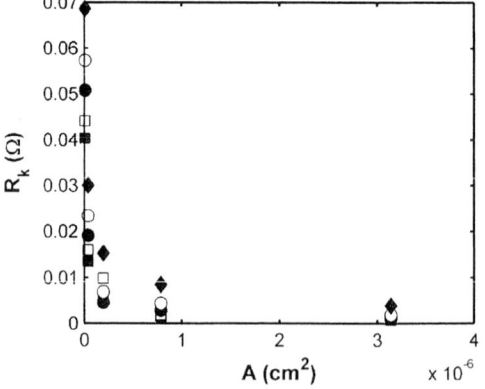

Fig. 3. Measured Kelvin resistance vs. contact size for given symmetric overlap area sizes for square contacts: $\delta = 0.2$ μm (■), $\delta = 0.5$ μm (□), $\delta = 1$ μm (●), $\delta = 2$ μm (○), $\delta = 5$ μm (◆).

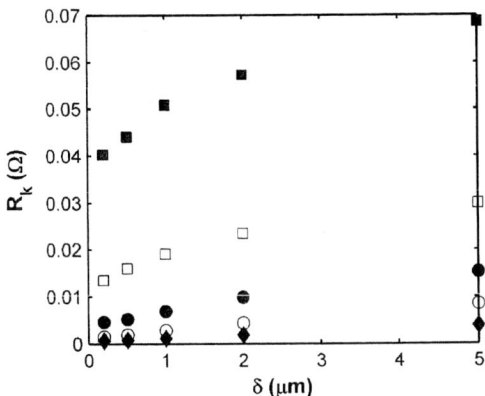

Fig. 4. Measured Kelvin resistance vs. symmetric overlap size for given square contact sizes: $L = 1$ μm (■), $L = 2$ μm (□), $L = 4.4$ μm (●), $L = 8.9$ μm (○), $L = 17.7$ μm (◆).

B. Measured Kelvin resistance R_k for square contacts with non-symmetric overlap areas for back and front metals

As expected for the non-symmetric overlaps for front and back metals, the R_k values were dependent on the direction of the forced current I. This is in contrast to the other structures with symmetric δ for back and front metals, where R_k was not direction dependent (Fig. 5). It is noteworthy that (4) is derived for extracting specific *metal-to-silicon* contact resistance, when only R_{sh} and δ of the diffusion layer are considered, since R_{sh} of the metal is much lower than that of even highly doped silicon. While measuring contact resistance between two materials with similar R_{sh}, δ and R_{sh} of both layers must be taken into account. Therefore, the R_k dependences on δ_B and δ_F were studied separately (Fig. 6a and 6b). It was demonstrated that R_k values increased with increasing δ_B or δ_F.

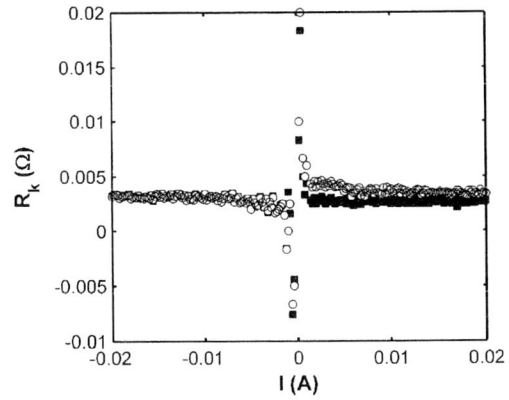

Fig. 5. Measured Kelvin resistance vs. forced current for symmetric (■) (i.e., $\delta_B = \delta_F$) and non-symmetric (i.e., $\delta_B \neq \delta_F$) (○) overlaps of square contacts.

(a)

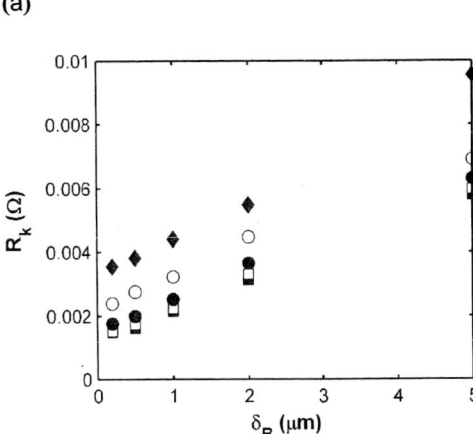

(b)

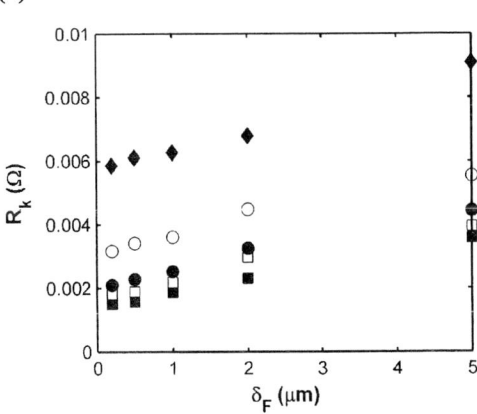

Fig. 6. Dependence of measured Kelvin resistance on overlap size δ_B **(a)** varying δ_F of 0.2 μm (■), 0.5 μm (□), 1 μm (●), 2 μm (○), 5 μm (◆) and on overlap size δ_F; **(b)** varying δ_B of 0.2 μm (■), 0.5 μm (□), 1 μm (●), 2 μm (○), 5 μm (◆) for given square contact size ($L = 8.9$ μm).

C. Measured Kelvin resistance R_k for round contacts

For round structures, the R_k data for different metal tap widths (V_{tap}) as a function of contact size *(A)* and overlap size *(δ)* are given in Fig. 7 and Fig. 8, respectively.

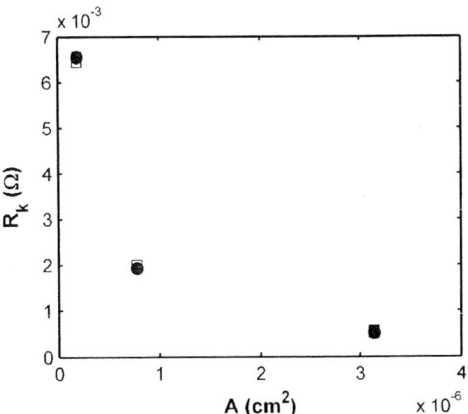

Fig. 7. Measured Kelvin resistance vs. contact size for given symmetric overlap area sizes for round contacts and various $V_{tap} = 2$ μm (■), $V_{tap} = 1.5$ μm (□), $V_{tap} = 1$ μm (●).

201

Similar to the square contacts, the R_k increased with increasing δ and decreased with increasing A, in agreement with the theory (4). The R_k was not dependent on V_{tap} (Fig. 7 and Fig. 8), proving validity of the measurements and supporting a correct definition of overlap size for round contacts. The R_k behavior for the non-symmetric overlaps was studied by varying δ_B and δ_F separately and revealed the same behavior as for the square contacts.

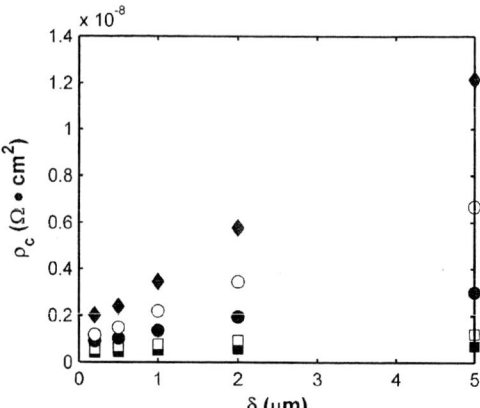

Fig. 9. Specific contact resistance obtained using 1D-approach vs. overlap size for given square contact sizes: $L = 1$ μm (■), $L = 2$ μm (□), $L = 4.4$ μm (●), $L = 8.9$ μm (○), $L = 17.7$ μm (◆).

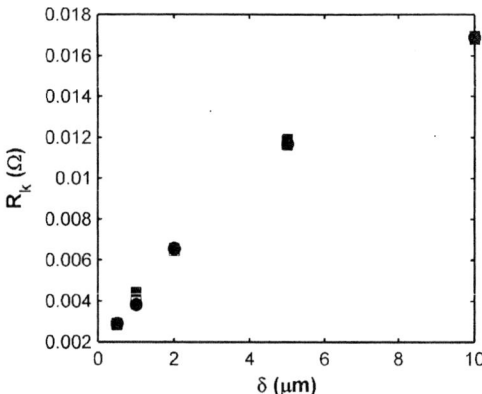

Fig. 8. Measured Kelvin resistance vs. symmetric overlap size for given round contact sizes and various $V_{tap} = 2$ μm (■), $V_{tap} = 1.5$ μm (□), $V_{tap} = 1$ μm (●).

D. Extraction of ρ_c using the analytical model of Schreyer and Saraswat

The specific contact resistance was extracted using both the 1D- and 2D- approximations for a variety of different contact and overlap sizes. The ρ_c values for square contacts with symmetric overlaps can be found in Fig. 9 and Fig. 10. The ρ_c values obtained using the 1D-approach (Fig. 9) were strongly dependent on the contact and overlap size. This supported the significance of applying the 2D-Model instead of the simple 1D-approximation, discussed earlier. The ρ_c values, extracted using the 2D-Model for the smallest contact sizes, were hardly dependent on the overlap dimensions and revealed similar values for different contact sizes. As the contact size increased, the disagreement with the model appeared, showing difference between the geometrical factor, calculated from (4) and the actual geometrical factor, which led to a clear dependence on δ (Fig. 10). For structures with non-symmetrical overlaps, the ρ_c values were extracted by varying δ_B and δ_F separately, using R_{sh} (4) of the corresponding metal layers. An example for δ_B can be found in Fig. 11, where the current direction was from the back to front metal. If the current direction is changed, the ρ_c extracted using δ_F (instead of δ_B) revealed the same values. In summary, it was found that ρ_c is determined by δ_B, if current enters from the back metal and by δ_F if the incoming current is from the front metal.

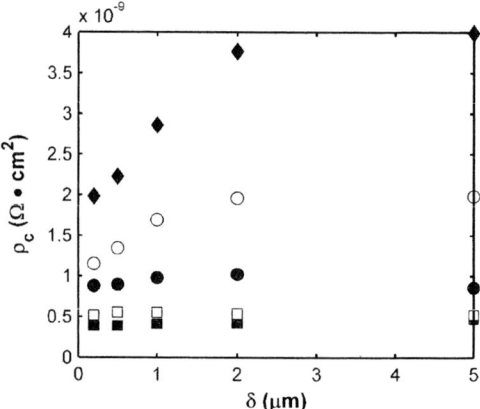

Fig. 10. Specific contact resistance obtained using 2D-approach vs. overlap size for given square contact sizes: $L = 1$ μm (■), $L = 2$ μm (□), $L = 4.4$ μm (●), $L = 8.9$ μm (○), $L = 17.7$ μm (◆).

For a given contact size, the ρ_c values obtained using the 1D-approach were also strongly dependent on the overlap size, in contrast to the values, extracted using the 2D-Model (Fig. 11). The latter was also observed for the round contacts with various V_{taps} (Fig. 12).

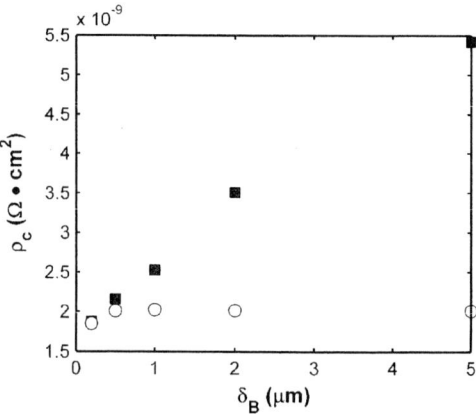

Fig. 11. Specific contact resistance obtained using 1D- (■) and 2D- (○) approach vs. δ_B for given $\delta_F = 2$ μm and square contact size ($L = 8.9$ μm).

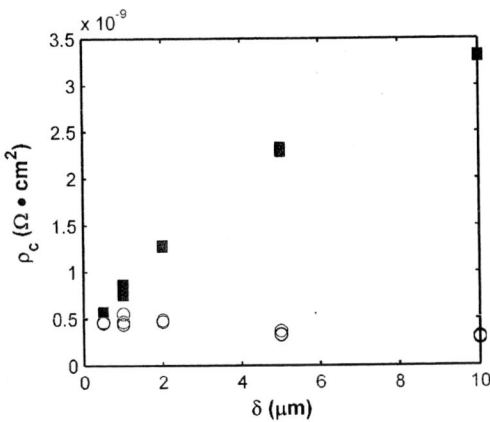

Fig. 12. Specific contact resistance obtained using 1D- (■) and 2D- (○) approach vs. overlap size for given round contact size.

E. Our approach to account for the actual current flow regions

A more accurate approach to extract the ρ_c value is the extrapolation of the measured dependence R_k versus δ to $\delta = 0$, and the calculation of ρ_c from (2) using the R_k value at $\delta = 0$ as the R_c. In this manner the model simplifications, assumed while deriving equation (4), can be ignored. However, for large contacts, the results were still dependent on the contact size (Fig. 13). An explanation of this observation is that the current, which contributes to V_{34}, can flow across a smaller area compared to the actual contact area A. As the contact size becomes larger, this effect enhances, causing significant differences while extracting ρ_c. To account for this effect, the potential distributions along 2 horizontal resistive layers, vertically separated by a resistive "contact", must be considered [10]. For the given test structures, such distributions can be described by the following pair of coupled differential equations:

$$\frac{\partial^2 V_B(x)}{\partial x^2} = \frac{R_{shB}}{\rho_{CG}}\left(V_B(x) - V_F(x)\right) \quad (5)$$

$$\frac{\partial^2 V_F(x)}{\partial x^2} = \frac{R_{shF}}{\rho_{CG}}\left(V_F(x) - V_B(x)\right), \quad (6)$$

where $V_F(x)$ and $V_B(x)$ are the potential distributions in the front and back metal layers, respectively, x is the coordinate along the contact length L, R_{shF} and R_{shB} are the sheet resistances of the front and back metals, respectively, and ρ_{CG} corresponds to the specific resistance caused by the properties and geometry of the contact. Applying the boundary conditions, appropriate for the particular geometry, and using ρ_c as a fitting parameter to obtain the corresponding R_k, the $V_{34}(x) = V_F(x) - V_B(x)$ (i.e., voltage difference distribution along the contact) can be calculated. It is important to note that the $V_{34}(x)$ dependence will indicate the actual current flow areas because the current can only flow from the back metal layer into

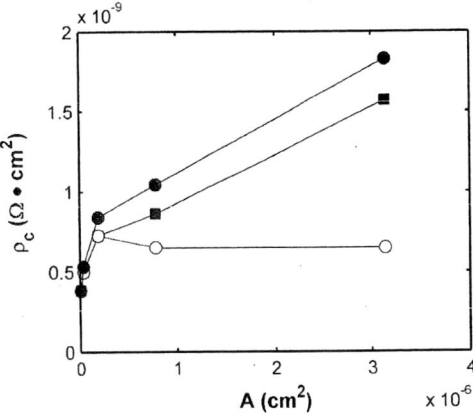

Fig. 13. Specific contact resistance vs. contact size obtained using Schreyer and Saraswat model and then extrapolated to $\delta = 0$ (●), the extrapolation of R_k to $\delta = 0$ (■) and the current-flow area correction (○).

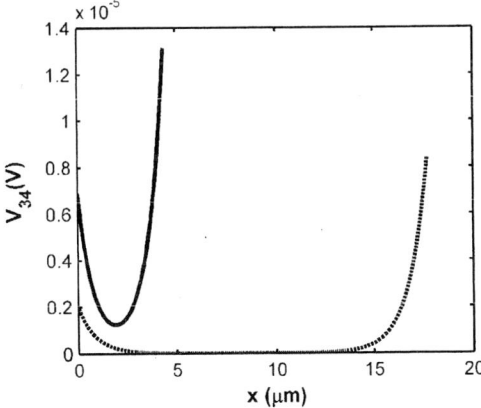

Fig. 14. Potential difference distribution $V_{34}(x) = V_F(x) - V_B(x)$ along the contact coordinate x for the contact length of 4.43 μm (solid line) and 17.72 μm (dashed line).

the front metal layer if $V_{34}(x) \neq 0$. It was shown that for the small contacts, the current flow area was identical to that of the contact, while for the larger contacts this area was much smaller than the designed contact area (Fig. 14). The demonstrated approach allowed to estimate the actual current-flow area size A_{cor} ($A_{cor} < A$) and therefore resulted in a corrected ρ_c. A comparison of the extraction methods is presented in Fig. 13. Our approach results in similar ρ_c values for various contact sizes. This obviously points to the importance of knowing the actual current flow distribution. The extracted ρ_c values were $(6.31 \pm 0.66) \cdot 10^{-10}$ $\Omega \cdot cm^2$.

The sheet resistances of both the back and front metal layers were measured using Van-der-Pauw structures, fabricated on the same wafers. The obtained values of 0.054 and 0.027 $\Omega/\square$ for the 0.675 μm-thick and 1.4 μm-thick metals, respectively, were in agreement with the corresponding thicknesses. Due to the fact that the two metals had different thicknesses, the potential distribution along the contact was non-symmetrical (Fig. 14). The estimated Kelvin resistance, determined by the two-metal stack of the

203

known geometry and the sheet resistances, matched with the measured R_k values, thus providing the minimum value of ρ_c to be accurately extracted from these CBKR structures.

CONCLUSIONS

A design and fabrication of various metal-to-metal CBKR structures has been realized. The structures included a large variety of contact geometries, i.e., various shapes and sizes for contact holes and overlap regions. The obtained Kelvin resistance, R_k, was in agreement with the analytical model proposed by Schreyer and Saraswat. This demonstrated the necessity to account for 2D current flow effects around the contact area while measuring low contact resistance values. However, as the calculated ρ_c values were still dependent on the contact size, we developed a new correction method to account for the actual current-flow areas through the contact. The approach allowed to obtain a potential difference distribution along the contact length and led to a physically-correct extraction of the ρ_c. The measured R_k values corresponded to the two-metal stack resistance, calculated from the given dimensions of the contact size and sheet resistances of the metals used. As a result, the minimum value to be accurately extracted from the CBKR structures was determined.

ACKNOWLEDGEMENTS

The authors would like to thank the DIMES Clean room staff of Delft University of Technology for processing the wafers.

REFERENCES

[1] S. Wolf and R. N. Tauber, Silicon Processing for the VLSI Era, ed., vol. 2. Sunset Beach: Lattice Press, 1990.

[2] D. K. Schroder, Semiconductor Material and Device Characterization, 3rd ed. New York: Wiley-Interscience/IEEE, 2006.

[3] W. M. Loh, S. E. Swirhun, E. Crabbe, K. Saraswat, and R. M. Swanson, "An Accurate Method to Extract Specific Contact Resistivity Using Cross-Bridge Kelvin Resistors," IEEE Electron Device Lett., vol. 6, pp. 441-443, 1985.

[4] T. A. Schreyer and K. C. Saraswat, "A Two-Dimensional Analytical Model of the Cross-Bridge Kelvin Resistor," IEEE Electron Device Lett., vol. 7, pp. 661-663, Dec. 1986.

[5] J. Santander, M. Lozano, A. Collado, M. Ullan, and E. Cabruja, "Accurate contact resistivity extraction on Kelvin structures with upper and lower resistive layers," IEEE Trans. Electron Devices, vol. 47, pp. 1431-1439, Jul. 2000.

[6] A. S. Holland, G. K. Reeves, and P. W. Leech, "Universal error corrections for finite semiconductor resistivity in Cross-Kelvin resistor test structures," IEEE Trans. Electron Devices, vol. 51, pp. 914-919, Jun. 2004.

[7] R. L. Gillenwater, M. J. Hafich, and G. Y. Robinson, "Extraction of the Minimum Specific Contact Resistivity Using Kelvin Resistors," IEEE Electron Device Lett., vol. 7, pp. 674-676, Dec. 1986.

[8] M. Finetti, A. Scorzoni, and G. Soncini, "Lateral Current Crowding Effects on Contact Resistance Measurements in 4 Terminal Resistor Test Patterns," IEEE Electron Device Lett., vol. 5, pp. 524-526, 1984.

[9] A. Scorzoni, M. Finetti, K. Grahn, I. Suni, and P. Cappelletti, "Current Crowding and Misalignment Effects as Sources of Error in Contact Resistivity Measurements. 1. Computer-Simulation of Conventional CER and CKR Structures," IEEE Trans. Electron Devices, vol. 34, pp. 525-531, Mar. 1987.

[10] D. B. Scott, W. R. Hunter, and H. Schichijo, "A Transmission-Line Model for Silicided Diffusions - Impact on the Performance of VLSI Circuits," IEEE Trans. Electron Devices, vol. 29, pp. 651-661, 1982.

Comb Capacitor Structures for Measurement of Post-processed Layers

D. Roy[1], J.H. Klootwijk[2], N.A.M. Verhaegh[1], H.H.A.J. Roosen[2], and R.A.M. Wolters[1]

[1] NXP Semiconductors, Eindhoven, The Netherlands
[2] Philips Research, Eindhoven, The Netherlands

ABSTRACT

We present a simple comb capacitive measurement structure to monitor the properties of post-processed layers. These measurement structures are easily fabricated in a single step in the last metallization layer of a standard IC process, while the post-processed layer in this article is formed over these comb structures by spray coating. The capacitive coupling of the structure on the substrate is modeled based on the electric field distribution around the structure. The change in composition of this post-processed layer is analyzed in terms of measured capacitance values.

INTRODUCTION

To meet the increasing demands and performance requirements microelectronic industries have to deliver technological solutions that include smaller feature dimensions, introduction of new materials, and novel processing methods. Another approach is to increase the functionality of the existing chip by addition of extra integrated components for sensory applications like humidity sensors or gas detectors and other micro devices [1, 2]. Adding functionalities on top of metal layers of a finished chip is called post-processing. For sensing applications the change in properties of a post-processed layer (which may or may not be a CMOS based process) with the measured parameter needs to be sensed and converted into a measurable parameter. In this article comb capacitive test structures are used to detect the presence of such a post-processed layer and to study its properties. Comb structures are preferred since they have a large area exposed to the layer and can be processed in the same metal layer as the last metallization step of a chip. Comb structures also have a higher flexibility in tuning the capacitor value by changing the dimensions of the comb structures. These structures with different dimensions are fabricated on silicon substrates. Prior to the application of the post processed layer these structures are electrically modeled to study the different capacitors in the network. Then the post processed layer is applied on the structures and is electrically evaluated by capacitance measurements. In this particular study used a security coating as the post processed layer. This layer prevents physical attacks on an IC. Moreover, it provides the implementation of a physically uncloneable function (PUF) given the wide range of measured capacitance values.

TEST STRUCTURE

The test structures studied in this article are comb capacitor structures as shown in Fig.1 (a). These structures have two interlinked finger like metallic combs of equal height and width. Across the two combs a capacitance can be established, the variation of which is used in the analysis of the post-processed layer. The use of these test structures for the measurement meets the following requirements or advantages:

1. These structures are fabricated in one single metal layer.
2. The test structures are made in the last metal layer of the IC process and the post-processed layer is applied on top of the metal.
3. Optimizations of the capacitance for different post-processed layers are easily possible by varying length, width or distance between combs of the structure.
4. These structures have a significant amount of fringe capacitive coupling that increases the variation of the measured capacitance values with changes in the post-processed layer.

The test structures used for measurement of post-processed layers and the layer itself are fabricated on boron doped silicon substrates having a resistivity of 700–1300 ohm-cm. These comb structures are isolated from the silicon substrate by a 1.5 μm thick insulating layer of thermal oxide grown at 1100°C. On top of this oxide layer a 1μm thick aluminum layer is formed by sputtering and is patterned by lithography to form comb capacitive structures. The different post-processed layers are made from a liquid chemical matrix of aluminum-meta-phosphate (AMP) with inclusion of of TiO_2 (dielectric particles) and TiN (conducting particles) [3]. Over this patterned aluminum layer a 50nm SiO_2 passivation layer is formed by plasma enhanced chemical vapor deposition (PECVD) at 400°C to prevent chemical reaction between the post-processed layer and the metallic comb structures. This post processed layer is formed on top of the structures with an automatic spray coating unit. The layer is annealed at 400°C for 30 minutes in N_2 atmosphere. This makes it hard and porous and it binds the included particles in the porous matrix resulting in good adhesion to the substrate.

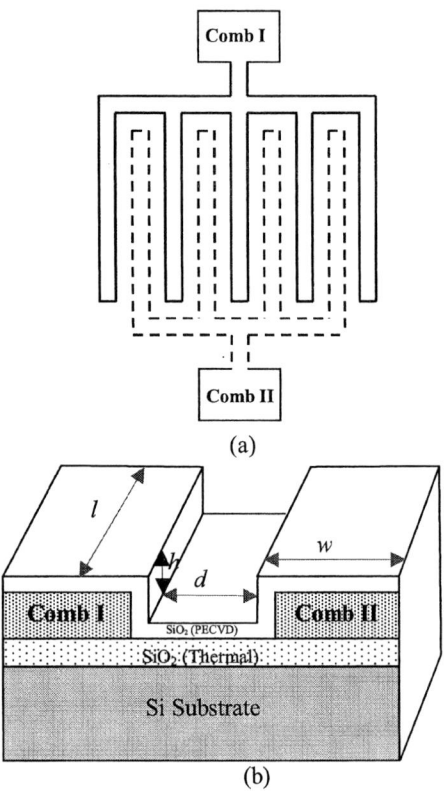

Figure 1: (a) Comb capacitor test structures; (b) parallel line approximated model for the comb structure on a Si substrate.

ELECTRICAL MODELING OF TEST STRUCTURE

The measured output from these test structure is a capacitance value. Detailed understanding of the measured capacitance involves determination of a capacitor model for these structures. This is done by capacitive measurements on the comb structures with different dimensions; width (w), length (l) and distance (d) between the combs. To simplify the capacitance calculations, interlinked comb structures are approximated to be two parallel lines as shown in figure 1(b). The height of the structures (h) is fixed to 1µm. Test structures with four different lengths are available, l = 1000µm, 2000µm, 5000µm and 10000µm. For each of these different lengths there are structures with different width and distance between the combs. The possible widths of the combs are 2 µm, 5 µm and 10 µm and the possible distance between the plates are 1 µm, 1.5 µm, 2 µm, 2.5 µm, 3 µm and 5 µm.

Based on the field distribution between the combs and the silicon substrate, the capacitors established between the two combs structures are modeled as shown in Fig 2. The capacitors formed between the two combs are lateral capacitor (C_l) (C_{s1}, C_{s2} and C_{s3} are capacitors due to SiO$_2$ passivation layer. The capacitance contribution due to this thin layer is neglected form the calculations) and lateral fringes (C_{f1}, C_{f2}). For comb structures with parallel vertical edges, the lateral capacitance per unit length is estimated by a parallel plate approximation with d as distance between plates. The fringe capacitors arise due to concentration of charges at the corners of metallic structures. These fringe capacitances are a function of the width of the plates as well as the separation between the plates and its value dependence can be modeled as a polynomial of d[4]. This fringe capacitance changes from zero to infinity as distance between the combs is varied form infinity to zero. For a fixed separation the charge density induced on each plate due to a finite potential on the other plate falls gradually with the increasing distance between each other. This gradually becomes insensitive to further increase on width. Another group of capacitors is the overlap capacitors (C_{a1}, C_{a2}) and overlap fringes (C_{f3}-C_{f6}) from the corners and the vertical faces of each comb to the silicon substrate. These capacitors can be seen as a MOS capacitor due to the stacking of metallic combs on silicon separated by an oxide layer. The overlap capacitance values cannot be calculated numerically since the substrate is not held at a fixed potential.

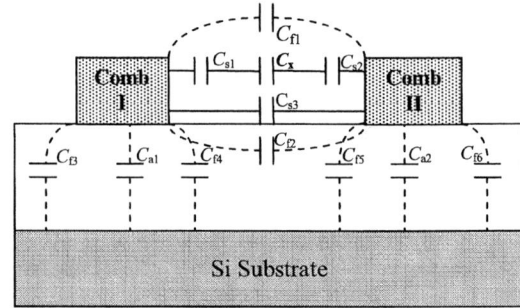

Figure 2: Capacitor model for comb structure on Si Substrate.

To model the capacitors in the network measurements were done by varying the width and distance between the combs for each structure for a fixed length. These capacitance measurements were carried out at 25ºC with an HP4275A multi frequency LCR analyzer. The measured capacitance (C_e) includes overlap (C_a), lateral (C_l) and fringe capacitances (C_{f1},C_{f2}). The capacitor C_a is the combination of overlap capacitors C_{a1}, C_{a2}, Overlap fringes C_{f3}-C_{f6} and the capacitance contribution due to the substrate. By comparing the capacitance network of the comb structure in Fig 2 to a parallel resistance- capacitance measurement network; the measured capacitance is approximated [5] to

$$C_e = C_l + C_{f1} + C_{f2} + \frac{C_a}{2} \qquad (1)$$

The approximation used in deriving this equation is that all capacitors in the network are ideal and impedance offered by the substrate is zero. In this capacitor model presented, only the value of lateral capacitor is well established. To separate the contribution of different capacitors from the measured capacitance value, measurements are carried out by varying the distance between the combs and by changing width of the combs. In Fig 3(a) the capacitance is plotted with the reciprocal of distance (d) between the combs. As the distance between combs is large the lateral components C_l, C_{f1} and C_{f2} becomes negligible and only the overlap component C_a remains. These can be obtained from the intercept of the curve with the capacitance axis. The value of C_a increases with the width of the combs due to an increase in the overlap capacitance values C_{a1} and C_{a2}.

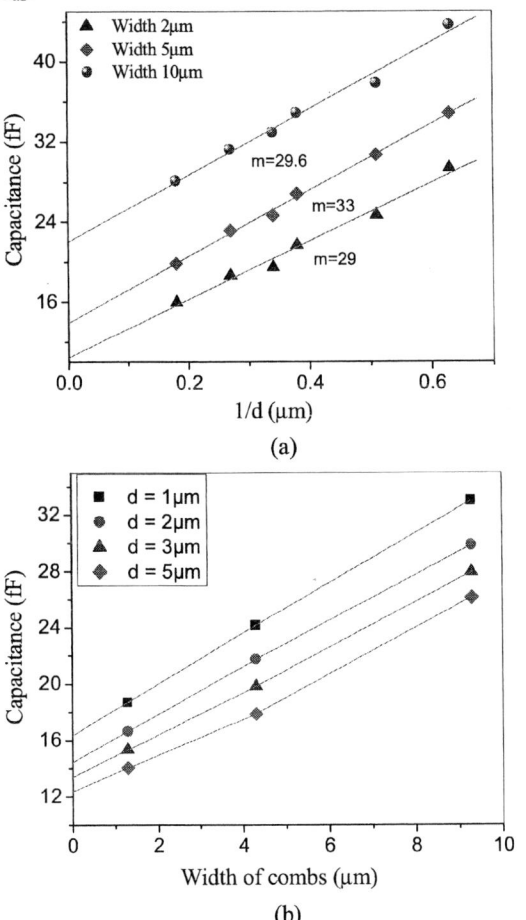

(a)

(b)

Figure 3: (a) Capacitance with reciprocal of distance between the combs; (b) capacitance with width of the combs.

Similarly in Fig 3(b) the capacitance is plotted with the width of the combs and the curve is extrapolated to zero width. From the intercept the lateral component C_l, part of C_{f1} and C_{f2} and the overlap component C_{f4} and C_{f5} including the substrate capacitances are estimated. If this intercept capacitance value in Fig. 3(b) is plotted with the inverse of the distance the lateral components can be eliminated and only the overlap components C_{f4} and C_{f5} including the substrate capacitance remains. By this analysis method all different capacitors in the network can be calculated. The application of a post-processed layer on the comb structures acts like a dielectric layer for the lateral capacitors of the network. The result is a change in the value of these lateral capacitors depending on the dielectric constant of the layer applied. This changes the measured value of capacitance making the comb structures ideal for measurement of a post-processed layer.

MEASUREMENTS

The change in measured capacitance values with and without a post processed is shown in Fig 4. Capacitance values increase with a post-processed layer having an effective dielectric constant of 32. This increase is mainly due to the increased contribution of the lateral capacitance. The capacitance increase with increasing length of the combs is due to the increase in lateral area (increase in fringes is not that pronounced compared with increase in lateral value so the curve still remains linear).

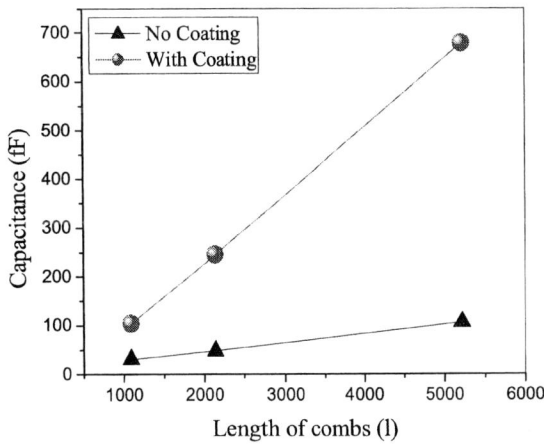

Figure 4: Capacitance with length of the structures.

The comb capacitance is not only sensitive to the variation in dielectric constant of the layer but also to the variation in the thickness of the post-processed layer. Figure 5 shows the variation in the capacitance values with thickness of the post-processed layer having a relative dielectric constant of 4 on top of the structures. The capacitance increases with thickness due to the increase in

207

the lateral fringe coupling, which saturates. Further increase in layer thickness has no effect on lateral fringes.

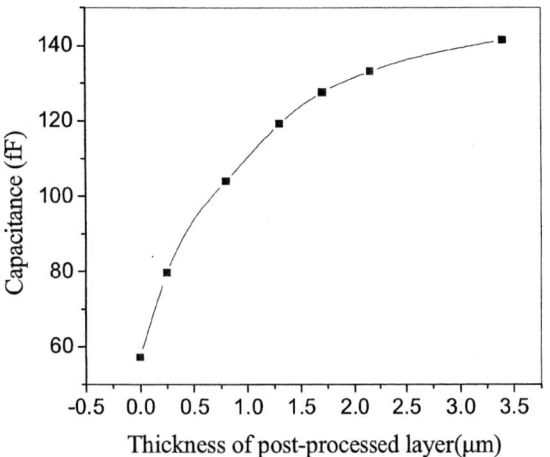

Figure 5: Variation in comb capacitance with thickness of post-processed layer.

Fig 6 shows the plot of the measured capacitance value of a large number (600) of identical comb structures spread over a 6 inch silicon wafer for three different situations. The first case is comb capacitor measurements without a processed layer, while for the other two situations the structures have a homogeneous and a non-homogeneous post processed layer on top. The mixing of the uniform sized TiO_2 particles in an aluminum phosphate matrix results in a homogeneous precursor solution for the homogeneous layer. The addition of conducting TiN particles of varying size to this homogeneous precursor solution results in a non-homogeneous precursor solution. The size of TiO_2 particles is 300nm while the size of TiN particles varies from 500 nm to 3 µm. These precursor solutions are then spray coated and processed over the comb capacitive structures. The comb capacitance measurements without a post-processed layer show the lowest mean value of 36fF with a standard deviation of 5fF. The spread in capacitance value is due to the variation in the fabrication process and measurement accuracy. Application of a homogeneous layer with a relatively high dielectric constant of 32, results in a mean capacitance value of 212fF with a standard deviation of 43fF. This increase in mean capacitance value is due to the increase in lateral capacitance and fringes by a relatively higher dielectric constant layer over the comb structures, the overlap capacitance remains unaffected. The spread in capacitance value is also due to variation in the layer thickness due to the spray coating process. As shown in Fig 6 the non-homogeneous layers have higher capacitance values and a higher spread. The average capacitance value for this layer is 1125fF with a standard deviation of 405fF.

The mechanism attributed to this increase in capacitance is first, the change in lateral coupling due to the presence of conducting TiN particles in the region between combs. This can effectively be seen as a reduction in distance between combs and there by increasing capacitance value.

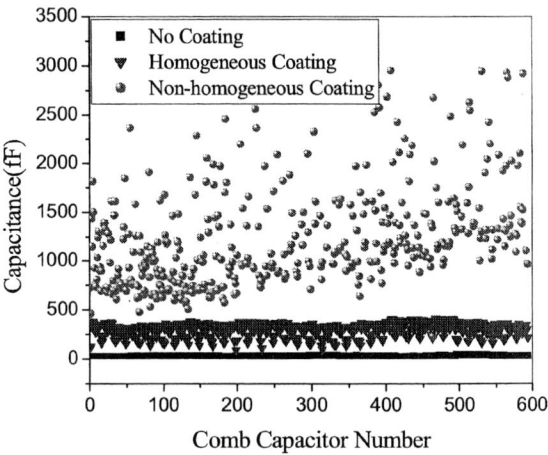

Figure 6: Capacitance values for different post-processed layers over the comb structures.

Fig 7(a) shows a SEM image of conducting particles in the dielectric layer. The presence of these conducting particles enhances the lateral capacitance value by forming a capacitive network between themselves and to the combs as depicted in the SEM image. There is also an additional coupling between the conducting particles and the Si substrate. The resulting capacitive network with conducting particles in the vicinity of comb structures is schematically shown in Fig 7(b). Each of these conducting TiN particles creates an additional node in the equivalent capacitance network. The large standard deviation of the measured capacitance values can be understood from geometrical considerations. The size range of the TiN particles is 0.5-3 µm, the thickness of the coating is 3-5 µm and the distance between the metal combs (d) of figure 1(b) is 1.5-2 µm for the measured structure (not in the SEM image). These TiN particles are incorporated in the dielectric layer in the lateral and vertical direction. This means that they will be distributed in position and size, leading to an unpredictable lateral and overlap coupling for each structure.

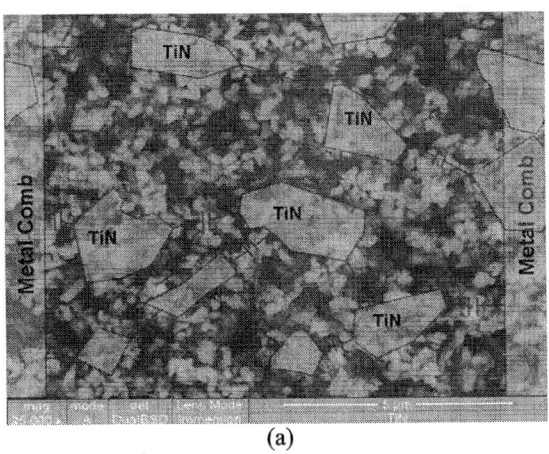

(a)

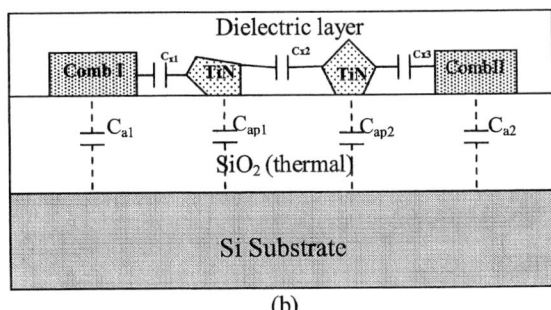

(b)

Figure 7: Non-homogeneous layer on comb structures. (a) SEM image of the layer with the capacitance network formed between the TiN particle and the combs; (b) Capacitive coupling formed to the substrate from these conducting particles.

CONCLUSIONS

In summary comb structures are used as an effective tool to measure the presence of a post-processed layer and to study the properties of this layer. By varying the dimensions of this structure a model is described for the comb capacitors on silicon substrate. This model shows the different capacitive couplings through the post-processed layer, which significantly contributes to the measured capacitance values. Measurement of a security coating with these structures show that any change in the layer properties (e.g. dielectric constant and in-homogeneity) reveals itself as a significant difference in the measured capacitance values. The inclusion of conducting (TiN) particles in the layer modifies the capacitance network. This results in higher capacitance values due to the decrease in effective dielectric distance and in a wide capacitance distribution across the wafer due to local variation in the position and size of these TiN particles.

REFERENCE

1. J. Schmitz, "Adding functionality to microchips by wafer post-processing," *Nuclear Instruments and Methods in Physics Research A*, vol. 567, pp. 142-149, 2007.

2. W. Smetana and W. Wiedermann, "Comb capacitive structures as humidity sensors," *Sensors and actuators*, Vol. 11, pp. 329-337, 1987.

3. P.E. de Jongh, P. van Tilborg, and H.J. Wondergem, "Wet-Chemical Formation of Aluminophosphates," *J. Sol-Gel Sci. and Technol.* Vol. 31, pp. 241, 2004.

4. U.Choudhury,A.Sangiovanni, "Automatic generation of analytical models for interconnect capacitances", *IEEE transactions on computer aided design of integrated circuits and systems,* vol.14, no.4, pp-470-480, (1995).

5. Deepu Roy, "Security coating as Physical Uncloneable Function", Master Thesis, June 2007, Technical University of Delft, pp 64-65.

Test structure for characterizing metal thickness in damascene CMP technology

Alain Toffoli [1], Sylvain Maitrejean [1], Jean Duport de Pontcharra [1], Francois de Crecy [1], David Bouchu [1], Lucile Arnaud [1], Fabien Boulanger [1]

CEA LETI-Minatec, 17 rue des Martyrs, 38054 Grenoble Cedex 9, France. (alain.toffoli@cea.fr)

ABSTRACT

In the damascene processed tracks, the electrical extraction of the metal thickness is reached through a complex analysis. This is due to processing defects such as dishing and erosion, and linewidths decreasing. This work introduces a new test structure, coupled with the Temperature Coefficient of Resistance (TCR) method [1] [2]. This allows an accurate characterization of metal thickness. At the same time, parameters such as sheet resistance and resistivity become easier to extract for statistical processing analysis.

1. INTRODUCTION

The continuous scaling necessary to produce Ultra Large Scale Integrated circuits induced a continuous improvement of the technology.

The Cu Damascene process [3], using Cu as conducting material, and low k dielectric as inter level dielectric (ILD), is now largely used for multi levels interconnect tracks production, in nano-scale dimensions. Cu is very attractive, because of its higher electrical conductivity, and a superior electro-migration performance. Unfortunately its atoms easily diffuse through silicon, and affect dielectric properties as well as transistors functionality; but an adjusted thin barrier layer stops it efficiently. When Cu is combined with low k dielectric, it allows reducing RC time delay and power consumption [4].

The Chemical Mechanical Polishing (CMP) [5] combines abrasive nano-particulates and mechanical polishing actions. It is used at the end of each interconnect level, in order to remove completely the excess of Cu and barrier material, delivering flat surfaces for the next level, and retaining the target thickness for both inter-metal dielectric and metal features.

The CMP process complexity, coupled with its drawbacks, like dishing and erosion [6], create conditions that prevent basic metal thickness measurements. This is the reason why the extraction of metal thickness in these technologies requires adapted structures and methods.

Therefore we have designed a new test structure, coupled to the TCR method for parametric tests. A complete study of its performance has been carried out.

2. PROCESS CHARACTERIZATION REQUIREMENTS

A particularity of the CMP results in a high number of precautions necessary to take against dishing and erosion, this thanks to both process adjustments and circuits design rules.

Dishing occurs when a larger quantity of metal is removed in the centre of a track, than at its edges with the ILD, as shown in Fig. 1. This is due to the preferential metal etching of the CMP, in case of wider track dimensions.

Erosion occurs when the dielectric is also polished away as shown in Fig. 2, this is due to the polish rate adjustment during the over polish stage. It occurs more with a high density of narrow metal and ILD lines.

Fig. 1: Dishing   Fig. 2: Erosion

Designing a new test structure for thickness extraction needs to take into account these effects.

3. TEST STRUCTURES

The proposed test structure is displayed in Fig. 3. It consists of a metal device (light blue or darker) allowing a four Kelvin connection. This one is surrounded by the ILD (light green or lighter), inside which metal dummies (gradated blue) are uniformly distributed. The advantage of this structure is that ILD anti-dummies (gradated green) are also uniformly distributed inside the metal area. This generates an adjustable ILD / Metal areas ratio, inside this material. This design contributes to reduce dishing and erosion effects; it also gives the opportunity of a precise metal thickness extraction, similar to the dimensions obtained with narrow lines.

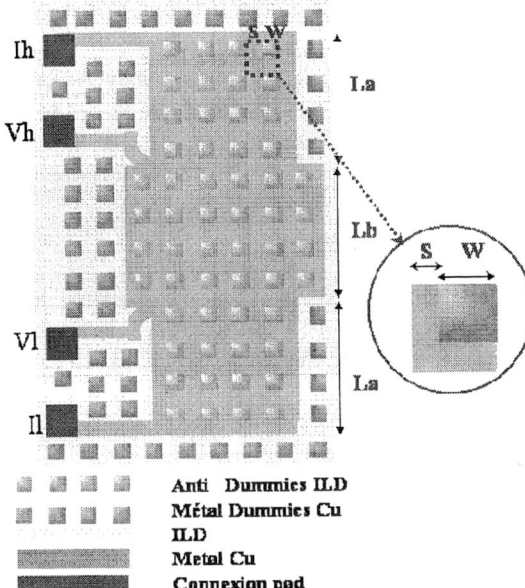

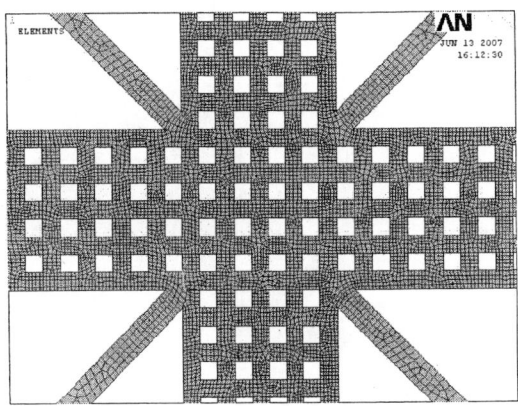

Fig. 4: simulated view of device CDRM6

Device	W	S	La	Lb	ILD/Cu areas ratio	α
CRDM1	0.3	0.3	239	2.7	0.33	0.8432
CRDM2	0.5	0.5	238	4.5	0.33	0.8419
CRDM3	1.0	1.0	235	9.0	0.33	0.8419
CRDM4	2.0	2.0	230	18.0	0.33	0.8419
CRDM5	5.0	5.0	220	45.0	0.33	0.8412
CRDM6	0.5	0.5	238	4.50	0.33	0.9177
CRDM7	0.58	0.42	238	4.42	0.5	1.1059
CRDM8	0.66	0.34	238	4.34	0.75	1.3859
CRDM9	0.70	0.30	238	4.30	1	1.5836

Table 1: geometric devices parameters (W, S, La and Lb dimensions are in µm) and extracted value from ANSYS™.

Fig. 3: New test structure. The inset represents a zoom on a unit Metal / ILD areas

4. TEST STRUCTURES CHARACTERISTICS

Each device is measured like a resistor, forcing current I between Ih and Il, and measuring voltage between Vh and Vl, then resistance R reads:

$$R = \frac{Vh - Vl}{I} = \frac{\alpha \rho}{e} \qquad (1)$$

where ρ is the copper resistivity, e the metal thickness and α a geometric form factor.

Nine structures have been designed in a test pattern with different ILD anti-dummies (W) and metal (S) experimental dimensions as shown in the inset of fig. 3.

In order to obtain a homogeneous current line flow inside the device, the necessary condition (2) has been applied

$$La \gg Lb \qquad (2)$$

La and Lb are defined on Fig.3.

A finite element modeling has been performed with ANSYS™, to calculate the α geometric form factor of Eq (1) for the various experimental dimensions. Fig. 4 shows an example of a simulated device for the structure CDRM6 referenced in table 1.

Table 1 summarizes the variation of these dimensions used for the design of test structures CDRM1 to CDRM9. It also gives the values of the ILD/Cu areas ratios, a key coefficient in CMP process, and the α factor, used for parameter extraction.

The structures CDRM1 to CDRM5 have ILD anti-dummies only between Vh and Vl connections; on the other hand the structures CDRM6 to CDRM9 have ILD anti-dummies also in the current connections areas.

5. MEASUREMENT TECHNIQUE AND EXPERIMENTAL RESULTS

The resistance between Vh and Vl is measured between T1=30°C and T2=80°C. With TCR method [1], it follows that:

$$\frac{1}{\alpha} \frac{dR}{dT} = \frac{1}{e} \frac{d\rho}{dT} \qquad (3)$$

where T is the device temperature. The resistance measurements and the thickness extraction have been carried out on 25 dies on several wafers. In addition four SEM cross sections have been completed for verifications.

Fig. 5 to Fig.8 represent SEM cross sections of CRDM6 to CRDM9 devices. The graph in Fig. 9 displays the evolution of metal thickness versus the device dimensions, both for electrical and SEM measurements. First we observe a good correlation between electrical and physical values.

Fig.5 SEM Cross section device CDRM6

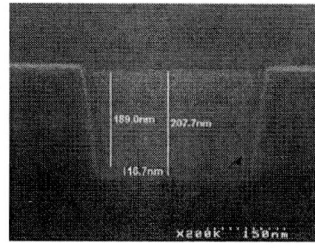

We found that the wider the lines are (devices CDRM1 to CDRM5), the bigger is the dishing effect, as expected. This results in low thickness values for the wider lines. On the contrary, for narrow lines (devices CDRM6 to CDRM9) the electrical thickness measurements converge in an asymptotic curve.

The second part of graphs in Fig. 9 (devices CDRM6 to CDRM9) and graph in Fig. 10 display the evolution of metal thickness versus the device dimensions and the ILD/Metal areas ratio. We see that dishing effect decreases also for narrow lines coupled with ILD/metal areas ratios close to 1.

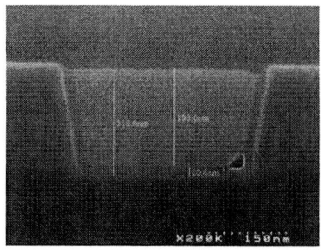

Fig.6 SEM Cross section device CDRM7

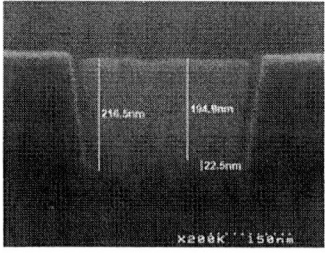

Fig.7 SEM Cross section device CDRM8

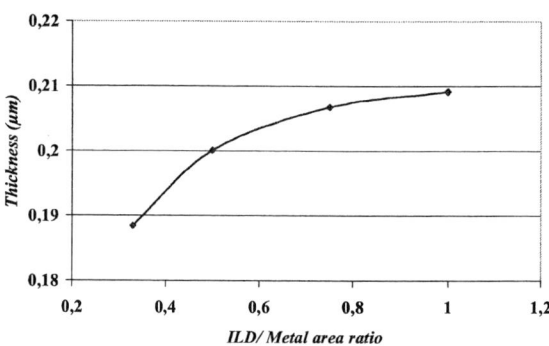

Fig.10: Evolution of average metal thickness, determined by electrical tests, versus ILD/metal areas ratio

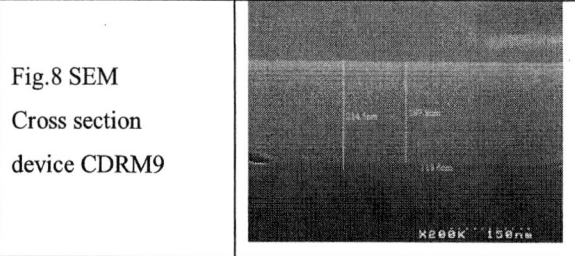

Fig.8 SEM Cross section device CDRM9

In conclusion, the design of devices with a line width lower than 0.5µm, and an ILD/Metal areas ratio bigger that 0.5, allows optimizing the thickness extraction, reducing parasitic effects of CMP.

Automatically, the sheet resistance and the 30°C and 80°C resistivity are the others parameters extracted from the electrical measurements.

The Fig.11 displays the evolution of the sheet resistance versus the thickness. As expected, the sheet resistance decreases as the average metal thickness increases.

The Fig.12 shows the resistivity homogeneity versus the thickness and line widths. This is the consequence of the enough large copper line widths, used here to prevent resistivity increase due to narrow line widths.

An example of wafer level thickness and resistivity uniformity are respectively given in figures 13 and 14, thanks to automatic test mode, which outputs wafer cartographies.

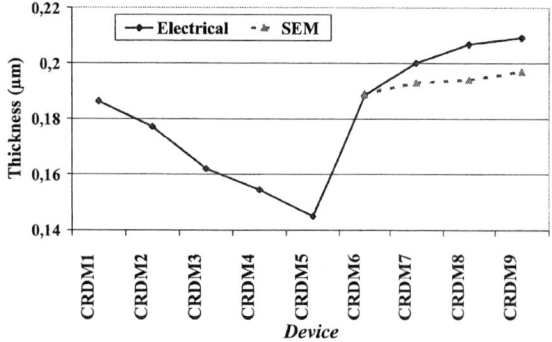

Fig. 9: Evolution of average metal thickness determined by electrical tests and SEM cross sections versus devices geometric parameters

212

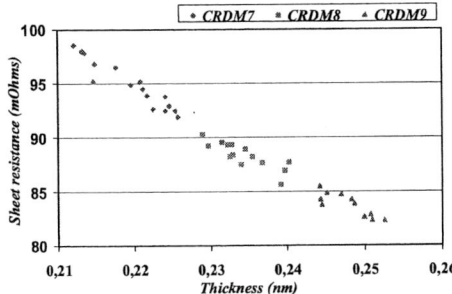

Fig. 11: Sheet resistance versus thickness

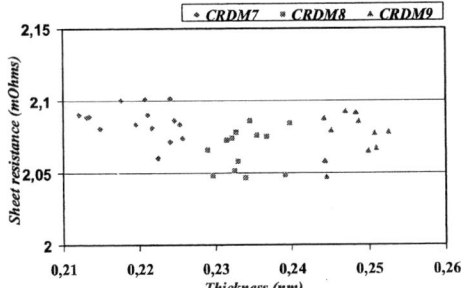

Fig.12: 30°C Resistivity versus thickness

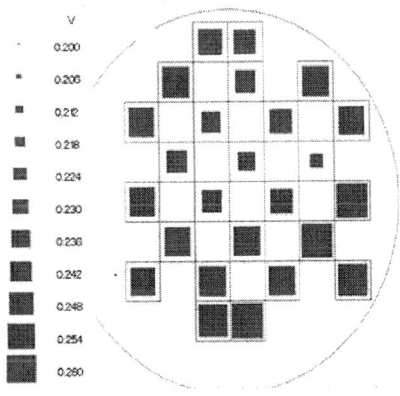

Fig.13: Wafer thickness cartography

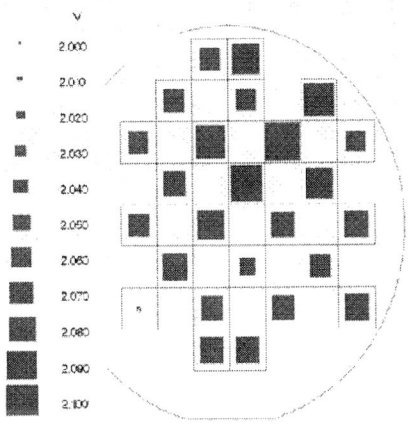

Figure 14: wafer 30°C Resistivity cartography

6. CONCLUSION

A new test structure has been designed, coupled with TCR method measurements. We have tested it considering several metal and ILD sizes, and several ILD/Metal areas ratios. This work gives significant adapted rules: line width smaller that 0.5µm, and ILD/Metal areas ratio bigger that 0.5 are required. This allows extracting metal thickness, resistivity and sheet resistance; reducing processing defects such as dishing, erosion and linewidths decreasing. Structure and method can be used with confidence and accuracy in statistic mode. A future work will consist in coupling this structure with narrow lines; In this case the complete metal lines parameters will be described.

References:

[1] F Warkusz J. Phys. D: Appl. Phys., Vol. 11, 1978.

[2] J.F. Guillaumond et al., Proc. of IITC 2003 conf., p132.

[3] D.C. Edelstein, "Advantages of Copper Interconnects" 1995 VMIC (1995).

[4] P. Singer, "Tantalum, Copper and Damascene: the future of interconnects", Semiconductor international, vol. 21, no. 6, pp. 90-98, June 1998.

[5] M. Fayolle et al., Proc. of IITC 2004 conf., p208

[6] J.M. Steigerwald et al., J Elec. Soc., vol 14, p.2842, 1994.

2008 IEEE Conference on Microelectronic Test Structures, March 24-27, Edinburgh, UK

9.4

Test Structure Definition for Dummy Metal Filling Strategy Dedicated to Advanced Integrated RF Inductors

Carine Pastore[1,2], Frédéric Gianesello[1], Daniel Gloria[1], Emmanuelle Serret[1], and Philippe Benech[2].

[1]STMicroelectronics, 850 rue Jean-Monnet, 38 926 Crolles Cedex, France.
[2]IMEP, 3, rue Parvis Louis Néel, BP 257, 38 016 Grenoble Cedex 1, France.

ABSTRACT

A complete strategy to manage dummy fills inside a large spectrum of integrated RF inductors realized in a 0.13 µm CMOS technology using a Dual Damascene Copper Back End Of Line (BEOL) is presented here. Thanks to the developed test structures, their RF characterization, and a Design Of Experiment (DOE) modeling analysis, it has been possible to determine the right metal fill density to insert inside inductors in order to be compliant with digital metal density rules without degrading their electrical performances.

INTRODUCTION

In today's advanced IC processes based on Damascene Copper technology, dummy metal fills are required to meet metal density rules and guarantee circuit integrity. This is a direct consequence of the Chemical Mechanical Polishing (CMP) process use which is actually sensitive to the overall layout density [1] and can create topographical and electrical defects called erosion in high density areas, and dishing in wide lines, see in Fig. 1, [2-3].

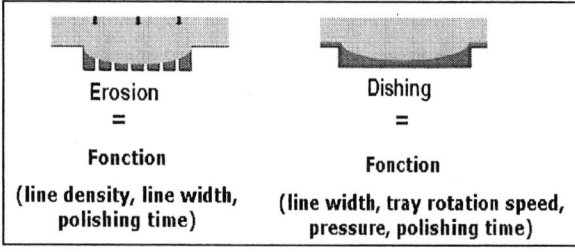

Fig. 1: Definition of erosion and dishing after Copper CMP.

In order to avoid these kinds of over-polishing and provide a surface planarity as uniform as possible, more or less severe design rules are applied to all BEOL structures [4], in particular for integrated inductors. Actually, dedicated metal density rules are set in order not to decrease the electrical performances of these key RF components. In most technologies, the inductor centre is wide and kept metallization free, and as coils are made of top metal levels in most inductor layouts, the space under the coils is lower metal level free too, as shown in Fig. 2. Then, specific design rules are applied around the inductor to match metal density requirements. However, these requirements are chip area consuming and difficult to apply in Advanced processes.

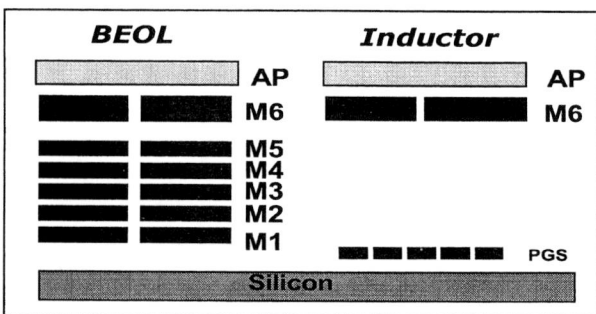

Fig. 2. BEOL and inductor architecture descriptions.

The main motivation of this paper is to use dedicated test structures, DOE modeling [5] and RF characterizations in order to provide a relevant and efficient dummy metal filling strategy inside various inductors integrated in an Advanced Copper BEOL, in respect with the digital density rules, and without degrading the RF device performances.

INDUCTOR DESCRIPTION

A: Manufacturing Process

Octagonal symmetric inductors with a Patterned Ground Shield (PGS) as described in [6] were used in this study. This set of inductors was fabricated in an industrial STMicroelectronics 0.13 µm node CMOS technology with a Dual Damascene Copper Back End on a conventional 10 Ω. cm silicon wafer.

The technology consists in six metallization levels, an aluminum capping layer AP, dielectric layers between all conductive layers with a dielectric constant k equal to 4 and a top passivation layer, see in Fig. 2.

B: Reference Inductors

In order to provide the most accurate dummy metal filling method for most RF applications, in comparison with earlier work [7], a large spectrum of integrated inductors in terms of geometrical and electrical parameters was studied, as given in Table 1.

Inductor1 is made of a unique coil built at the upper two metal levels (M5-M6) stacked with the aluminum capping, see in Fig. 2. It is characterized by a high Q-factor and a low inductance L_s value, as shown in Table 1.

Inductor2 is made of two turns and wide conductors. Its coils are built using the last metal level (M6) stacked with the aluminum capping layer, see in Fig. 2, and present a localized underpass with the stack M4-M5. This device is

978-1-4244-1801-5/08/$25.00 ©2008 IEEE

214

characterized by a high Q-factor and a medium inductance value L_s, as shown in Table 1.

Inductor3 and Inductor4 are multi-turn inductors with narrow conductors. Their coils are built using the upper three metal levels (M4-M5-M6) without the aluminum capping layer, and present a continuous underpass with the stack M4-M5, and an upperpass M6, see in Fig. 2. This device targets medium inductance L_s value and compact size, as shown in Table 1.

Table 1. Electrical and geometrical parameters of the reference test structures.

	Inductor1	Inductor2	Inductor3	Inductor4
Q_{max}	14.04 @ 13.25GHz	13.48 @ 4.5 GHz	5.07 @ 2.3 GHz	5.72 @ 1.1 GHz
L_s [nH] @ 100 MHz	0.68	1.11	9.32	18.32
Cut-off F_c [GHz]	30.50	9.75	4.10	1.90
Serial R_s [Ω] @ 100 MHz	1.93	1.00	17.54	13.58
Radius [µm]	110	62.68	15	35
Coil width/spacing [µm]	5	34/20	2/0.6	5/0.6
Number of turns	1	2	12	11

C: Inductors with Dummy Fill-Cells

The objective of this work is to define a way to manage fill metals inside this set of inductors without degrading RF performances, and being compliant with digital Design Rule Control. In order to evaluate the effect on the device electrical characteristics, squared dummy metals were additionally inserted in the centre of the inductor coils from M3 to M6.
The variable parameters are the density D, and the distance d1 from the inductor loops, and the dummy configuration: aligned A or crossed C dummy fills, see in Fig. 3.

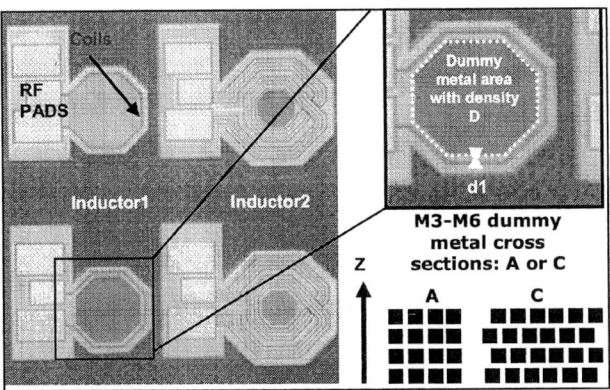

Fig. 3. Used RF pads, Inductor1 (on the top: D=25% inside, d1=1 µm, A; at the bottom: D=80% inside, d1=1 µm, A) and Inductor2 (on the top: D=52.5%, d1=1 µm, A; at the bottom: D=52.5%, d1=5.5 µm, A) images showing the considered DOE input factors.

A large density range from 25% up to 80% in comparison with [7] is provided by the squared dummy fill size for a fixed 0.46 µm spacing value. The used dummy fill widths are 0.46 µm, 1.21 µm and 3.9 µm. For the distance d1 range, it starts from 1 µm up to 10 µm with a middle point at 5.5 µm.

To perform this novel evaluation, only twelve samples for each kind of inductors (Inductor1, 2, 3 and 4) were chosen using the DOE methodology with a D-Optimal matrix and employing D, d1 and the type of dummy metal stacks (A or C) as input factors in Statgraphics software.

MEASUREMENTS AND INDUCTOR PARAMETER EXTRACTION

A: Measurements

One-port-S-parameter measurements up to 50 GHz were performed using an Agilent HP8510C VNA and Cascade Microtech Infinity GSG RF probes on the achieved inductors including the reference structures and those where fill metals were inserted, see in Fig. 3. De-embedding was performed using a dedicated open structure to remove pad contribution [8-9].
In order to determine the influence of dummy fills on the inductor RF performances, three main parameters have been extracted: the quality factor Q (1), the coil inductance L_s, and the coil resistance R_s values, (2) and (3), respectively:

$$Q = \frac{\text{Im}(Z_{11})}{\text{Re}(Z_{11})} \quad (1)$$

$$L_s = \frac{\text{Im}(Z_{11})}{2 \cdot \pi \cdot frequency} \quad (2)$$

$$R_s = \text{Re}(Z_{11}) \quad (3)$$

Cut off frequency F_c has also been extracted when Q=0.
No significant variation has been observed on inductance L_s, as shown in Fig. 4 and resistance R_s values see in Fig. 5. Consequently, only results on quality Q-factor are presented hereafter.

B: Measurement Results for High-Q Inductors

Concerning Inductor1, measured quality Q-factors of the device with and without dummy fills are plotted in Figure 4. It is observed that if aggressive design rules are used (D=80% and d1=1 µm, C), the impact on Q factor is high with a 30 % reduction on Q_{max} value and a 15 % decrease on the Q peak frequency because of an increase of parasitic capacitance [10]. In the opposite, if relaxed design rules are applied (D=25% and d1=1 µm, A), the impact on inductance performance is moderated with less than 5% on Q_{max} value and without any impact on the Q peak frequency.

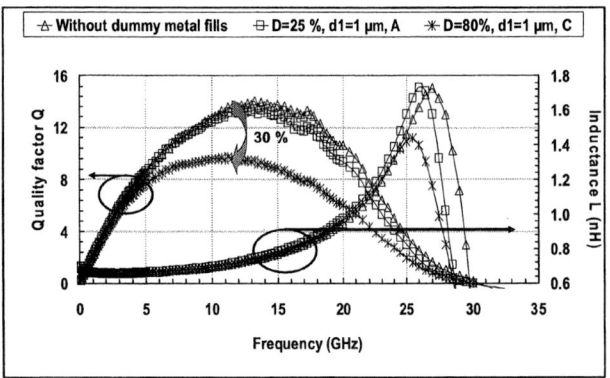

Fig. 4 Measured quality Q-factor of Inductor1 with and without dummy cells.

The same conclusions are observed for the measured quality Q-factor of Inductor2 with and without dummy fills, see in Fig. 5, but with a lower impact. Actually, if aggressive design rules are used (D=80% and d1=1 μm, A), the maximum value Q_{max} is reduced by 8.6 % whereas if relaxed design rules are employed (D=25% and d1=1 μm, C), the impact on Q_{max} value is very low, less than 1%.

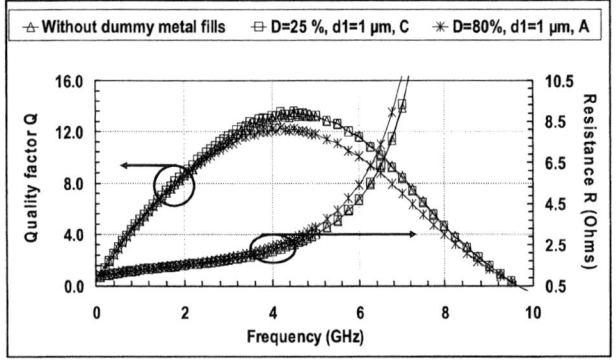

Fig. 5 Measured quality Q-factor of Inductor2 with and without dummy cells.

C: Measurement Results for Moderate-Q Multiturn Inductor

Concerning Inductor3, measured quality Q-factors with and without dummy fills are plotted in Fig. 6. It shows that if aggressive design rules are used (D=80% and d1=1 μm, A), the maximum value Q_{max} is lowered by 3.9 % because of an increase of parasitic capacitance [10]. In the opposite, if relaxed design rules are employed (D=25% and d1=1 μm, A), the impact on inductance performance is low, less than 1% on Q_{max} value.

The same conclusions are observed for the measured quality Q-factor of Inductor4 with and without dummy fills but with a lower impact, as shown in Fig. 7. If aggressive design rules are used (D=80% and d1=1 μm, C), the effect on Q factor is less than 1% on Q_{max} value. If relaxed design rules are applied (D=52.5 % and d_1=10 μm, A), the impact on inductance performance is very low with a reduction of 1.6 % on Q_{max}.

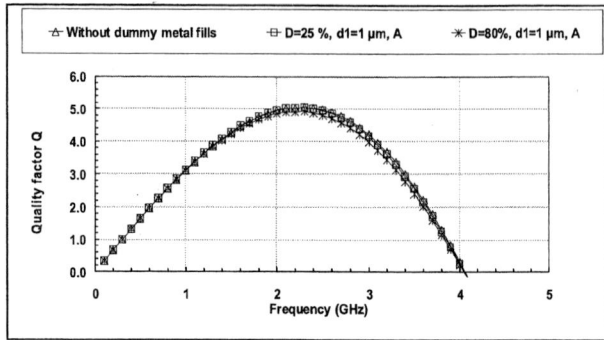

Fig. 6 Measured quality Q-factor of Inductor3 with and without dummy cells.

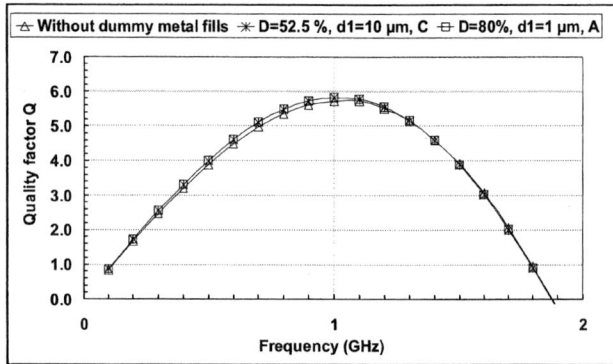

Fig. 7 Measured quality Q-factor of Inductor4 with and without dummy cells.

DOE MODELING AND DOMINANT INPUT FACTORS

As it has been demonstrated in the previous subsections, the sensitivity of inductor to dummy metal filling is structure dependant. Actually, high-Q inductors are more impacted by metal fills whereas moderate Q inductors seem to be less sensitive because of the high turn-to-turn parasitic capacitance.

A: Modeling Results for High Q-inductors

Concerning Inductor1, the standardized pareto chart for Q_{max}, given in Fig. 8, with an adjusted R-squared equal to 99%, indicates that the first significant parameter among the considered input factors is the dummy fill density D. However, the distance d1 and its combination with the density D also have a strong influence on the Q-factor. The bigger the distance is, the lower the impact on Q_{max} value is. Moreover, the DOE analysis reveals that the stack configuration (aligned A or crossed C fill metals) is not a critical parameter.

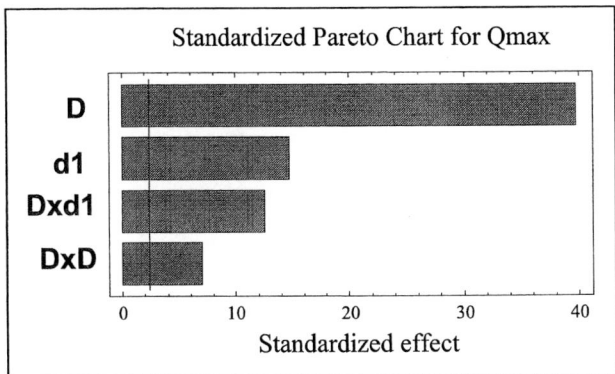

Fig. 8.　Pareto diagram of Inductor1 with adjusted R^2=99%.

For a low density (D=25%) the distances d1 has no effect on the Q factor, Fig. 8. For a high density (D=80%), the best performances are observed for the highest distance (d1=10 μm), as it is shown in the estimated response surface graph, see in Fig.9. This illustrates the importance of fringe capacitance between dummy cells and coils for high fill metal density.

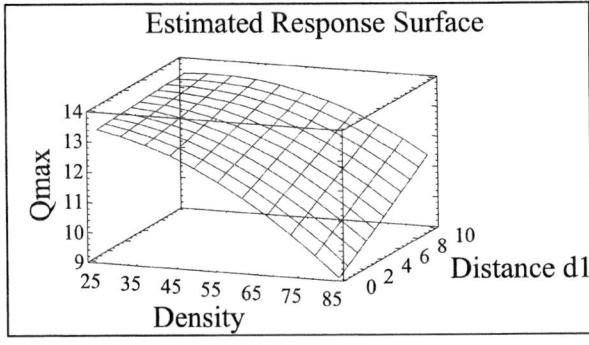

Fig. 9.　Estimated response surface graph for Inductor1.

Concerning Inductor2, the low adjusted R-squared equal to 26%, indicates that the DOE modeling is not relevant and shows that the studied parameters are not significant for the electrical device performance degradation, see in Fig. 10. Investigations for this structure are still on going.

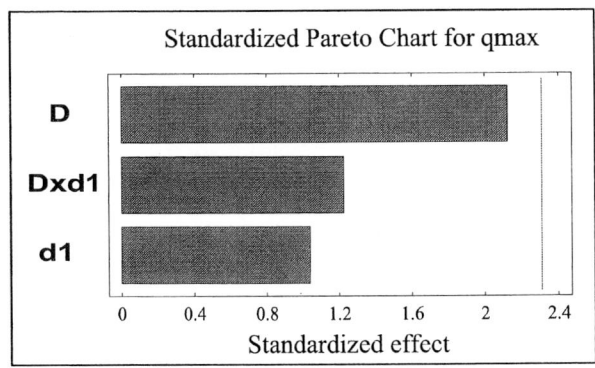

Fig. 10.　Pareto diagram of Inductor2 with adjusted R^2=26%..

B: Modeling Results for Moderate-Q Multiturn Inductor

Concerning Inductor3, the standardized pareto chart, for Q_{max}, given in Fig. 11, with an adjusted R-squared equal to 80%, indicates that the first significant parameter among the considered input factors is the dummy fill density D. However, the distance d1 and its combination with the density D also have a strong influence on the Q-factor. The bigger the distance is, the lower the impact on Q_{max} value is. Moreover, the DOE analysis reveals that the stack configuration (aligned A or crossed C fill metals) is not a critical parameter.

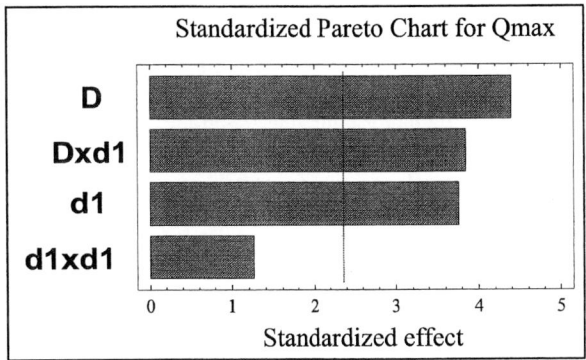

Fig. 11.　Pareto diagram of Inductor3 with adjusted R^2=80%.

For a low density (D=25%) the distances d1 has no effect on the Q factor, Fig. 8. For a high density (D=80%), the best performances are observed for the highest distance (d1=10 μm), as it is shown in the estimated response surface graph, see in Fig.12.

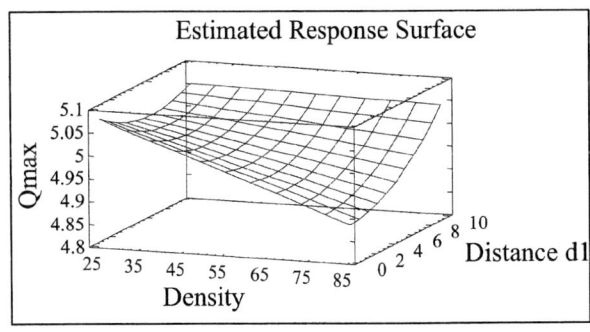

Fig. 12.　Estimated response surface graph for Inductor3.

Concerning Inductor4, the standardized pareto chart for Q_{max} with an adjusted R-squared equal to 64%, indicates that the first significant parameter among the considered input factors is the dummy fill density D. However, as a first observation, the stack configuration (aligned A or crossed C fill metals) seems to play a low role in Q_{max} degradation, as illustrated in Fig. 13.

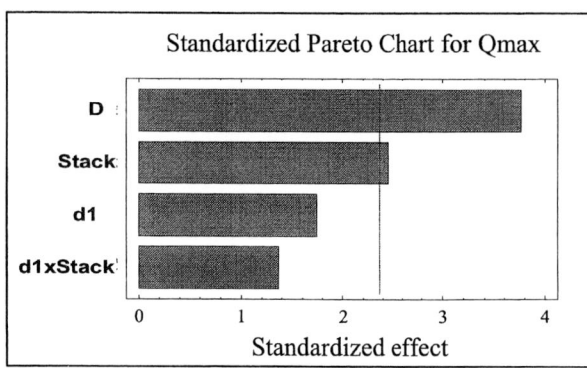

Fig. 13. Pareto diagram of Inductor4 with adjusted R^2=64%.

However, if we consider the type of stack separately in the modeling part, the DOE analysis reveals that in both cases (A or C configuration), the most significant parameter is the density D as shown in Fig. 14 and Fig. 15.

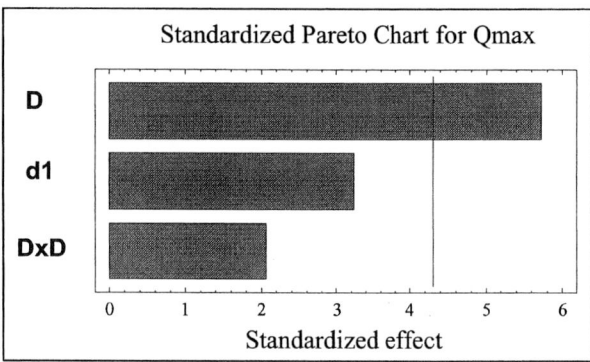

Fig. 14. Pareto diagram of Inductor4 with adjusted R^2=91%, stack configuration A.

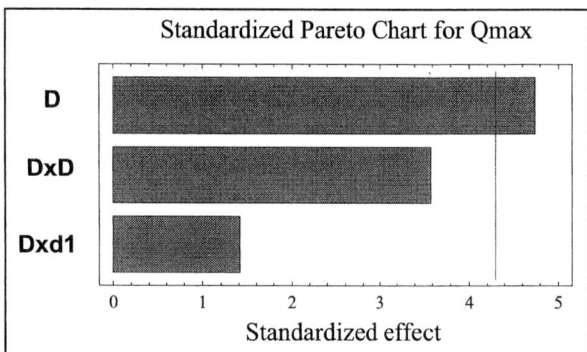

Fig. 15. Pareto diagram of Inductor4 with adjusted R^2=86%, stack configuration C.

In both cases, for a low density (D=25%) the distances d1 has no effect on the Q factor. For a high density (D=80%), the best performances are observed for the highest distance d1=10 µm, as it is shown in the estimated response surface graph, see in Fig. 16 and Fig. 17.

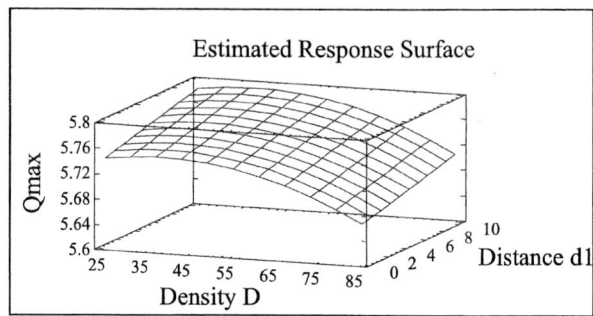

Fig. 16. Estimated response surface graph for Inductor4, stack configuration A.

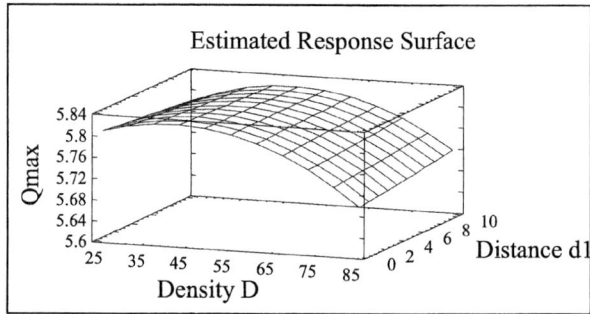

Fig. 17. Estimated response surface graph for Inductor4, stack configuration C.

DISCUSSION

Thanks to the DOE modeling, an upper limit for density filling has been first determined (~25%) in a minimal number of experiments and fulfill metal density requirements without decreasing the electrical performances of integrated inductors. The density D and the distance d1 are firstly recognized as the most significant parameters in terms of RF performance inductor degradation whereas the stack configuration (A or C) is not a critical parameter.

CONCLUSION

For the first time, a complete dummy filling strategy based on dedicated test structures, DOE analysis and RF measurements, using STMicroelectronics 0.13 µm bulk CMOS technology, has been proposed to evaluate the impact on RF performances of a large spectrum of integrated inductors. The DOE modelling confirms that it is possible to introduce dummy cells (up to 25%) into the coil centre, controlling the distance between the inserted fill metals and the coils, without degrading the electrical RF performances. Thanks to this new approach, it seems now possible to reduce the number of design rules dedicated to inductors, by introducing dummy fills inside the inductors which would enable inductor layout to fulfil digital density rules, keeping their RF performances constant.

REFERENCES

[1] S Kordic, "Fixing Today's Issues with Copper Electroplating and Copper CMP", SEMICON Europa, 2000, pp. 50-59.

[2] T. Park, and al., "Pattern and Process Dependencies in Copper Damascene Chemical Mechanical polishing Processes, VMIC, 1998, pp.437-442.

[3] S. Kordic, "Failure Analysis and characterization methods for Cu-CMP", CMP-MIC, 2005, pp.557-586.

[4] A. B. Kahng, G. Robins, A. Singh, and A. Zelikovsky, "Filling algorithms and analyses for layout density control", IEEE Trans. On Computer-Aided design of integrated Circuits and Systems, vol. 18, no. 4, 1999, pp. 445-462.

[5] A. Dean and D. Voss, "Design and analysis of experiments", Springer-Verlag New York, Inc., 1999.

[6] C. P. Yue, S. S. Wong, "On-Chip Spiral Inductors with Patterned Ground Shields for Si-Based RF IC's", IEEE Journal of Solid-State Circuits, vol. 33, No. 5, 1998, pp.743-752.

[7] L. F. Tiemeijer, R. J. Havens, H. J. Pranger, "Physics-Based Wideband Predictive Compact Model for Inductors with High Amounts of Dummy metal fills", IEEE Transactions on Microwave theory and Techniques, Vol. 54, No. 8, 2006, pp. 3378-3386.

[8] T. E, Kolding, "On-wafer Calibration Techniques for giga-hertz CMOS measurements", Proceeding of the IEEE International Conference on Microelectronic Test Structures, 1999, pp. 105-110.

[9] T. E Kolding, and al, "Ground-shielded measuring technique for accurate on-wafer characterization of RF CMOS devices", Proceeding of the IEEE International Conference on Microelectronic Test Structures, 2000, pp. 246-251.

[10] Lan Nan, and al.,"Experimental characterization of the effect of metal dummy fills on spiral inductors", IEEE Radio Frequency Integrated Circuits Symposium, 2007, pp. 307-310

SESSION 10
Matching II – Mechanisms

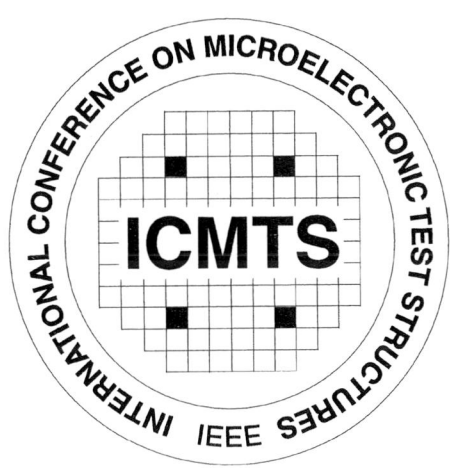

March 26, 10:50–12:10

Co-Chairs: Mark Poulter, *National Semiconductor, USA*

Hugues Brut, *STMicroelectronics, France*

2008 IEEE Conference on Microelectronic Test Structures, March 24-27, Edinburgh, UK

10.1

Fully considered layout variation analysis and compact modeling of MOSFETs and its application to circuit simulation

[†]Takuji Tanaka, [*]Akira Satoh, Mitsuru Yamaji, Osamu Yamasaki, [*]Hiroshi Suzuki,
Tsuyoshi Sakata, [‡]Yoshio Inoue, Masaru Ito, Seiichiro Yamaguchi, and Hiroshi Arimoto
FUJITSU Ltd., [*]FUJITSU Laboratories Ltd., and [‡]FUJITSU VLSI Ltd.
50 Fuchigami, Akiruno, Tokyo 197-0833 Japan
Email: tanaka.takuji@jp.fujitsu.com

Abstract— We have developed a total system of circuit design to treat dependency of MOSFET electric characteristics on layout patterns. Our new methodology with two-step multivariate analysis realizes highly reliable compact modeling, and its application to SPICE simulation significantly improves accuracy of circuit modeling. Our system is a powerful tool of design for manufacturing in 65 nm technology node and beyond.

I. Introduction

In circuit design in recent nano-scaled technology nodes, we never can ignore the variation of MOSFET electric characteristics on layout patterns (Fig. 1): i.e. active and STI shape [1,2], gate numbers and pitch [1,3,4], contact numbers and positions [5,6], due to variation of mechanical stress and controllability in manufacturing. On the other hand, conventional compact modelings (e.g. Berkeley Short Channel IGFET Model 4, BSIM4 [7]) have not supported such a complex layout parameters yet. Some previous modeling requires sophisticated process/device or lithography simulation [2, 3, 6] with expensive CPU cost. Moreover, we have to properly treat huge number of layout parameters and correctly model those effects on the electric characteristics. To solve the problem, practical modeling of layout variation which is easily applicable to EDA (electrical design automation) tools with low CPU cost is desired.

In this paper, we propose fully-considered layout variation analysis (LVA) and compact modeling with a new modeling methodology. We discuss detail of our modeling: measurement, modeling, application to circuit simulation, and validity of the system.

II. Basic Strategy of Layout Dependent Modeling

Figure 2 illustrates basic strategy of our modeling. We developed an EDA tool that can extract layout parameters from physical layout given by GDSII and converts them to modulation of SPICE model parameters by polynomial approximation. The modulated model parameters can apply to conventional circuit simulator taking layout dependence into account. Our goal of this work is to construct a practical and reliable model of parameter conversion for the tool.

We propose parameter conversion procedure from layout parameters to model parameters by way of electric parameters with two-step multivariate analysis. The purposes

are as follows: Interpolation of SPICE model parameters sometimes gives discrete results. Electric parameters with interpolation and extrapolation give much realistic and stable data than SPICE model parameters. Physical parameters converted from electric parameters are better to grasp what happened in the MOSFETs with layout difference than model parameters. We do not make an approach via process, device or lithography simulation [2,3,6] since their validity is limited by accuracy of those modeling and it is always too difficult to catch up newly found physical effects. On the other hand, our approach need only new test patterns for the new effects. We believe our model takes much advantage of black-boxed straight-forward direct conversion from layout parameters to model parameters.

III. Measurement of Test Patterns

We prepared test patterns specialized to measure layout dependence of I–V. We generated hundreds of variation of layout patterns by applying table of orthogonal arrays with difference of dozens of layout parameters. Figures 1 and 3 show examples of layout parameters and generated layout patterns, respectively. In addition, we prepared some monitors to analyze more detailed dependence on active shape, gate pitch and contact layouts. Those patterns are set in the 16×16 transistor matrix per a chip as similar to Ref. 8. We measured I–V characteristic of typically 5120 transistors per one monitor pattern to avoid errors due to random and inter/intra-chip variations.

IV. Modeling

We converted from measured I–V data to several electric parameters (e.g. V_{thl}, V_{ths}, $g_{m\,\text{max}}$, I_{on}, ...) for the intermediate parameter set between step I and step II (Fig. 2). The set of electric parameters is more strongly related than raw I–V data to physical phenomena such as low field mobility μ, channel impurity density N_{ch}, gate length/width $L_{\text{gate}}/W_{\text{gate}}$, DIBL (drain induced barrier lowering), saturation velocity v_{sat}, etc. It helps us to easily understand what happened in the MOSFETs with layout difference. We selected a set of four BSIM4 model parameters for instance parameters in SPICE simulation. It will be necessary and sufficient to represent several physical phenomena that is strongly affect electric properties: $VTH0$, $ETA0$, $U0$, and

978-1-4244-1801-5/08/$25.00 ©2008 IEEE

223

VSAT mainly covers variations of (N_{ch}, L_{gate}, W_{gate}), DIBL, μ, and v_{sat}, respectively.

We extracted polynomial regression equation by multivariate analysis at the both steps I and II. We achieved good fitting at almost all layout patterns (Figs. 4, 5) and averaged residual significantly decreased to 15 mV in V_{th} and 5–6 % in I_{on} (Fig. 6). It is found that at least five independent variables in the intermediate electric parameter set are required for good fitting in all instance parameters (Fig. 7).

V. RESULT AND DISCUSSION

A. Newly found layout dependence

We show here some example of our result. Dependence on layout patterns along the W_{gate} direction such as crank-shaped active and neighbor active (Fig. 8) has not been reported in the previous works [1–6] and not been considered in conventional compact models yet. 20 % difference of I_{on} was measured at the maximum due to those layouts (Figs. 9, 10) and our model well reproduces those properties. Figure 11 shows whole I_d-V_g and I_d-V_d curves well fit in both H- and I-type thanks to four instance parameters in SPICE model.

B. SPICE simulation with layout variation analysis

We applied the layout variation analysis to SPICE simulation (Figs. 12–14). We found significant improvement of accuracy. In delay time of various gates (Fig. 14), SPICE simulation with layout variation analysis decreases the error with the order of 10 %.

VI. CONCLUSION

We have successfully developed a total system of layout analysis, compact modeling, and circuit simulation to treat dependency of MOSFET electric characteristics on layout patterns: i.e. active/STI patterns, gate numbers and pitch, contact numbers and positions. Its application to SPICE simulation significantly improves accuracy of circuit modeling. Our newly developed system is a powerful tool of design for manufacturing in 65 nm technology node and beyond.

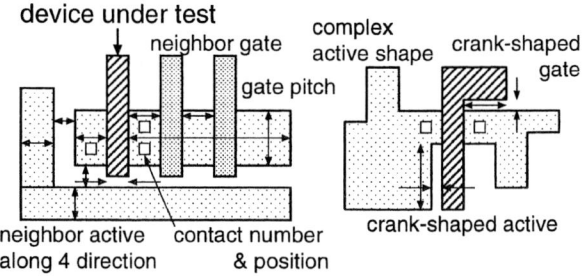

Fig. 1. Schematic illustration of MOSFET layout variation in test patterns of our measurement. Some examples of layout parameters are also shown by arrows.

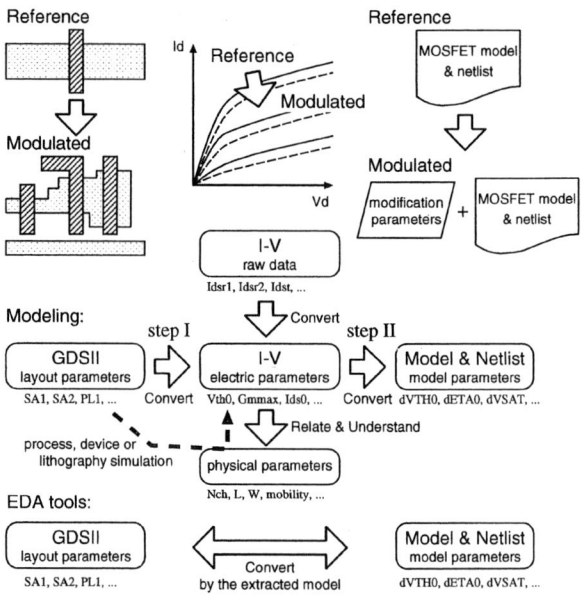

Fig. 2. Schematic illustration of the basic strategy for compact modeling of layout dependent MOSFET characteristics and those application to EDA tools. We make an approach not via process/device/lithography simulations but via straight-forward multivariate analysis with two steps.

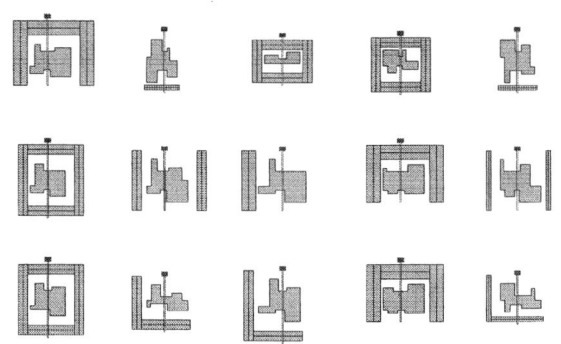

Fig. 3. Example of various layout of gate and active in test patterns concerning active shape and neighbor active shape.

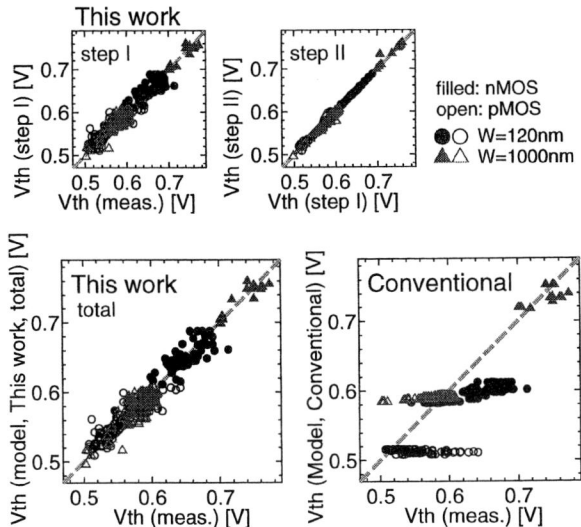

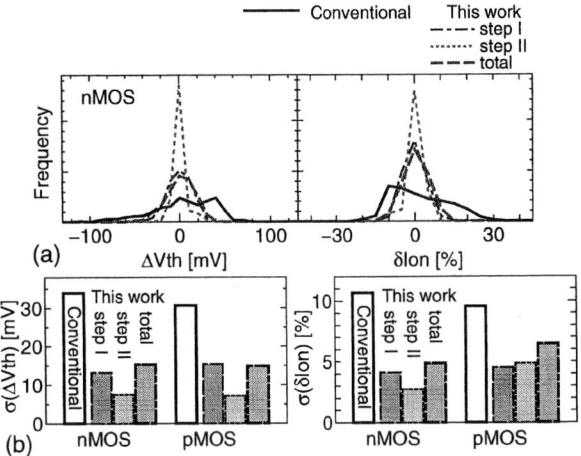

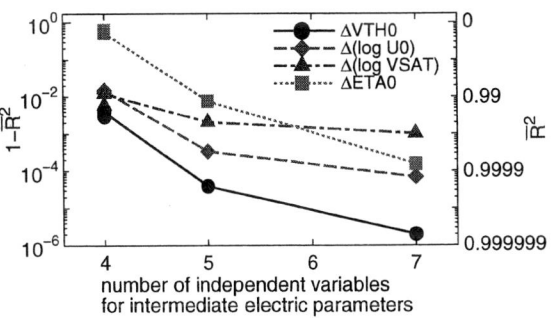

Fig. 4. Correlation plot between measured and modeled V_{th} among different layout patterns corresponding to Fig. 5. The results with (This work) and without (Conventional) layout variation analysis are compared.

Fig. 6. (a) Frequency distribution and (b) standard deviation of residual of modeled V_{th}, I_{on} among various layout patterns. The results of SPICE simulation without (Conventional) and with (This work) layout variation analysis are compared.

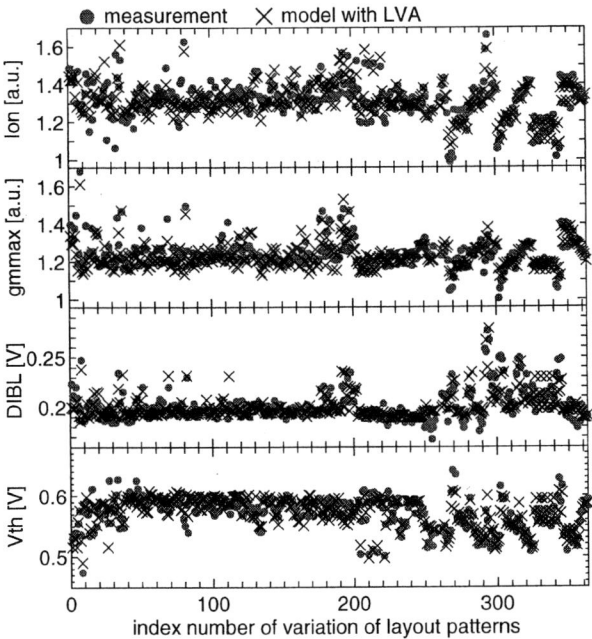

Fig. 5. Measured and modeled electric parameters (I_{on}, $g_{m\,max}$, DIBL, V_{th}) versus MOSFET index number of variation of layout patterns which include Figs. 1, 3, and 8.

Fig. 7. Coefficient of determination $\overline{R}^2$ at regression of step II comparing different sets of intermediate electric parameters as a function of number of independent variables.

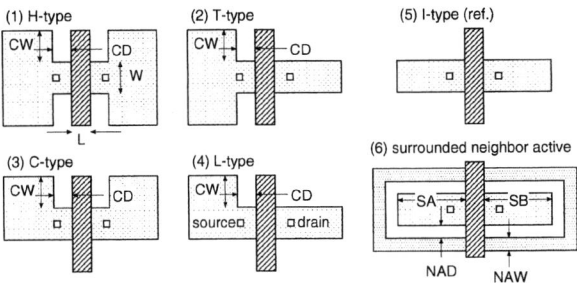

Fig. 8. Schematic illustrations of several layout patterns under test: (1)–(4) four types of crank-shaped active, (5) reference, and (6) surrounded neighbor active. Dependence on these layout patterns has not been reported in previous works [1–6] and not been considered in conventional compact models yet.

225

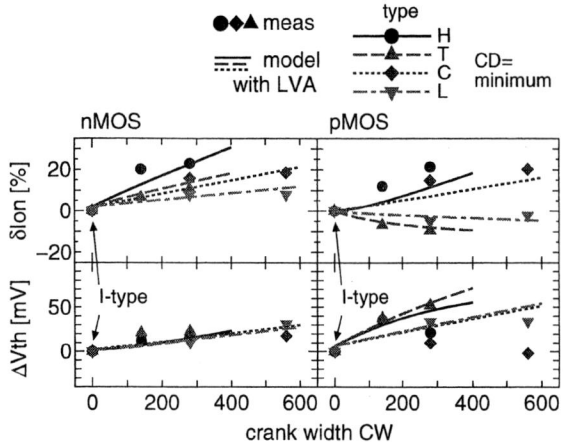

Fig. 9. V_{th}, I_{on} relative to that of reference as a function of crank-shaped active width. The parameters are shown in Fig. 8(1)–(5).

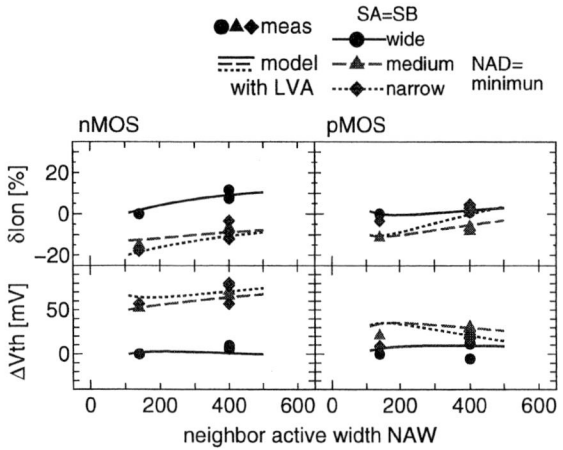

Fig. 10. V_{th}, I_{on} relative to that of reference as a function of neighbor active width. The parameters are shown in Fig. 8(6).

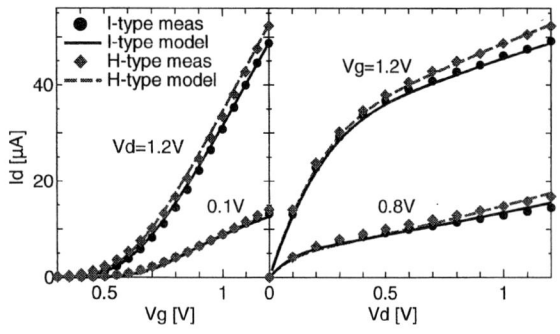

Fig. 11. I–V characteristics of MOSFETs with/without crank-shaped active comparing measurement and SPICE simulation with layout variation analysis.

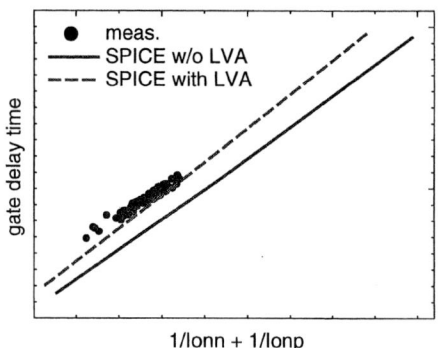

Fig. 12. Gate delay of a ring oscillator pattern versus $1/I_{on\,n} + 1/I_{on\,p}$ comparing measurement and SPICE simulation with/without layout variation analysis.

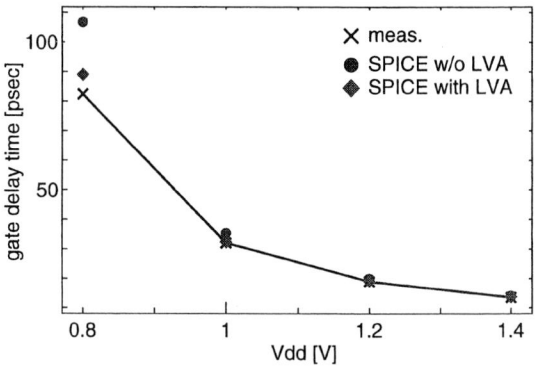

Fig. 13. Gate delay of a ring oscillator pattern versus supply voltage V_{dd} comparing measurement and SPICE simulation with/without layout variation analysis.

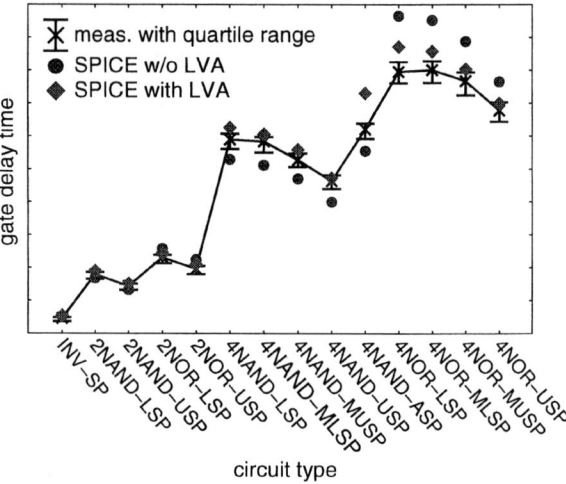

Fig. 14. Gate delay time of various ring oscillator patterns comparing measurement and SPICE simulation with/without layout variation analysis. The measurement is shown by median and quartile range.

226

REFERENCES

[1] R. A. Bianchi, G. Bouche, and O. Roux-dit-Buisson, "Accurate modeling of trench isolation induced mechanical stress effects on MOSFET electrical performance,", IEEE Electron Devices Meeting (IEDM) Tech. Dig., pp. 117–120, 2002.

[2] H. Tsuno, K. Anzai, M. Matsumura, S. Minami, A. Honjo, H. Koike, Y. Hiura, A. Takeo, W. Fu, Y. Fukuzaki, M. Kanno, H. Ansai and N. Nagashima, "Advanced Analysis and Modeling of MOSFET Characteristic Fluctuation Caused by Layout Variation", Symp. on VLSI Technol. pp. 204–205, 2007.

[3] Munkang Choi, L. Milor, L. Capodieci, "Simulation of the circuit performance impact of lithography in nanoscale semiconductor manufacturing", Int. Conf. on Simulation of Semiconductor Processes and Devices (SISPAD), pp. 219–222, 2003.

[4] A. Oishi, O. Fujii, T. Yokoyama, K. Ota, T. Sanuki, H. Inokuma, K. Eda, T. Idaka, H. Miyajima, S. Iwasa, H. Yamasaki, K. Oouchi, K. Matsuo, H. Nagano, T. Komoda, Y. Okayama, T. Matsumoto, K. Fukasaku, T. Shimizu, K. Miyano, T. Suzuki, K. Yahashi, A. Horiuchi, Y. Takegawa, K. Saki, S. Mori, K. Ohno, I. Mizushima, M. Saito, M. Iwai, S. Yamada, N. Nagashima and F. Matsuoka, "High Performance CMOSFET Technology for 45nm Generation and Scalability of Stress-Induced Mobility Enhancemt", IEEE Electron Devices Meeting (IEDM) Tech. Dig., pp. 239–243, 2005.

[5] S. M. Cea, M. Armstrong, C. Auth, T. Ghani, M. D. Giles, T. Hoffmann, R. Kotlyar, P. Matagne, K. Mistry, R. Nagisetty, B. Obradovic, R. Shaheed, L. Shifren, M. Stettler, S. Tyagi, X. Wang, C. Weber, and K. Zawadzki, "Front end stress modeling for advanced logic technologies,", IEEE Electron Devices Meeting (IEDM) Tech. Dig., pp. 963–966, 2004.

[6] R. Liebmann, M. Nawaz and K. H. Bach, "Efficient 2D approximation for Layout-dependent Relaxation of Etch Stop Liner Stress due to Contact Holes", Int. Conf. on Simulation of Semiconductor Processes and Devices (SISPAD), pp. 173–175, 2006.

[7] BSIM4.2.1 MOSFET Model User's manual

[8] S. Ohkawa, M. Aoki and H. Masuda, "Analysis and characterization of device variations in an LSI chip using an integrated device matrix array", Int. Conf. on Microelec. Test Struct. (ICMTS) pp.70–75, 2003.

On-Mask Mismatch Resistor Structures for the Characterisation of Maskmaking Capability

S. Smith*, A.Tsiamis*, M. McCallum†, A.C. Hourd‡, J.T.M. Stevenson*, A.J. Walton* and S. Enderling*

*Institute of Integrated Micro and Nano Systems, School of Engineering and Electronics,
Scottish Microelectronics Centre, The University of Edinburgh, EH9 3JF, UK   Email: Stewart.Smith@ed.ac.uk

†Nikon Precision Europe GmbH, Appleton Place, Appleton Parkway, Livingston, West Lothian, EH54 7EZ, UK

‡Compugraphics International Ltd., Eastfield Industrial Estate, Glenrothes, Fife, KY7 4NT, UK

Abstract—This paper presents results from the use of electrical measurements to investigate dimensional mismatch in an advanced photomask process. Test structures consisting of matched pairs of Kelvin resistors have been measured and the results analysed to obtain information about the capability of the mask making process. The mask plate used in this work has an array of over 350 sets of mismatch test structures, providing an unprecedented volume of data. A second mask, which was underexposed, has allowed the relationship between mismatch and the location of nearby mask writing pattern boundaries to be investigated for the first time.

I. INTRODUCTION

The present work is part of a continuing collaborative project investigating the use of electrical test patterns for the dimensional characterisation of advanced photomask plates. Previous publications have described the use of sheet resistance and electrical linewidth test structures, which can be directly probed on-mask, for the characterisation of alternating aperture phase shifting masks [1]–[7], corner serif optical proximity correction (OPC) features [8], iso-dense mark-space features [9] and dimensional mismatch [10]. Results presented in these papers have demonstrated the ability of the on-mask electrical structures to provide faster and more repeatable measurements of critical dimension (CD) than traditional optical or CD-SEM techniques.

This paper is a continuation of the work published in [10] investigating on-mask, dimensional mismatch test structures. This revealed previously unobserved systematic width offsets between pairs of closely spaced resistors but the source of this was unclear. There were no obvious patterns related to processes such as spin developing or etching and so it was concluded that the electron-beam mask writing tool was the most likely source. In order to further investigate these effects, new masks have been fabricated and initial results are presented in this paper.

II. TEST STRUCTURES

The test structures employed in this study have been previously described in detail in [10] and are shown schematically in in Fig. 1.

They consist of pairs of Kelvin connected bridge resistors, 600μm long with a nominal width of 0.5μm. The resistors are separated by 30μm and each mismatch block has two pairs of

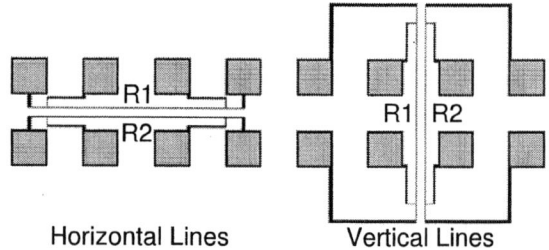

Fig. 1.   Layout of mismatch bridge resistor test structures

structures, one pair configured horizontally and one vertically. This means that offsets can be measured along two axes. The test mask (MSN7520) used in this work is shown in Fig. 2. It differs from the previous test plates, with the full available area being taken up with a regular array of mismatch test structures with a 6mm pitch.

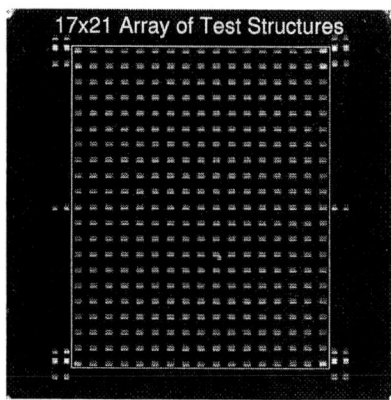

Fig. 2.   Scanned image of chrome on quartz test mask MSN7520

This array of 357 blocks of matching test structures provides a considerably more comprehensive data set than the previous test masks that included these structures. Mask MSN7520 is a "vanilla" plate, fabricated without the standard GHOST correction process [11] which equalises the background dose from the e-beam mask writing system. GHOST correction is performed by applying a second exposure with the inverse

978-1-4244-1801-5/08/$25.00 ©2008 IEEE

of the main pattern at a lower dose, which helps to correct for iso-dense bias effects. The effect of the procedure is demonstrated in [12], which compares results from on-mask electrical linewidth test structures on GHOSTed and unGHOSTed photomasks.

A second photomask (MSN7544) with the same pattern as MSN7520 was fabricated using the GHOST process but a problem with an unexpectedly low exposure dose meant that it was unsuitable for its initial purpose. However, this turned out to be serendipitous as the underexposed plate clearly shows the boundaries between the e-beam patterning segments. In a previous paper on this subject [10], it was suggested that these boundaries were a possible source for the observed pattern of misalignment. As mask MSN7544 was fabricated using the same lithography job file as MSN7520 it provides a unique opportunity for further investigation.

III. MEASUREMENTS

The test structures are measured by passing a current (I) between the outer pads and measuring the voltage (V) developed between the voltage taps connected to the inner pads, as shown in Fig. 3.

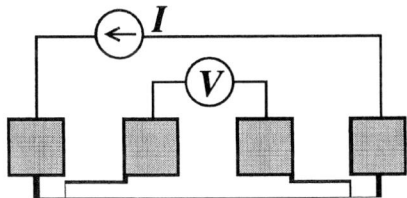

Fig. 3. Measurement of a Kelvin bridge resistor forming one half of a mismatch test feature

The resistance ($R = V/I$) of each line is then calculated and used to estimate the electrical width of the feature. This requires a value for the sheet resistance (R_S) of the chrome layer and, as the mismatch blocks do not include test structures to measure this, an average value for the whole mask is used. Results from previous masks from the same manufacturer and using mask blanks from the same supplier, have suggested that the sheet resistance of the chrome layer is around $22.5\Omega/\square$ with a variation across the mask of less than 1%. The short range variation of R_S is likely to be significantly lower and so we can be confident that two closely spaced lines are fabricated from material with the same sheet resistivity. Assuming that $R_S = 22.5\Omega/\square$ means that the calculated linewidths are approximations but this does not affect the mismatch figure $\Delta W/W(\%)$ which is a relative value. The linewidth offset ΔW is the width of resistor R2 subtracted from the width of R1, for both the vertical and horizontal lines shown in Fig. 1.

IV. RESULTS

Histograms of $\Delta W/W$ for the mask wide array of mismatch structures can be seen in Fig. 4a, for horizontal lines, and Fig. 4b for vertical lines.

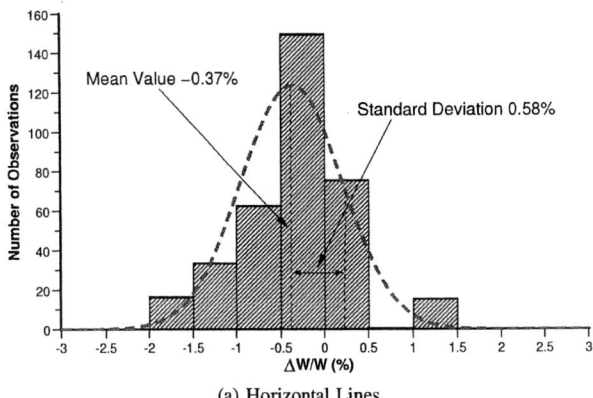

(a) Horizontal Lines

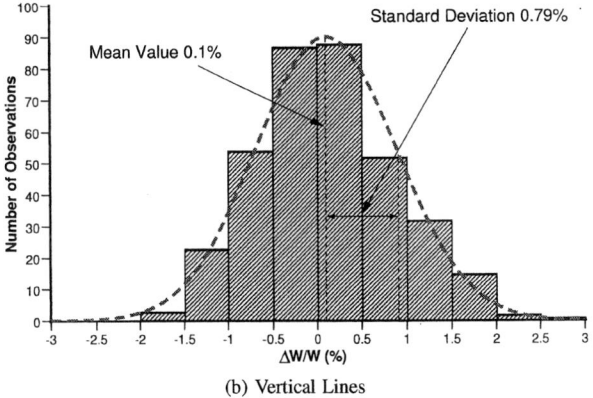

(b) Vertical Lines

Fig. 4. Histograms of CD mismatch results for MSN7244 fitted with normal distributions

Although both histograms have been fitted with normal distributions the mismatch figures, especially those from the horizontal lines, do not show the expected random statistical variation. As with mask MSN5757 in [10] this mask has been processed without using GHOST correction and as with that mask the mean value of the mismatch data is within 1σ of zero for both horizontal and vertical lines.

In order to determine if there are any patterns in the data related to the position of the structures on the mask the mismatch data is plotted against either the vertical or horizontal position in Fig. 5 and Fig. 6.

Fig. 5b shows that the results for the horizontal lines are strongly grouped by row or vertical position. However there is no clear pattern to this or agreement between rows. The results from the vertical lines show a much more random distribution but there is a suggestion of a downward trend in $\Delta W/W$ from the bottom to the top of the mask. This is highlighted in Fig. 6b which shows a linear fit to the data. However, this trend is relatively small compared to the range of variation in the data and so may not be significant.

Fig. 7 shows the linewidth offset (ΔW) in X and Y for each block of matching structures as a vector plot. Due to the fact that the linewidth values are calculated from an assumed value of sheet resistance these results should be

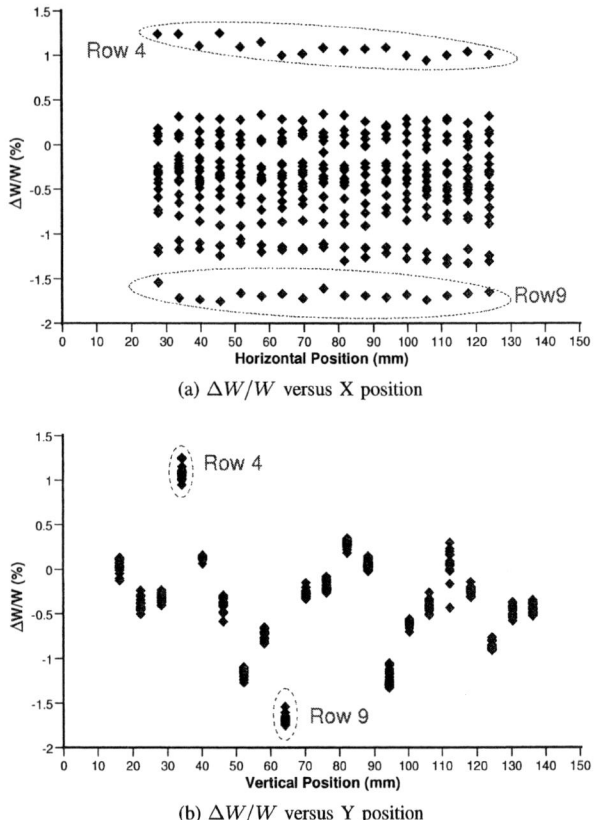

(a) $\Delta W/W$ versus X position

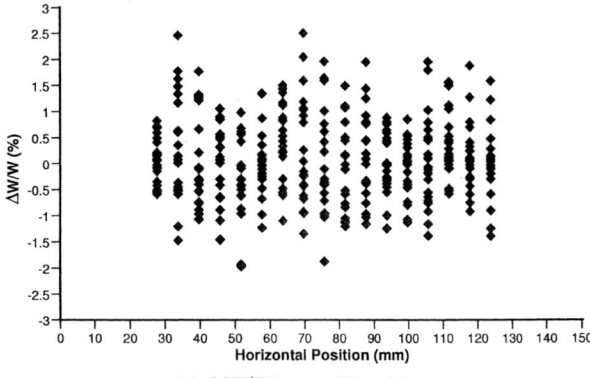

(a) $\Delta W/W$ versus X position

(b) $\Delta W/W$ versus Y position

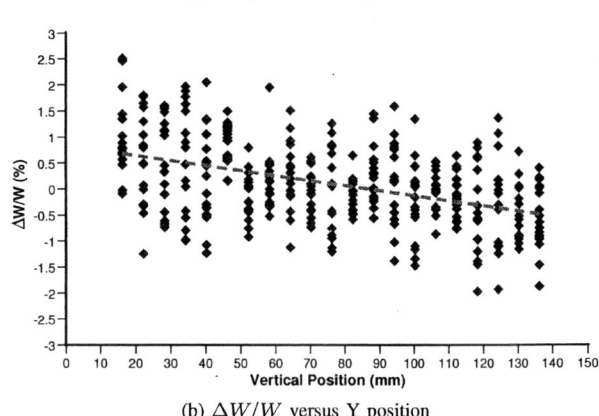

(b) $\Delta W/W$ versus Y position

Fig. 5. CD mismatch figures as a function of position for horizontal lines

Fig. 6. CD mismatch figures as a function of position for vertical lines

interpreted as indicative rather than as absolute values of linewidth offset. This plot shows that the agreement in the vertical component, taken from the horizontal lines, within each row. However, it also suggests there is some sort of pattern to the horizontal component that was masked by the way the data was represented in Fig. 6.

V. DISCUSSION

In a previous paper [10] it was suggested that a possible source for some of the obvious patterns of mismatch observed on masks MSN5757 and MSN6659 was the position of the boundaries between data segments in the mask writing tool. The mask data is sent to the mask writing tool in blocks which are 20971.52μm long and 655.36μm high. The original understanding of the manner in which the pattern is written was that the mask is moved horizontally while the electron beam is scanned vertically, as shown in Fig. 8.

At the end of each block a new set of data is loaded into the tool and the next segment in the row is written. When the edge of the plate is reached the mask is moved back to the start of the next row in a similar manner to the raster scan of a television. This suggests that if there are any issues with differences between adjacent blocks they are more likely to happen at the horizontal boundaries and will therefore affect the horizontal lines more than the vertical lines. It was proposed that this could be the source of the characteristic

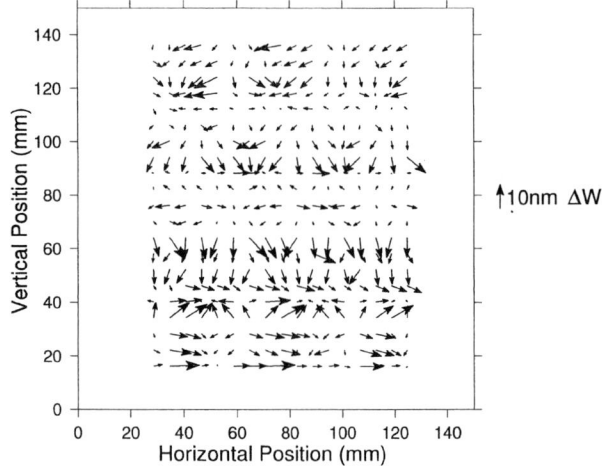

Fig. 7. Vector plot showing linewidth bias for MSN7520

agreement within a row for the horizontal lines, as shown in Fig. 5b. The highest and lowest values of mismatch are highlighted in this graph and these originate from row 4 and row 9.

As was described in section II mask MSN7544 was underexposed which served to highlight the actual positions of the mask writing boundaries. Microscope images of horizontal

230

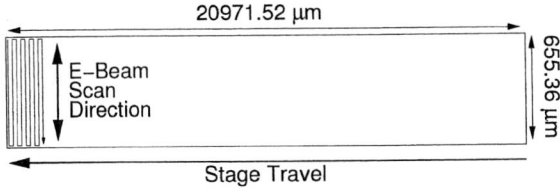

Fig. 8. Diagram showing the e-beam raster writing method

test structures in rows 4 and 9, which show the write segment boundaries, can be seen in Fig. 9. It is clear from these images that the vertical height of the write segments is not the same as is suggested in Fig. 8 and is in fact around 160μm. The system that controls the mask writing tool takes the segmented data and presents it to a rasterising engine, which re-segments the data to optimise the write operation. Although it is possible to define the dimensions of the data segments in the file sent to this system it is, unfortunately, practically impossible to determine exactly how this will be finally written by the lithography tool following rasterisation.

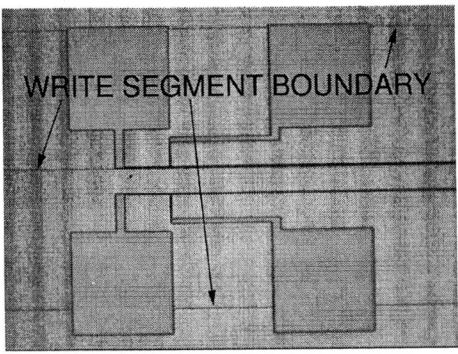

(a) Row 4

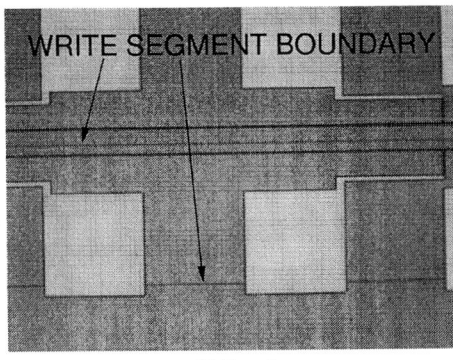

(b) Row 9

Fig. 9. Microscope images of mask MSN7544 showing write segment boundaries

Both row 4 and row 9 have a horizontal segment boundary between the two horizontal lines and, in the case of row 4, it is within a few microns of one of the lines. Consequently it could be concluded that this might be the cause of the observed mismatch. However, rows 1 and 17 also have a segment

boundary which runs between the horizontal test structures and they have a mean mismatch which is close to zero. Similar images were taken of vertical test structures in each column of the array but only column 9 was observed to have a boundary running between the lines. There is no obvious consequence of this in the data in Fig. 6.

VI. CONCLUSION

A new test mask (MSN7520) with an array of on-mask electrical test structures designed to measure dimensional mismatch has been fabricated and tested. The test structures consist of matched pairs of Kelvin resistor structures orientated either vertically or horizontally. Unlike previous designs, this mask plate features a large, regular array of matching test structures, which covers the entire printable area. This provides a much larger data set than previous masks without gaps caused by the inclusion of other types of test structure.

The test structures have been measured electrically and the results analysed. As with previous test masks [10] it is clear that the mismatch does not display a random, statistical distribution. This is especially true for the horizontally oriented lines where there is very little variation in the mismatch between structures on the same row of the test array. The vertical lines appear at first glance to be randomly distributed but a pattern or process footprint is revealed in a vector plot showing the linewidth offsets. The range of mismatch values for this mask is less than ±2% for the horizontal lines and less than ±3% for the vertical lines. This is comparable with the unGHOSTed mask in [10]. The higher standard deviations observed here are probably due to the complete array of structures. On previous masks, rows with high mismatches may only have had a small number of test structure blocks. It should be remembered that the mismatch is not randomly distributed.

A second test mask was prepared using the same pattern but was underexposed, which allowed the boundaries between blocks of data generated by the mask writing tool to be clearly observed. These boundaries could be the cause of some of the observed mismatch, especially where they fall between the two measured structures. Photomicrographs of MSN7544 show that the rows where the largest mismatches were observed for the horizontal structures do have writing segment boundaries between the two measured lines. However, there are two other rows on the mask where this also occurs and these both have an average ΔW/W which is close to zero. The vertical structures do not seem to be affected in any way by the location of the segment boundaries. One column of structures has a boundary which falls between the vertical measured lines but there is no obvious consequence of this that can be observed in the data.

ACKNOWLEDGMENT

The authors would like to acknowledge the support of the Edinburgh Research Partnership in Engineering and Mathematics and the associated Institute of Integrated Systems, Nikon Precision Europe GmbH and Compugraphics International Ltd.

REFERENCES

[1] S. Smith, M. McCallum, A. J. Walton, and J. T. M. Stevenson, "Electrical CD Characterisation of Binary and Alternating Aperture Phase Shifting Masks," in *Proceedings of IEEE International Conference on Microelectronic Test Structures*, Cork, Ireland, April 2002, pp. 7–12.

[2] S. Smith, M. McCallum, A. J. Walton, J. T. M. Stevenson, and A. Lissimore, "Comparison of Electrical and SEM CD Measurements on Binary and Alternating Aperture Phase-Shifting Masks," *IEEE Transactions on Semiconductor Manufacturing*, vol. 16, no. 2, pp. 266–272, May 2003.

[3] S. Smith, M. McCallum, A. J. Walton, J. T. M. Stevenson, P. D. Harris, A. W. S. Ross, A. C. Hourd, and L. Jiang, "Test Structures for CD and Overlay Metrology on Alternating Aperture Phase-Shifting Masks," in *Proceedings of IEEE International Conference on Microelectronic Test Structures*, Awaji Yumebutai International Conf. Center, Japan, March 2004, pp. 29–34.

[4] M. McCallum, S. Smith, A. C. Hourd, A. J. Walton, and J. T. M. Stevenson, "Cost Effective Overlay and CD Metrology of Phase-Shifting Masks," in *Proceedings of the SPIE, Vol. 5567: 24th Annual BACUS Symposium on Photomask Technology*, 2004, pp. 596–603.

[5] S. Smith, M. McCallum, A. J. Walton, J. T. M. Stevenson, P. D. Harris, A. W. S. Ross, and L. Jiang, "On-Mask CD and Overlay Test Structures for Alternating Aperture Phase Shift Lithography," *IEEE Transactions on Semiconductor Manufacturing*, vol. 18, no. 2, pp. 238–245, May 2005.

[6] S. Smith, A. Walton, M. McCallum, A. C. Hourd, J. T. M. Stevenson, A. W. S. Ross, and L. Jiang, "Improved Test Structures for the Electrical Measurement of Feature Size on an Alternating Aperture Phase-Shifting Mask," in *Proceedings of IEEE International Conference on Microelectronic Test Structures*, K.U. Leuven, Leuven, Belgium, April 2005, pp. 17–22.

[7] S. Smith, A. Tsiamis, M. McCallum, A. C. Hourd, J. T. M. Stevenson, and A. Walton, "Comparison of Optical and Electrical Measurement Techniques for CD Metrology on Alternating Aperture Phase-Shifting Masks," in *Proceedings of IEEE International Conference on Microelectronic Test Structures*, Austin, Texas, USA, March 2006, pp. 119–123.

[8] A. Tsiamis, S. Smith, M. McCallum, A. C. Hourd, O. Toublan, A. J. Walton, and J. T. M. Stevenson, "Electrical Test Structures for the Characterisation of Optical Proximity Correction," in *Proceedings of the SPIE, Vol. 6533: The 23rd European Mask and Lithography Conference (EMLC 2007)*, Grenoble, France, January 2007, p. 65330M.

[9] A. Tsiamis, S. Smith, M. McCallum, A. C. Hourd, O. Toublan, J. T. M. Stevenson, and A. J. Walton, "Development of Electrical On-Mask CD Test Structures Based on Optical Metrology Features," in *Proceedings of IEEE International Conference on Microelectronic Test Structures*, Tokyo, Japan, March 2007, pp. 171–176.

[10] S. Smith, A. Tsiamis, M. McCallum, A. C. Hourd, J. T. M. Stevenson, and A. J. Walton, "Electrical Measurement of On-Mask Mismatch Resistor Structures," in *Proceedings of IEEE International Conference on Microelectronic Test Structures*, Tokyo, Japan, March 2007, pp. 3–8.

[11] G. Owen and P. Rissman, "Proximity Effect Correction for Electron Beam Lithography by Equalization of Background Dose," *Journal of Applied Physics*, vol. 54, no. 6, pp. 3573–3581, 1983.

[12] A. Tsiamis, S. Smith, M. McCallum, A. C. Hourd, O. Toublan, J. T. M. Stevenson, and A. J. Walton, "Investigation of Electrical and Optical CD Measurement Techniques for the Characterisation of On-Mask GHOST Proximity Corrected Features," in *To be presented at IEEE International Conference on Microelectronic Test Structure 2008*, March 2008.

Physics and Modeling of Transistor Matching Degradation under Matched External Stress

Xiaoju Wu, Zhenwu Chen, Praful Madhani
(xjwu@ti.com)

Analog Technology Development, MS 364, Texas Instruments, Dallas, TX 75243, USA

ABSTRACT

In this paper, we report detailed studies on transistor matching reliability under external stresses such as NBTI. Transistor threshold voltage VT mismatching as a function of stress voltage, ambient temperature as well as stress time has been established based on device physics and statistics. A closed form equation for current mismatching under stress has also been obtained using a correlated mobility-shift and VT-shift model. Methods to extract relevant parameters in the models have been established. The models have been validated experimentally and been used to predict the transistor matching shift under normal operating conditions for a 40V high performance precision analog technology. It has been found that transistor matching degrades ~10% during the 10 years lifetime of products when operating at VGS=15V and 100C. This suggests ~10% over design for critical matched transistors at time zero in order to achieve a 10 year reliability requirement.

I. INTRODUCTION

Differential pair and current mirror are essential building blocks in analog circuit design. Final product performances are determined by transistor threshold voltage and current matching in these blocks [1]. More importantly, the transistor matching must be maintained during product lifetime in order to guarantee its long term reliability. On the other hand, transistors in these precision blocks are under constant stresses such as channel hot carrier (CHC) and negative bias temperature instability (NBTI). Transistor parameter shift and its impact to circuit performance under NBTI and CHC stresses have been studied extensively [2, 3] for many years. However, matching behaviors among transistors under *matched stresses* (all matched transistors are biased at same conditions) have remained largely unexplored. Transistor matching stability was examined and found no matching drift after 1000 hours burn-in at 125C under well-matched DC biasing [4]. Transistor VT and current mismatching under NBTI stress was investigated assuming uncorrelated VT and mobility shifts [5]. In this paper, we present detailed studies on transistor matching behavior under matched stress.

Closed-form expressions for VT and ID matching coefficients as function of stress strength, ambient temperature and stress time have been obtained based on physical model including the correlation effect between mobility and threshold voltage shifts. The model has been validated experimentally and applied to a 40V high performance precision analog technology.

In general, a transistor under external stresses experiences charge and quantum state generation at the silicon-oxide interface. NBTI involves positive fixed charge and interface state generations when a PMOS transistor is biased in the inversion region. The charge generation process is essentially independent lateral electrical field and thus exhibits very weak channel length dependence. Charge generation under matched NBTI stresses is area related and occurs in the region under the transistor gate [5]. In contrast to the NBTI, charge generation under CHC stress condition exhibits strong channel length dependence and is localized at the edge of a transistor close to the drain side. Even under perfect matched or balanced stress conditions, electrical parameters shift among matched transistors are different due to the fact that the charge generation process is physically stochastic. It is anticipated that the stress induced transistor matching becomes worse as the charge generation accumulates.

II. THRESHOLD VOLTAGE MISMATCHING UNDER MATCHED NBTI STRESS

A. VT Mismatching Model

Assuming Ni(t) is stress induced interface charges per unit area at time t, transistor VT as a function of time can be expressed as VT(t) = VT(0) + q Ni(t)/Cox, where VT(0), q and Cox are the VT value at time zero, electron charge and gate capacitance per unit area, respectively. Notice that interface charge generation is independent of charge in the depletion region and follows the Poisson distribution statistically. The variance of interface charge Ni(t) is given by Var[Ni(t)] = Ni(t)/(WL) for a transistor with active area of WL. Then, the variance of the threshold voltage at time t can be expressed by equation (2.1) accordingly.

$$Var[VT(t)] = Var[VT(0)] + (q/Cox)^2 Ni(t)/WL \quad (2.1)$$

Since VT shift $\delta VT(t)$ and interface charge $Ni(t)$ are related by $\delta VT(t) = qNi(t)/Cox$, (2.1) may be rewritten as $Var[VT(t)] = Var[VT(0)] + (q/Cox) \delta VT(t)/WL$. For a matching pair, transistor VT matching can be expressed by $\sigma^2[VT(t)] = \sigma^2[VT(0)] + 2(q/Cox) \delta VT(t)/WL$ after simple algebra manipulation. Transistor matching at time zero is expressed by the well-known equation $\sigma^2[VT(0)] = A^2vt(0)/WL$, where $Avt(0)$ is the VT matching slope at time zero. Combing all expressions together, we finally obtain

$$\sigma^2[VT(t)] = A^2vt(t)/WL \quad (2.2)$$

$$A^2vt(t) = A^2vt(0) + 2\alpha(q/Cox).\delta VT(t) \quad (2.3)$$

where parameter α is introduced to accommodate random charge spatial distribution and must be determined experimentally.

Equations (2.2) and (2.3) show that transistor matching still follows the well-known Pelgrom law [6] (parameter matching ~ 1/WL) under matched NBTI stress, although the matching slope degrades due to the interface charge generation. The time zero matching slope Avt(0) can be extracted during the standard matching characterization using different transistor sizes. Typically, $\delta VT(t)$ under NBTI type of stress is modeled by $\delta VT(t) = Ae^{\beta Vs} e^{-Ea/KT} t^n$, where parameters A, n, Ea and β are model parameters and can be determined during standard NBTI tests under different voltages, temperatures and stress times. Determination of parameter α involves transistor matching characterization at different stress conditions and is very time consuming. Accelerating stress at higher voltage and elevated temperature is used for test time reduction. Overstress of the transistors during the tests should be avoided since it leads to a model over estimating mismatching of transistor.

B. Results and Discussions

The device matching degradation model has been applied to a 40V high performance analog technology [7]. Figure 1 shows the VT shift $\delta VT(t)$ as a function of time under different gate stress voltages at 150C for 40V high voltage PMOS. Figure 2 shows silicon and modeling results of transistor matching slope as function of time during the high gate voltage stress test at 150C. It demonstrates that the physical model is capable of fitting experimental values with proper adjustment of parameter. Once the parameters are determined and validated, the model has been used to predicate the transistor VT matching degradation under nominal operation conditions. Figure 3 and Figure 4 show the modeling predications of transistor

VT matching slope as function of time at various of temperatures when the device operates at VGS=30V and VGS=15V, respectively. Avt(t) degradation is enhanced at higher operating temperature and higher gate voltage as evidenced by the model. Results shows ~10% (VG=15V) and ~20% (VG=30V) matching degradation during 10 year product lifetime when operating at temperature of 100C. To ensure the long term product reliability, proper transistor over design must be applied accordingly to compensate the matching deterioration for precision matching pair.

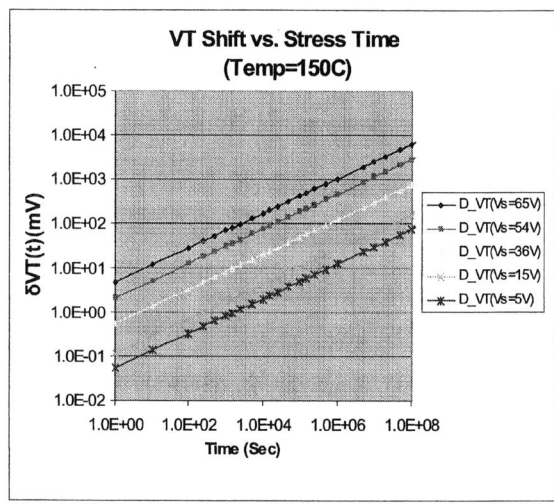

Fig. 1: VT shift $\delta VT(t)$ as function of stress times under different stress voltages at 150C. The model is established using equation $\delta VT(t) = A.e^{\beta Vs}. e^{-Ea/KT} t^n$ under various stress conditions.

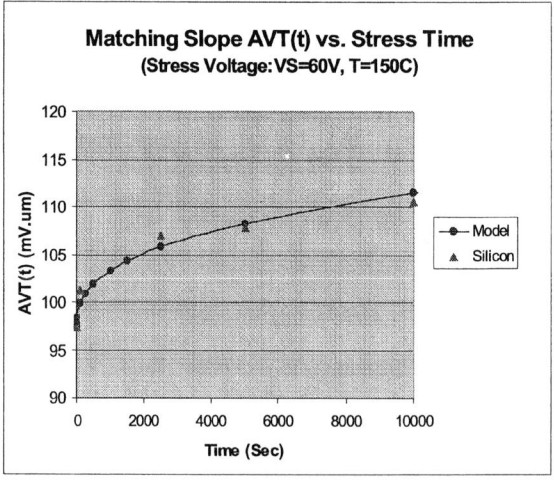

Fig. 2: High voltage PMOS VT matching slope as a function of stress time under accelerating stress voltage VS=60V at 150C. Equation (2.3) is used to fit the silicon results.

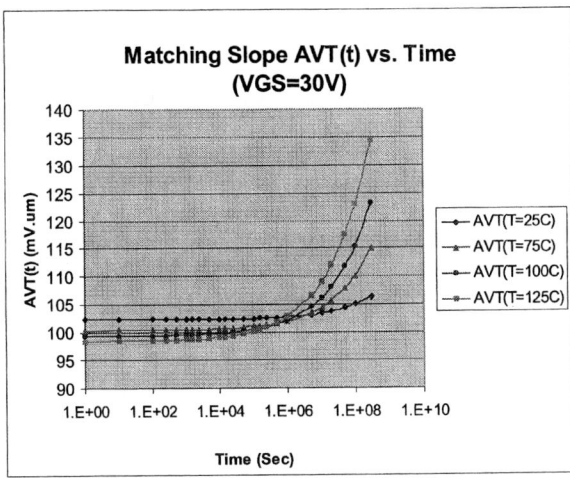

Fig. 3: High voltage PMOS VT matching slope degradation as a function of time at different temperatures under DC biasing condition VDS=30V. Avt(t) degrades by about 20% at 100C.

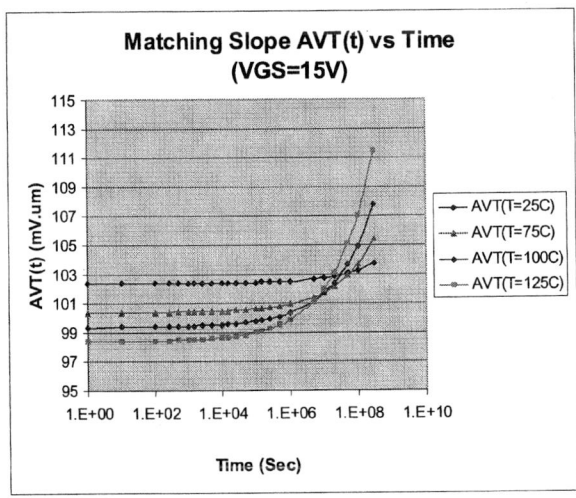

Fig. 4 High voltage PMOS VT matching slope degradation as a function of time at different temperatures under DC biasing condition VDS=15V. Avt(t) degrades by about 10% at 100C.

III. CURRENT MISMATCHING UNDER MATCHED NBTI STRESS

A. Current Mismatching Model

Charge generation at silicon-oxide interface not only causes threshold voltage VT shift $\delta VT(t) = qNi(t)/Cox$ as discussed in the previous section, but also degrades carrier mobility as well. The correlation between VT-shift and mobility-shift has important implication in calculating the variance of current shift due to NBTI stress. In general, the current change due to mobility and VT shifts under gate voltage VGS is

given by

$$\delta Id(t)/Id = \delta\mu(t)/\mu_0 - \gamma\, \delta VT(t)/(VGS-VT) \qquad (3.1)$$

Where $\delta Id(t)/Id$ and $\delta\mu(t)/\mu$ are current and mobility shifts respectively and γ is number between 1 and 2 depending on device operation region. According to the Mathiessens's scattering probability summation rule, mobility at time t is given by $1/\mu(t) = 1/\mu0 + 1/\mu s(t)$, where $1/\mu s(t) = q\, S\, Ni(t)$ is mobility contributed by interface charge scattering and S is scattering rate at the interface [8]. Mobility shift can thus be expressed as

$$\delta\mu(t)/\mu(t) = -q\,\mu0\,S\,Ni(t) = -q\,\mu0\,Cox\,S\,\delta VT(t) \quad (3.2)$$

Combining (3.1) and (3.2) and assuming scatting rate S is inversely proportional to channel carrier density, we finally obtain current shift at time t under NBTI stress

$$\delta Id(t)/Id = -\lambda\,/(VGS-VT)\,\delta VT(t) \qquad (3.3)$$

Where λ is phenomenological constant and must be determined experimentally. Using equation (3.3) and statistical theorem, a close form current matching equations are obtained as follows

$$\sigma^2[\Delta ID(t)/ID] = A^2id(t)/WL \qquad (3.4)$$

$$Aid^2(t) = A^2id(0)\,[ID(0)/ID(t)]^2 + B^2id(t) \qquad (3.5)$$

$$B^2id(t) = 2q/Cox\,\lambda^2\,\delta VT(t)/[VGS-VT(t)]^2 \qquad (3.6)$$

Where Aid(t) and Aid(0) are current matching slopes at time t and time zero, respectively. Combed with NBTI VT shift $\delta VT(t)$, (3.4), (3.5) and (3.6) can be used to model current mismatching as a function of time under any operation condition. It is evident that the current matching also follows ~ 1/WL law albeit the slope is increased by the external stress. Bid(t) in (3.5) represents additional current mismatching caused by interface charge generation during the stress. Constant λ can be determined from equation (3.3) using standard NBTI characterization method by monitoring the current shift as a function of time. Another alternative to determine λ is fitting the current mismatching results using equations (3.4), (3.5) and (3.6). The later involves extensive current mismatching measurements at different stress conditions.

B. Result and Discussions

Figure 5 shows both measured and modeling results of the current shift $\delta Id(t)/Id$ as a function of time under stress voltage of 60V at 150C at two different operating regimes. In both cases, the model fits the silicon data very well. Theoretical and experimental current matching results as function of time at same stress condition is plotted in Figure 6.

Good agreement between silicon data and modeling results has also been demonstrated. Measurements must be conducted carefully to average out equipment noise during the characterizations. This results may be used to determine parameter used to determined λ in using equation (3.3). Again, overstress may lead to overestimation of λ and thus current mismatching slope.

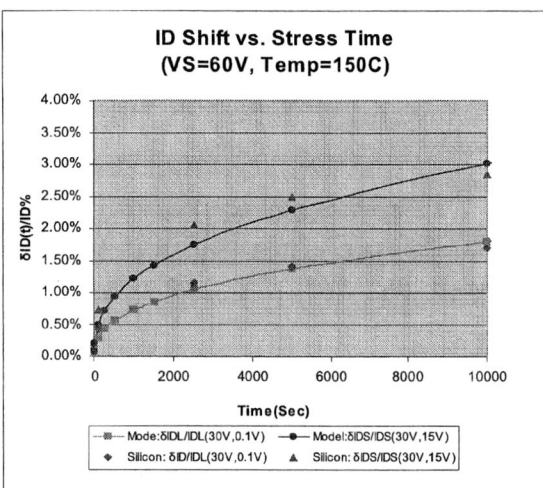

Fig. 5: Driving current shift $\delta Id(t)/Id$ as function of stress times under stress voltages VS=60V at 150C. The post-stress current is measured at different DC operation regimes.

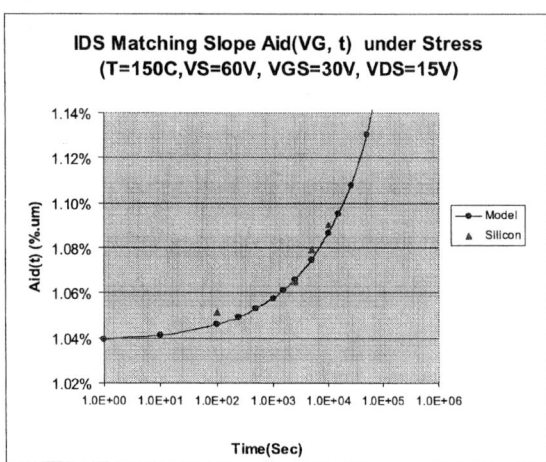

Fig. 6: High voltage PMOS current matching slope as a function of stress time under accelerating stress voltage VS=60V at 150C. The post-stress current mismatching is measured at VGS=30V and VDS=15V.

Figure 7 shows how the current matching slope evolves as a function of time at different temperatures under operating condition VGS=30V. The results show that the current matching deteriorates faster at higher temperature due to enhanced interface charge generation. Current matching as a function of time at different operating conditions is demonstrated

in Figure 8. The results indicate that the current matching degrades at lower speed at intermediate voltage (VGS =10V and 15V), while opposite is observed at other (higher VGS=30V and lower VGS=5V) voltages. Higher matching degradation rate at VGS=30V is attributed to the fast charge generation, while the same effect under low VGS is due to larger augmentation factor $\sim 1/(VGS-VT)^2$ in the current mismatching equation. Similar to the VT matching, the current matching model also implies 10%~20% over design in precision current process blocks to ensure product reliability.

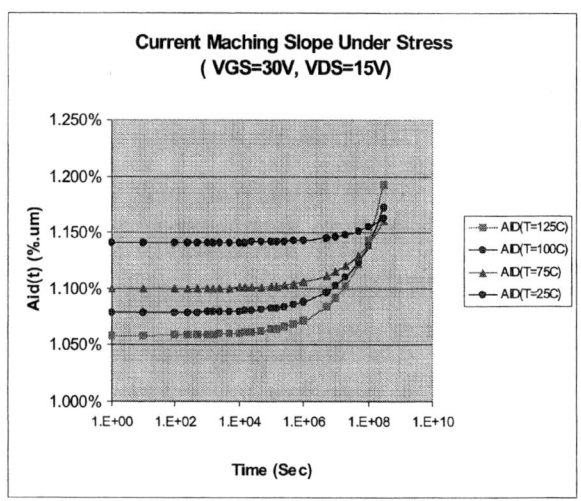

Fig. 7: High voltage PMOS current matching slope degradation as a function of time at different temperatures under DC biasing condition VDS=30V.

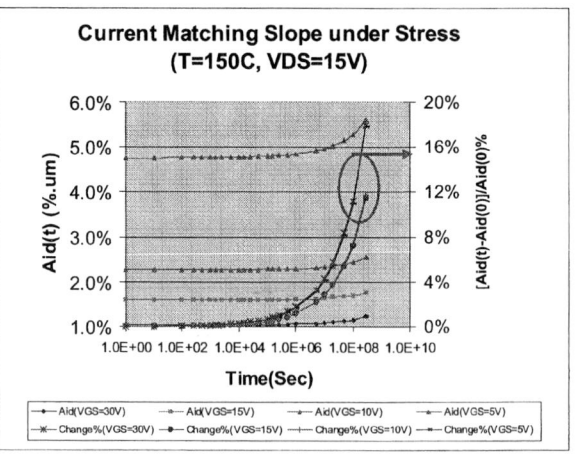

Fig. 8: High voltage PMOS VT matching slope degradation as a function of time at different DC biasing conditions at 150C. Left axis shows Aid(t) degradation relative to Aid(0). Avt(t) degrades by about 10%~20% depending DC operating point.

IV. SUMMARY

Transistor matching degradation as a function time under different operating voltages and ambient temperatures have been developed based on charge generation at the silicon-oxide interface using device physics and statistics. The models can be used to predicate transistor matching degradation under device normal operating conditions. The transistor matching slopes Avt(t) and Aid(t) are demonstrated to closely relate to the interface charge density Ni(t) = Cox/q δVT(t). Methods to extract parameters in the model have also been described. To finalize the matching model, three characterizations must be conducted. First, mismatching characterization at different temperature and biasing conditions is conducted to determine Avt(0) and Aid(0) before applying any external stresses. Then, parameter shifts δVT(t) and δId(t)/Id(t) models are developed using accelerating stress with different strengths, temperatures and time. This is typically done during the standard reliability test in the technology development. Finally, matching slopes Avt(t) and Aid(t) are measured at different stress voltage, temperatures and times for various device sizes. The models and parameter extraction methodology have been applied to a 40V high performance analog technology. It has been demonstrated that the models fit silicon results well at different operation conditions and temperatures. The results suggest 10%~20% over design for precisely matched transistors in order to ensure 10 yeas lifetime for product operating at VGS between 15V-30V at 100C temperature.

ACKNOWLEDGMENT

The authors like to express their sincerely thanks to ATD management for supporting of this work. The authors also like to thank Mr. Robert Higgins for many discussions on NBTI measurements.

REFERENCE

[1] K. R. Lakshmikumar et al., "Characterization and Modeling of Mismatch in MOS Transistors for Precision Analog Design", IEEE J. of Solid-State Circuits, Dec. 1986, Vol. 21, pp.1057-1066.

[2] M. Agostinelli et al., "PMOS NBTI-Induced Circuit Mismatching in Advanced Technology", Proceeding of IRPS, April 2004, pp. 171-175.

[3] R. Thewes et al., "Device Reliability in Analog CMOS Application", IEDM Tech. Dig., Dec. 1999, pp 81-84.

[4] M. Christopher et al., "Mismatching Drift: a Reliability Issue for Analog MOS Circuits", Processing of IRPS, April 1992, pp. 82-84.

[5] S. E. Rauch III, "The Statistics of NBTI-Induced VT and β Mismatching Shift in pMOSFETs ", IEEE Tans. Electron Devices, Dec. 2002, Vol. 2, pp.89-93.

[6] M. Pelgrom et al., "Matching Properties of MOS Transistor", IEEE J. of Solid-State Circuits, Oct. 1989, Vol. 24, pp.1433-1439.

[7] Xiaoju Wu et al., "Thin-Gate CMOS and Super-Thick Gate DECMOS Integration in 0^0 on axis <100> starting Wafer: Process Challenges and Solutions", Proceeding of ISSM, Oct, 2006, pp.439-442.

[8] L. Tewksnury III, PhD Thesis, Massachusetts Institute of Technology (1992).

2008 IEEE Conference on Microelectronic Test Structures, March 24-27, Edinburgh, UK   10.4

Influence of STI stress on drain current matching in advanced CMOS

Nicole Wils[1], Hans Tuinhout[1], Maurice Meijer[2]

[1] NXP-TSMC Research Center; [2] NXP Semiconductor Research

High Tech Campus 37, 5656AE Eindhoven, The Netherlands

Abstract

Using a dedicated set of – asymmetrically designed – matched pair test structures and a data analysis technique based on so-called mismatch sweeps, we answer some important questions in the discussions on variability in advanced CMOS technologies.

Introduction

Shallow Trench Isolation (STI) induces mechanical stress and consequently has an influence on MOS transistor performance [1,2,3]. This results in a deterministic drain current mismatch between devices with identical width and length but different (active) layout. But does this systematic offset also have an influence on the stochastic fluctuations and thereby increase overall device performance spread specifications (variability)? So far, only limited results about the influence of STI stress on random fluctuations have been published [3,4]. Using a new approach based on a collection of asymmetrically designed matched pair test structures and data analysis through mismatch sweeps, we show that STI induced strain indeed causes significant offsets for NMOS devices, but that there is no significant influence on random fluctuations.

Test structures and analysis technique

Test structure layout

This study is based on a new set of 65 nm CMOS technology node matched pair test structures, fabricated using the (former) C2A multi-project-wafer service. The layout of our basic matched pair test structure and probe pad configuration is shown in Figure 1. Ultimate care was taken to assure that all possible (undesired) effects due to asymmetric metallization layout and unequal access resistances are avoided. For the same reason, CMP tiling was performed manually in the direct vicinity of the transistor pair. Our test module (Figure 2) includes a full (width and length) range of basic NMOS and PMOS matched pairs. Furthermore, a selection of matched pair test structures with W/L=2/0.5 was designed with a deliberate dissimilarity between T_1 and T_2. For the STI related matched pairs (Figure 3), we designed pairs where the distance (d) from the poly gate to the edge of the STI was kept constant at 0.175 μm for T_1, while it was varied from 2.0 to 0.175 μm for T_2. Layout details of these structures are given in Table 1. Since the amount of stress induced by the STI will increase with decreasing distance d [5], comparison of device characteristics between T_1 and T_2 will indicate the influence of STI stress on device performance. As with any properly designed matched pair, the effects of parametric wafer gradients and other process spreads are mitigated due to the relatively close spacing. This allows for independent assessment of microscopic device architecture fluctuations and systematic mismatch effects due to intentional layout asymmetries. We also compared W/L=1/0.06 NMOS devices with different distances from the gate edge to the STI. We did this by combining two T_1's of two matched pair structures (E and F in table 1), which were separated by a few millimeters in our test module. We could combine transistors from two structures, because we found that the contribution of parametric wafer gradients over such distances is negligible. This important observation is substantiated by the example in the following subsection.

Analysis: mismatch sweeps

For quick qualitative as well as proper quantitative evaluation of our 65 nm CMOS mismatch performances we measure complete linear region (V_{ds}=50 mV) I_d-V_{gs} curves simultaneously on both transistors of each pair. Subsequently we calculate the relative drain current mismatch ($\Delta I_d/I_d$) for each bias point of each pair. When this is done for a population of pairs, we obtain a collection of curves (mismatch sweeps) as depicted in Figure 4a. Next, the median (μ) and standard deviation (σ) of the relative drain current mismatch ($\Delta I_d/I_d$) can be determined as function of gate voltage (Figure 4b). The advantage of these so-called mismatch median and fluctuation sweeps is that - at a glance - they reveal mismatch characteristics (offsets as well as fluctuations) in different operating regions, based on measured drain currents. This as opposed to basing judgments on derivative parameters, such as V_t and β. If there is no systematic mismatch, the median (μ) should be close to zero (or at least significantly smaller than the standard deviation). The typical behavior for the mismatch fluctuation standard deviations (σ) is that they are relatively high at low gate bias (σ_$\Delta I_d/I_d$'s larger than 50% are no exception) and then drop to much lower values in strong inversion. The levels of the mismatch sweeps in sub-threshold are indicative for the amount of systematic threshold voltage shift (μ_ΔV_t) and the amount of random V_t mismatch fluctuation (σ_ΔV_t), while the levels in strong

978-1-4244-1801-5/08/$25.00 ©2008 IEEE   238 inversion (at the highest gate voltages) are indicative for the amount of systematic mobility (trans-conductance) shift and the amount of random mobility mismatch fluctuation respectively.

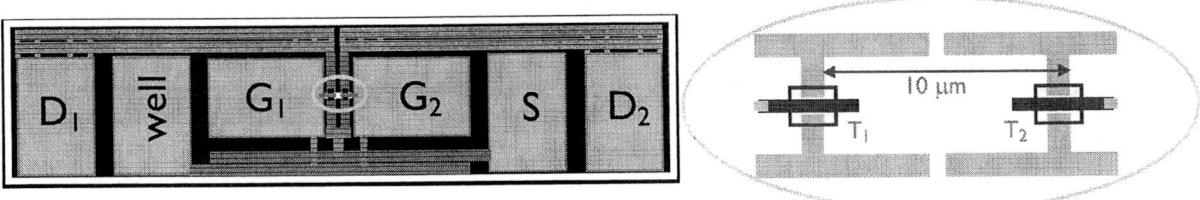

Figure 1: Ideal MOSFET matched pair test structure layout (left). Right: enlarged view of the actual matched pair.

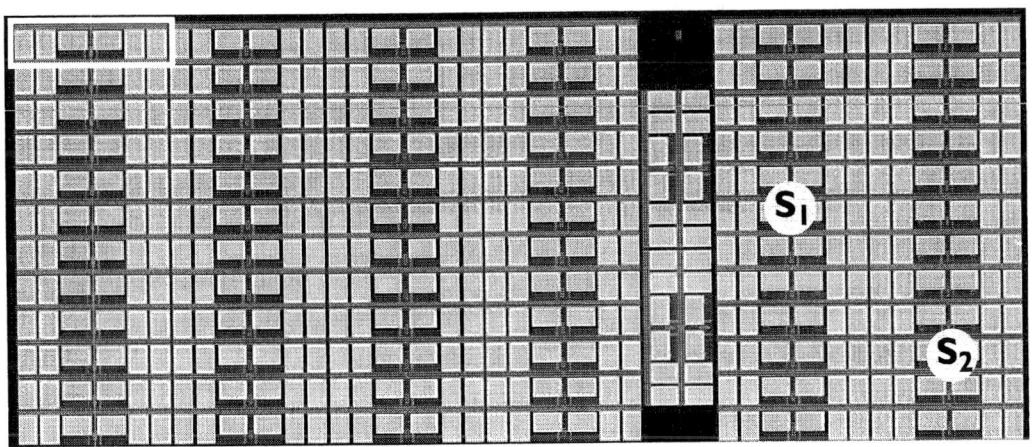

Figure 2: All matched pair structures in the test module have the same configuration. S_1 and S_2 are identical matched pairs separated 1 mm from each other.

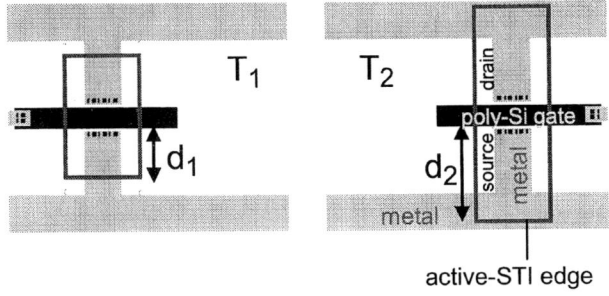

Figure 3: Dedicated matched pair test structures were used to test the influence of STI stress. The actual values of d_1 and d_2 can be seen in table 1.

Table 1: Layout details

structure	W [µm]	L [µm]	d_1 [µm]	d_2 [µm]
A	2.0	0.5	0.175	2.0
B	2.0	0.5	0.175	0.525
C	2.0	0.5	0.175	0.350
D	2.0	0.5	0.175	0.175
E	1.0	0.06	0.175	0.175
F	1.0	0.06	2.0	2.0

In structures A, B and C, d_1 differs from d_2. Structure D is a reference ($d_1=d_2$) taken from the standard matched pairs set. For W/L=1/0.06 devices the T_1's of structures E and F were compared.

239

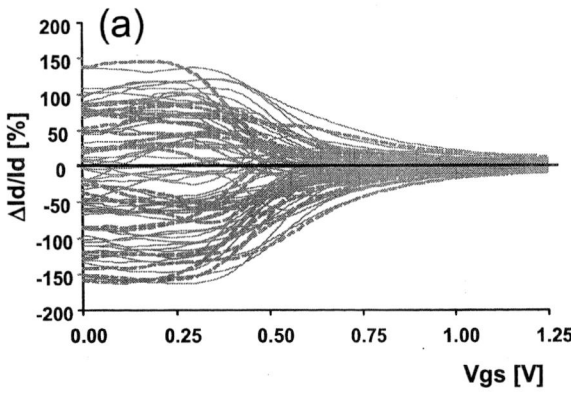

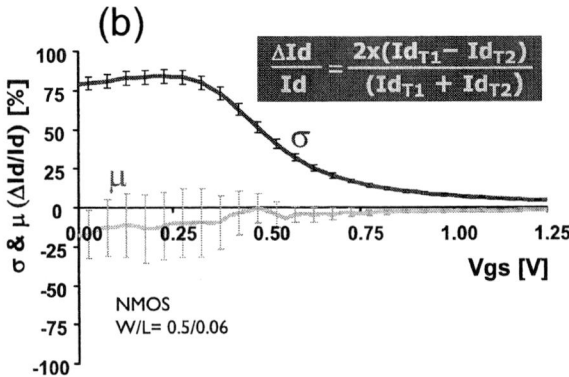

Figure 4: Relative drain current mismatch as a function of gate voltage for a population of 71 matched pair devices (a); the resulting mismatch median ($\mu\_\Delta I_d/I_d$) sweep and mismatch fluctuation ($\sigma\_\Delta I_d/I_d$) sweep are shown in (b). Some 1σ uncertainty bars (from bootstrapping) have been added.

Figure 5 provides a good example of how mismatch sweeps can be used to assess the relevance and significance of mismatch differences between different populations of pairs. In this case we compare the drain current mismatch sweeps (median and fluctuations) of a standard (0.5/0.06 NMOS) matched pair (S_1) to the mismatch sweeps of pairs that were constructed using a combination of transistors of two pairs (S_1 and S_2) separated by approximately 1 mm distance. Module S_2 is an exact copy of S_1. The positions of S_1 and S_2 are indicated in figure 2. From figure 5 it should be clear that there is no significant difference in drain current mismatch (μ nor σ) between the two cases. This means that for the wafers we measured it is allowed to combine transistors of spatially separated pairs (at least up to a few millimeters) for mismatch studies, without having to worry about possible distance effects (parametric gradients). This is relevant for some of the short channel transistor mismatch results presented in the following sections.

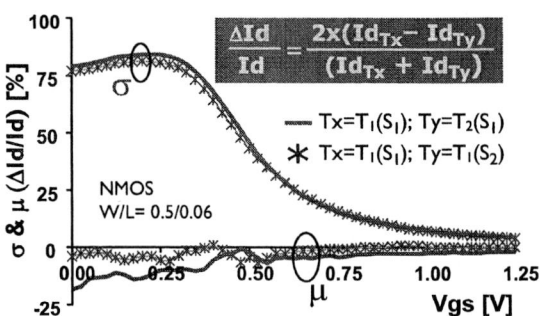

Figure 5: Mismatch sweeps for NMOS devices, derived from real matched pairs (solid lines) or from a combination of two spatially separated (1mm), but identical transistors (asterisks).

Results

Systematic offsets

NMOS devices

Figure 6 shows I_d-V_{gs} characteristics (in this case medians of the drain currents for populations of 31 NMOS devices with W/L=2/0.5) for transistor T_2 of structures A, B, C and D. As expected, it is observed that a larger distance to the STI edge results in a higher drain current for the complete range of gate voltages. This is further substantiated in Figure 7, where the mismatch median sweeps of the relative drain current (between T_1 and T_2 of structures A, B, C and D) are shown. Note that the reference matched pair (structure D, d_1 =d_2) shows no (statistically significant) systematic current mismatch over the entire bias range. For structure A (d_T_2=2.0 μm and d_T_1=0.175 μm) the relative current mismatch varies between 12% for high V_{gs} to about 30% in sub-threshold. In other words: the STI stress causes an offset in mobility as well as a moderate offset in threshold voltage (V_t). The higher the difference in stress between T_1 and T_2, the larger the offsets. The latter is also illustrated in Figure 8, where the median of the V_t difference ($\mu\_\Delta V_t$) and relative current factor difference ($\mu\_\Delta\beta/\beta$) between T_1 and T_2 are shown for structures A, B, C and D. In this case the transistor parameters V_t and β were calculated from the original linear region I_d-V_{gs} curves using fixed overdrive three-point-extractions [6].

To investigate the influence of STI induced strain on short channel devices we combined transistors T_1 of structures E and F (table 1). The drain current mismatch median sweep between E and F is presented in Figure 9. For comparison, also the sweep for the 2/0.5 NMOS devices has been included. Clearly, short devices (L=0.06 μm) show a much higher systematic drain current mismatch in sub-threshold than the longer ones (L=0.5 μm). This implies that for short NMOS devices the increase in threshold voltage with increasing STI induced strain is much higher than for long devices. The median increase in threshold voltage between F and E was 25 mV. This increase in threshold voltage with decreasing distance was also found by Sheu et.al. [2] and has been explained by strain induced retardation of dopant diffusion.

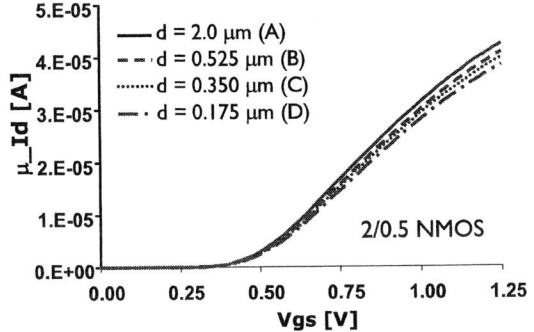

Figure 6: Median drain current (Vds = 50 mV) as function of gate voltage for 31 NMOS devices.

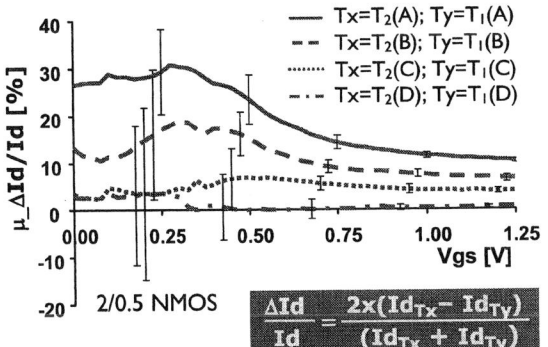

Figure 7: Relative drain current mismatch median sweeps for 31 NMOS matched pairs (A, B, C and D). For some points, 3σ uncertainty bars have been added.

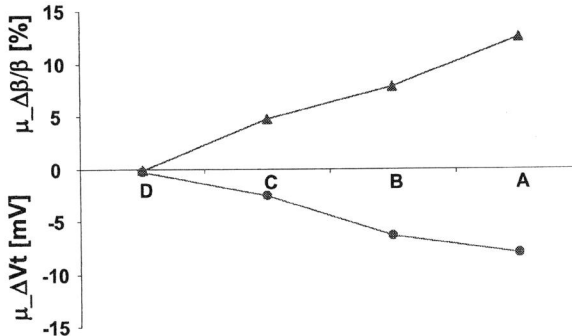

Figure 8: Median of threshold voltage and current gain difference between T_1 and T_2 for structures A, B, C and D (2/0.5 NMOS). The larger the difference in gate-STI distance between the two transistors, the higher the mismatch.

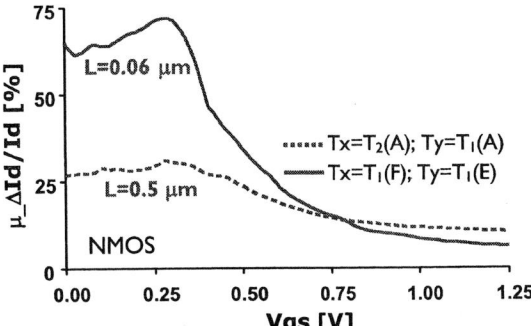

Figure 9: Relative drain current mismatch median sweeps for long and short NMOS devices. In sub-threshold the offset for the short channel device is significantly higher.

PMOS devices

On our wafers we found no significant influence of STI stress on the drain current of the PMOS devices, as is illustrated by the mismatch median sweeps in Figure 10. It is known that a compressive stress (as induced by the STI) will generally increase hole mobility. The fact that we don't see any influence on the p-channel drain current, is probably caused by the fact that 'twisted notch' wafers (<100> notch direction) are used for this 65 nm technology node. On such substrates, the hole mobility is almost unaffected along the [100] and [010] current flow directions [7].

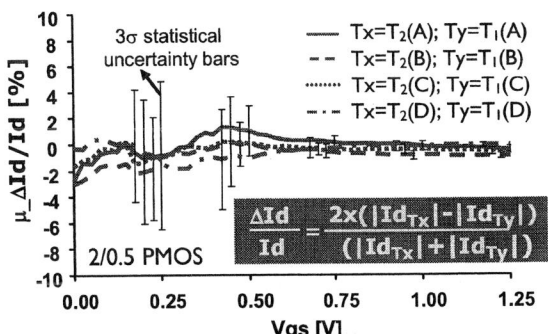

Figure 10: Mismatch median sweeps for 71 PMOS matched pairs (structures A, B, C and D).

Random fluctuations

From the previous sections it is clear that STI induced strain can create significant offsets in NMOS drain current. The main question however remains: do these offsets have an influence on random mismatch fluctuations? Figures 11 and 12 show the mismatch fluctuation sweeps of the relative drain current for structures A, B, C and D, for NMOS and PMOS respectively. Clearly, there is no significant impact of STI induced strain on the random mismatch fluctuations! The same was found for the short NMOS devices (Figure 13). Once again it should be noted that any trends that one may be tempted to see in these figures are generally not monotonous and certainly not statistically significant. To elucidate this, 3σ uncertainty bars (from bootstrapping), indicating the statistical uncertainty, have been added to figures 11 and 12.

The fact that the mismatch fluctuation standard deviations are practically equal for all populations implies that the STI stresses (differences) are constant over the wafer.

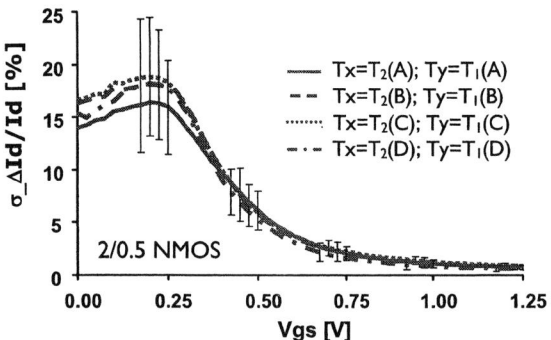

Figure 11: Relative drain current mismatch fluctuation sweeps for 31 NMOS matched pairs. There is no significant difference between the results of structures A, B, C and D. For some points, 3σ bars have been added to indicate statistical uncertainty.

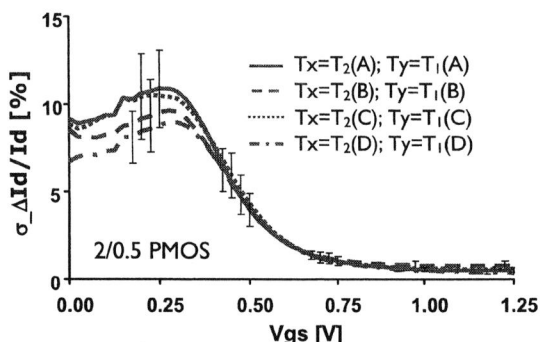

Figure 12: Relative drain current mismatch fluctuation sweeps for 71 PMOS matched pairs. There is no significant difference between the results of structures A, B, C and D. For some points, 3σ uncertainty bars have been added.

Conclusions

We describe matched pair test structures and a so-called mismatch sweep analysis technique, using intentionally dissimilar pairs. We confirm and quantify the impact of STI stress on the drain currents and threshold voltages of advanced NMOS devices. Clear differences are found between NMOS and PMOS devices and a strong impact of channel length on threshold voltage offset is demonstrated. However, we also show that there is no significant influence of increased stress on mismatch fluctuations. One can therefore conclude that systematic STI induced strain differences, which can well be modeled as systematic

parameter shifts in advanced compact models [8], do not contribute to the overall process variability range, once they are quantified and monitored through appropriate LVS (layout versus schematics) checks. The fact that the mismatch fluctuation standard deviations are not affected is good news in the sense that no additional (time consuming) statistical simulations are required to accommodate for (apparently insignificant) microscopic stress fluctuations. Furthermore, the lack of any distance effects up to distances of several millimeters also facilitates modeling and simulations of variability substantially. Such conclusions are very valuable contributions to the often hectic and controversial discussions on variability in advanced CMOS technologies.

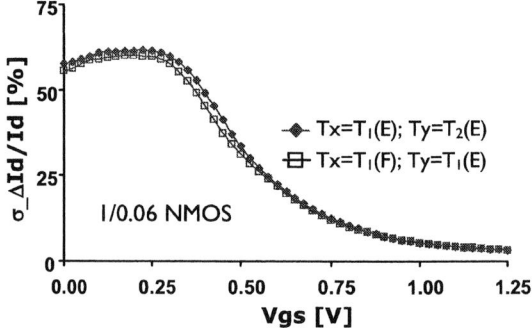

Figure 13: Relative drain current mismatch fluctuation sweeps for short channel NMOS devices. The systematic offset caused by dissimilarities in STI stress has no significant influence on the random fluctuations.

References

[1] Gregory Scott et al., "NMOS Drive Current Reduction Caused by Transistor layout and Trench Isolation Induced Stress"; IEDM 1999 Conference Proceedings, pp. 827-830

[2] Y. Sheu et al.," Modelling Mechanical Stress Effect on Dopant Diffusion in Scaled MOSFETs"; Transactions on Elec-tron Devices, vol. 52, no.1, January 2005, pp. 30-38

[3] P.G. Drennan et al., "Implications of Proximity Effects for Analog Design"; CICC 2006 Conference Proceedings, pp.169-176

[4] Tan et al., "Effect of Shallow Trench Induced Stress on CMOS Transistor Mismatch", ICSE 2004 Conference Pro-ceedings, pp.189-192

[5] Miyamoto et al., "Impact of Reducing STI-Induced Stress on Layout Dependence of MOSFET Characteristics"; Transac-tions on Electron Devices, vol. 51, no.3, March 2004

[6] J.A. Croon et al., "A comparison of extraction techniques for threshold voltage mismatch", ICMTS 2002 Conference Proceedings, vol.15, pp.235-239

[7] Dragan Manic, "Drift in Silicon Integrated Sensors and Circuits due to Thermo-Mechanical Stresses", Series in Micro systems, Volume 8, Hartung-Gorre (2000)

[8] web page: http://www.eigroup.org/cmc/

AUTHOR INDEX

A

Agarwal, K. 17
Allen, R.A. 35
Allport, P.P. 101
Ammo, H. 8
Ancient, R. 117
Annamalai, U. 166
Aota, H. 107
Argoud, F. 52
Arima, T. 8
Arimoto, H. 223
Armstrong, B.M. 65, 101
Arnaud, L. 210
Arriordaz, A. 166
Aziza, H. 52

B

Bain, M. 65
Baine, P. 65
Bauwens, F. 46
Beach, E. 11
Benech, P. 214
Bhushan, M. 147
Bhuva, B.L. 142
Bordez, S. 90
Borot, B. 52
Bouchu, D. 210
Boulanger, F. 210
Bowcock, T. 101
Brun, J. 117
Brut, H. 160
Burrows, J. 131

C

Casse, G. 101
Castellon, L. 43
Cathignol, A. 90
Chagawa, T. 86
Chang, L. 3
Chappaz, C. 43
Chen, W. 180
Chen, Z. 233
Cheng, J. 131
Cheng, W. 155
Constant, A. 137

Cornils, M. 23
Cote, W. 56
Craig, M. 56
Crean, G.M. 180
de Crecy, F. 210
Cresswell, M.W. 35
Cros, A. 160
Cummins, G. 111

D

Das, K.K. 147
Delecourt, H-X. 46
Depoutot, F. 117
Dieudonné, F. 137
Dixson, R.G. 35
Dolin, A. 175
Dubois-Bonvalot, B. 117
Dunare, C.C. 123
Duport de Pontcharra, J. ....
.... 210

E

Enderling, S. 228

F

Ferrante, W. 56
Fleury, D. 160
Franiatte, R. 117
Frank, D.J. 3
French, W. 96
Fukishima, M. 185

G

Gabrys, A. 96
Gadlage, M.J. 142
Galloway, K.F. 142
Gamble, H.S. 65, 101
Gasasira, A. 56
Gautheron, B. 137
Ghibaudo, G. 90, 160
Giansello, F. 214
Gillon, R. 46
Gloria, D. 214

Goswami, K. 71
Greig, W. 96
Gundlach, A.M. 123

H

Haworth, L. 80
Hayes, J. 17
Hegsted, D. 46
Henning, A. 56
Hess, C. 131
Heuting, R.J.E. 190
Holman, W.T. 142
Hourd, A.C. 29, 35, 228

I

Inoue, Y. 223
Inui, C. 185
Ito, M. 223
Iwata, H. 76

J

Ji, B.L. 3

K

Karthikeyan, M. 56
Ketchen, M.B. 3,147
Keur, W. 190
Khan, I. 11
Kinget, P. 175
Klee, M. 190
Klootwijk, J.H. 199, 205
Kolagunta, V. 166
Komuro, T. 155
Kovalgin, A.Y. 199
Kubota. M. 80
Kuroda, R. 155

L

Lauer, I. 3
Li, Y, ... 80
Liao, S. 65
Lin, H. 111,123
Lindley, R. 131

Lindorfer, P. 96
Liu, J. 190

M

Madhani, P. 11, 233
Maeda, H. 8
Maitrejean, S. 210
Massengill, L.W. 142
Mathewson, A. 117
Matsuda, T. 76
Mauczock, R. 190
McCallum, M. 29, 35, 228
McCarthy, K.G. 180
McNeill, D.W. 65
Medina, L. 56
Meijer, M. 238
Mennillo, S. 90
Merbeth, T. 56
Mita, Y. 80
Mizumura, A. 8
Mondal, S. 71
Montgomery, J.H. 101

N

Narasimham, B. 142
Nassif, S. 17
Née, D. 52
Ning, Z. 46
Nishida, A. 86
Nishiyama, E. 107
Nowodzinski, A. 117

O

O'Sullivan, J.A. 180
Ohmi, T. 155
Ohzone, T. 76
Orji, N.G. 35

P

Parkes, W. 80
Pastore, C. 214
Patra, D. 71
Paul, O. 23
Pearson, D.J. 3
Portal, J.M. 52
Potzick, J.E. 35

R

Reimann, K. 190
Revel, J.-F. 137
Rigaud, F. 52
Roosen, H.H.A.J. 205
Rosa, J. 137
Roy, D. 205
Ruddell, F.H. 101

S

Sakata, T. 223
Satoh, A. 223
Schmidt, J. 166
Schmitz, J. 190
Schrimpf, R.D. 142
Serret, E. 214
Shiling, E. 56
Shroff, M. 166
Sillon, N. 117
Smith, B. 166
Smith, S. 29, 35, 123, 228
Squcciarini, M. 131
Stavitski, N. 199
Steinmann, P. 11
Stellari, F. 3
Stevenson, J.T.M. ....
............29, 35, 123, 228
Suder, S.L. 101
Sugawa, S. 155
Sugiyama, Y. 76
Sural, S. 71
Suzuki, H. 223
Suzuki, T. 8
Swimmer, A. 131

T

Takakuwa, J. 76
Tanaka, T. 223
Terada, K. 86
Teramoto, A. 155
Thompson, J. 175
Tian, W. 11
Tiggelman, M.P.J. 190
Toffoli, A. 210
Van Torre, P. 46
Toublan, O. 29

R

Tsiamis, A. 29, 35, 228
Tsuji, K. 86
Tsunomura, T. 86
Tuinhout, H. 238
Tye, C. 155

U

Ueda, N. 107

V

Vast, J. 52
Vendrame, L. 90
Verhaegh, N.A.M. 205
Vlachakis, B. 46
de Vylder, E. 46

W

Walton, A.J. ....
..... 29, 35, 80, 111, 123, 228
Waltz, P. 43
Watabe, S. 155
Watanabe, H. 107
West, A.J. 71, 96
Wils, N. 238
Winters, S. 131
Witulski, A.F. 142
Wolters, R.A.M. 199, 205
Wu, X. 233

X

Xiang, J. 86

Y

Yamaguchi, S. 223
Yamaji, M. 223
Yamasaki, O. 223
Yu, S. 131

Z

van Zeijl, H.W. 199

CURRAN ASSOCIATES INC.
proceedings
.com

9781424418008